Computational Biology

Volume 22

The *Computational Biology* series publishes the very latest, high-quality research devoted to specific issues in computer-assisted analysis of biological data. The main emphasis is on current scientific developments and innovative techniques in computational biology (bioinformatics), bringing to light methods from mathematics, statistics and computer science that directly address biological problems currently under investigation.

The series offers publications that present the state-of-the-art regarding the problems in question; show computational biology/bioinformatics methods at work; and finally discuss anticipated demands regarding developments in future methodology. Titles can range from focused monographs, to undergraduate and graduate textbooks, and professional text/reference works.

More information about this series at http://www.springer.com/series/5769

Bir Bhanu · Prue Talbot
Editors

Video Bioinformatics

From Live Imaging to Knowledge

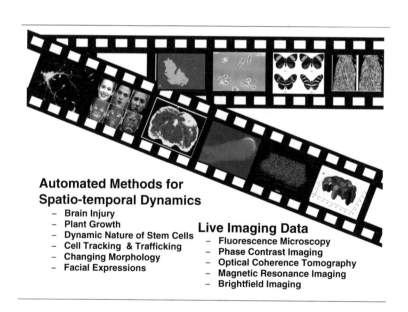

**Automated Methods for
Spatio-temporal Dynamics**
- – Brain Injury
- – Plant Growth
- – Dynamic Nature of Stem Cells
- – Cell Tracking & Trafficking
- – Changing Morphology
- – Facial Expressions

Live Imaging Data
- – Fluorescence Microscopy
- – Phase Contrast Imaging
- – Optical Coherence Tomography
- – Magnetic Resonance Imaging
- – Brightfield Imaging

 Springer

Editors
Bir Bhanu
University of California
Riverside, CA
USA

Prue Talbot
University of California
Riverside, CA
USA

ISSN 1568-2684
Computational Biology
ISBN 978-3-319-79526-3   ISBN 978-3-319-23724-4   (eBook)
DOI 10.1007/978-3-319-23724-4

Springer Cham Heidelberg New York Dordrecht London
© Springer International Publishing Switzerland 2015
© Springer International Publishing Switzerland (outside the USA) 2015 for Chapter 9
Softcover re-print of the Hardcover 1st edition 2015

Printed on acid-free paper

Springer International Publishing AG Switzerland is part of Springer Science+Business Media (www.springer.com)

Preface

The recent advances in high-throughput technologies for functional genomics and proteomics have revolutionized our understanding of living processes. However, these technologies, for the most part, are limited to a *snapshot analysis* of biological processes that are by nature *continuous and dynamic*. Modern visual microscopy enables video imaging of cellular and molecular dynamic events and provides unprecedented opportunities to understand how spatiotemporal dynamic processes work in a cellular and multicellular system. The application of these technologies is becoming a mainstay of the biological sciences worldwide. To gain a more mechanistic and systematic understanding of biological processes, we need to elucidate cellular and molecular dynamic processes and events.

Video Bioinformatics *as defined by the first author (BB) is concerned with the automated processing, analysis, understanding, data mining, visualization, query-based retrieval/storage of biological spatiotemporal events/data and knowledge extracted from microscopic videos. It integrates expertise from the life sciences, computer science and engineering to enable breakthrough capabilities in understanding continuous biological processes.* The video bioinformatics information related to spatiotemporal dynamics of specific molecules/cells and their interactions in conjunction with genome sequences are essential to understand how genomes create cells, how cells constitute organisms, and how errant cells cause disease.

Currently, new imaging instrumentation and devices perform *live video imaging* to image molecules and subcellular structures in living cells and collect biological videos for on-line/off-line processing. We can now see and study the complex molecular machinery responsible for the formation of new cells. Multiple imaging modalities can provide 2D to 5D (3D space, time, frequency/wavelength) data since we can image 2D/3D objects for seconds to months and at many different wavelengths. However, data processing and analysis (informatics) techniques for handling biological images/videos have lagged significantly and they are at their infancy. There are several reasons for this, such as the complexity of biological videos which are more challenging than the structured medical data, and the lack of interdisciplinary research at the intersection of life sciences and engineering and computer science.

We already are at a point where researchers are overwhelmed by myriads of high-quality videos without proper tools for their organization, analysis, and interpretation. This is the main reason why video data are currently underutilized. We believe that the next major advance in imaging of biological samples will come from advances in the automated analysis of multi-dimensional images. Having tools that enable processes to be studied rapidly and conveniently over time will, like Hooke's light microscope and Ruska's electron microscope, open up a new world of analysis to biologists, scientists, and engineers.

This interdisciplinary book on *Video Bioinformatics* presents computational techniques for the solution of biological problems of significant current interest such as 2D/3D live imaging, mild-traumatic brain injury, human embryonic stem cells, growth of pollen tubes, cell tracking, cell trafficking, etc. The analytical approaches presented here will enable the study of biological processes in 5D in large video sequences and databases. These computational techniques will provide greater sensitivity, objectivity, and repeatability of biological experiments. This will make it possible for massive volumes of video data to be analyzed efficiently, and many of the fundamental questions in life sciences and informatics be answered. The book provides examples of these challenges for video understanding of cell dynamics by developing innovative techniques. Multiple imaging modalities at varying spatial and temporal resolutions are used in conjunction with computational methods for video mining and knowledge discovery.

The book deals with many of the aspects of the video bioinformatics as defined above. Most of the chapters that follow represent the work that was completed as part of an NSF-funded IGERT program in Video Bioinformatics at the University of California in Riverside. Several of the chapters deal with work that keynote speakers presented at retreats sponsored by this program (Chaps. 14 and 16). Most other chapters are work done by IGERT Ph.D. fellows who were selected to participate in this program. The program emphasizes an interdisciplinary approach to data analysis with graduate students from engineering and life sciences being paired to work together as teams. These resulting chapters would likely never have been produced without cooperation between these two distinct disciplines and demonstrate the power of this type in interdisciplinary cooperation.

We appreciate the suggestions, support feedback and encouragement received from the IGERT faculty, IGERT fellows and NSF IGERT Program Directors Richard Tankersley, M.K. Ramasubramanian, Vikram Jaswal, Holly K. Given, and Carol Stoel. Authors would like to thank Dean Reza Abbaschian, Dean Joe Childers, Dallas Rabenstein, David Eastman, Victor Rodgers, Zhenbiao Yang, Vassilis Tsotras, Dimitri Morikis, Aaron Seitz, Jiayu Liao, David Carter, Jerry Schultz, Lisa Kohne, Bill Bingham, Mitch Boretz, Jhon Gonzalez, Michael Caputo, Michael Dang and Benjamin Davis for their support and help with the IGERT program. Authors would also like to thank Atena Zahedi for the sketch shown on the inside title page. Further, the authors would like to thank Simon Rees and Wayne Wheeler of Springer and Priyadarshini Senthilkumar (Scientific Publishing

Services) for their efforts related with the publication of this book. The first author (BB) would like to acknowledge the support from National Science Foundation grants DGE 0903667 video bioinformatics, CNS 1330110 distributed sensing, learning and control, IIS 0905671 video data mining, IIS 0915270 performance prediction, CCF 0727129 bio-inspired computation, and DBI 0641076 morphological databases. The second author (PT) would like to acknowledge support from the Tobacco Related Disease Research Program of California (18XT-0167; 19XT-0151; 20XT-0118; 22RT-0217), the California Institute of Regenerative Medicine (CL1-00508), and NIH (R01 DA036493; R21 DA037365).

Riverside, CA, USA   Bir Bhanu
June 2015   Prue Talbot

List of Video Links with QR Codes

Software for Using QR Codes:

In order to scan a QR Code you must install a QR Code reader app on your smartphone. You can download an app on Google Play (Android Market), Blackberry AppWorld, App Store (iOS/iPhone), or Windows Phone Marketplace.

- Most QR scanning apps are free. Any app that can read barcodes should be able to process QR codes.
- Recommended for Android: Google Goggles Version 1.9.4
- Recommended for iOS: Quick Scan—QR Code Reader Version 1.1.5

Hyperlinks and QR Codes for each Video Referenced in the Book

IGERT on Video Bioinformatics YouTube Channel:
   https://www.youtube.com/channel/UCX9tYZRm-mGwhg6866FWizA

   Chapter 6:
   Video 1: https://www.youtube.com/watch?v=3ylu3oHZEC4

   Video 2: https://www.youtube.com/watch?v=sfNhuK9JSxA

Chapter 8:
Video 1: https://www.youtube.com/watch?v=dQTkldB8lNk

Video 2: https://www.youtube.com/watch?v=NgJZ49xgev8

Video 3: https://www.youtube.com/watch?v=plb_CInFoEY

Video 4: https://www.youtube.com/watch?v=tHJ5JDcpkWY

Chapter 9:
Video 1: https://www.youtube.com/watch?v=wN1F_K_2vsg

Video 2: https://www.youtube.com/watch?v=INrVYlemvCY

Video 3: https://www.youtube.com/watch?v=xUcU3HpiMwI

Video 4: https://www.youtube.com/watch?v=RAiqZtK80Dw

Video 5: https://www.youtube.com/watch?v=hYayvIvuzOM

Video 6: https://www.youtube.com/watch?v=KLaFIhkovPI

Video 7: https://www.youtube.com/watch?v=Orw_lUs8KFI

Video 8: https://www.youtube.com/watch?v=M-asAiwiLnM

Video 9: https://www.youtube.com/watch?v=OHD0j74qbk0

Video 10: https://www.youtube.com/watch?v=aZlbfqfDDyg

Video 11: https://www.youtube.com/watch?v=OTBt1WBqqGw

Chapter 11:
Video 1: https://www.youtube.com/watch?v=l-Sb2iqzw1Y

Chapter 12:
Video 1: https://www.youtube.com/watch?v=jvdL7L4nQIs

Chapter 15:
Video 1: https://www.youtube.com/watch?v=pxg14SdvNbE

Video 2: https://www.youtube.com/watch?v=NnPHrGxJ7j0

Chapter 18:
Video 1: https://www.youtube.com/watch?v=6KZc6e_EuCg

Contents

List of Figures

List of Tables

Contributors

Stephen Ashwal Department of Pediatrics School of Medicine, Loma Linda University, Loma Linda, CA, USA

Bir Bhanu Center for Research in Intelligent Systems, University of California, Riverside, CA, USA

Anthony Bianchi Department of Electrical Engineering, University of California Riverside, Riverside, CA, USA

D.K. Binder School of Medicine, University of California, Riverside, CA, USA

Katherine A. Borkovich Department of Plant Pathology and Microbiology, University of California, Riverside, CA, USA

Ilva E. Cabrera Department of Plant Pathology and Microbiology, University of California, Riverside, CA, USA

Anirban Chakraborty Department of Electrical Engineering, University of California, Riverside, CA, USA

Hang Chang Lawrence Berkeley National Laboratory, Berkeley, CA, USA

Alberto C. Cruz Center for Research in Intelligent Systems, University of California, Riverside, CA, USA

Barbara Davis Department of Cell Biology and Neuroscience, UCR Stem Cell Center, University of California, Riverside, CA, USA

M.M. Eberle Department of Bioengineering, Materials Science and Engineering 243, University of California, Riverside, CA, USA

Ehsan T. Esfahani Department of Mechanical and Aerospace Engineering, University at Buffalo SUNY, Buffalo, NY, USA

Iryna Ethell Biomedical Sciences Department, University of California Riverside, Riverside, CA, USA

Linan Feng Center for Research in Intelligent Systems, Bourns College of Engineering, University of California at Riverside, Riverside, CA, USA

Gerald V. Fontenay Lawrence Berkeley National Laboratory, Berkeley, CA, USA

Nirmalya Ghosh Department of Pediatrics, School of Medicine, Loma Linda University, Loma Linda, CA, USA

Torsten Groesser Lawrence Berkeley National Laboratory, Berkeley, CA, USA

Benjamin X. Guan Center for Research in Intelligent Systems, University of California, Riverside, CA, USA

Ju Han Lawrence Berkeley National Laboratory, Berkeley, CA, USA

John Heraty Entomology Department, University of California at Riverside, Riverside, CA, USA

M.S. Hsu Translational Neuroscience Laboratory, University of California, Riverside, CA, USA

Sabrina C. Lin Department of Cell Biology and Neuroscience, UCR Stem Cell Center, University of California, Riverside, CA, USA

Min Liu Department of Electrical Engineering, University of California, Riverside, CA, USA

Nan Luo Department of Botany and Plant Sciences, University of California, Riverside, CA, USA

Devin W. McBride Department of Physiology and Pharmacology, Loma Linda University, Loma Linda, CA, USA

Amine Merouane University of Houston, Houston, TX, USA

Katya Mkrtchyan Department of Computer Science, University of California, Riverside, CA, USA

Arunachalam Narayanaswamy Google Inc., Mountain View, CA, USA

Andre Obenaus Department of Pediatrics, School of Medicine, Loma Linda University, Loma Linda, CA, USA

Vincent On Electrical and Computer Engineering Department, University of California Riverside, Riverside, CA, USA

B.H. Park Department of Bioengineering, Materials Science and Engineering 243, University of California, Riverside, CA, USA

Bahram Parvin Lawrence Berkeley National Laboratory, Berkeley, CA, USA

Rattapol Phandthong Department Cell, Molecular, and Developmental Biology, Department of Cell Biology and Neuroscience, UCR Stem Cell Center, University of California, Riverside, CA, USA

Janice Pluth Lawrence Berkeley National Laboratory, Berkeley, CA, USA

Natasha V. Raikhel Institute for Integrative Genome Biology, University of California, Riverside, CA, USA

C.L. Rodriguez Department of Bioengineering, Materials Science and Engineering 243, University of California, Riverside, CA, USA

Amit Roy-Chowdhury Department of Electrical Engineering, University of California, Riverside, CA, USA

Badrinath Roysam University of Houston, Houston, TX, USA

Somayeh B. Shafiei Department of Mechanical and Aerospace Engineering, University at Buffalo SUNY, Buffalo, NY, USA

J.I. Szu Translational Neuroscience Laboratory, University of California, Riverside, CA, USA

Prue Talbot Department of Cell Biology and Neuroscience, UCR Stem Cell Center, University of California, Riverside, CA, USA

Asongu L. Tambo Center for Research in Intelligent Systems, University of California, Riverside, CA, USA

Ninad S. Thakoor Center for Research in Intelligent Systems, University of California, Riverside, CA, USA

Nolan Ung Institute for Integrative Genome Biology, University of California, Riverside, CA, USA

Y. Wang Department of Bioengineering, Materials Science and Engineering 243, University of California, Riverside, CA, USA

Nikki Jo-Hao Weng Department of Cell Biology and Neuroscience, University of California, Riverside, CA, USA

Zhenbiao Yang Department of Botany and Plant Sciences, University of California, Riverside, CA, USA

Henry Yip Department of Cell Biology and Neuroscience, UCR Stem Cell Center, University of California, Riverside, CA, USA

Atena Zahedi Bioengineering Department, University of California Riverside, Riverside, CA, USA

Part I
Video Bioinformatics: An Introduction

Chapter 1
Live Imaging and Video Bioinformatics

Bir Bhanu and Prue Talbot

Abstract This chapter provides an overview of *live imaging* and *video bioinformatics*. It introduces the term of video bioinformatics and provides motivation for a deeper understanding of dynamic living processes in life sciences. It outlines computational challenges in understanding dynamic biological processes and a conceptual way of addressing them. The themes covered in the book range from oragnaismal dynamics to intercellular and intracellular dynamics with associated software systems. This chapter gives an overview of different parts of the book and a synopsis of each subsequent chapter.

1.1 Introduction

The recent advancements in high-throughput technologies for functional genomics and proteomics have revolutionized our understanding of living processes. However, these technologies, for the most part, are limited to a *snapshot analysis* of biological processes that are by nature continuous and dynamic. Biologists such as Lichtman (Univ. Washington) and Fraser (CalTech) suggested that bioinformatics based on static images is like learning about a sport by studying a scrapbook [1]. To determine the rules of American football one can examine 1000 snapshots taken at different times during 1000 games—but the rules of the game would probably remain utterly obscure and the role of the halftime marching band would be a mystery. Similarly, to gain a more mechanistic and systematic understanding of

B. Bhanu (✉)
Center for Research in Intelligent Systems, University of California,
Winston Chung Hall Suite 216, 900 University Ave., Riverside, CA 92507, USA
e-mail: bhanu@cris.ucr.edu

P. Talbot
Department of Cell Biology & Neuroscience, University of California,
2320 Spieth Hall, 900 University Ave., Riverside, CA 92507, USA
e-mail: talbot@ucr.edu

© Springer International Publishing Switzerland 2015
B. Bhanu and P. Talbot (eds.), *Video Bioinformatics*,
Computational Biology 22, DOI 10.1007/978-3-319-23724-4_1

biological processes, we need to elucidate cellular and molecular dynamic events (e.g., spatiotemporal changes in protein localization and intracellular signals).

One of the most exciting research developments has been the ability to image molecules and subcellular structures in living cells. Without harming a cell, we can now see and study the complex molecular machinery responsible for the formation of new cells. The imaging field is becoming more precise; for example, the resolution attainable by advanced techniques that break the diffraction limit is of the order of 1–30 nm [1, 2]. Multiple imaging modalities can provide 2D (x, y) to 5D $(x, y, z, t,$ wavelength) data since we can image 2D/3D objects for seconds to days to months and at many different wavelengths. This ability, combined with the power of genetics and novel methods for eliminating individual proteins, will answer questions that are centuries old.

To quote Murphy et al. [3], "…The unraveling of the molecular mechanisms of life is one of the most exciting scientific endeavors of the twenty-first century, and it seems not too daring to predict that, within the next decade, image data analysis will take over the role of gene sequence analysis as the number one informatics task in molecular and cellular biology."

The advances in modern visual microscopy coupled with high-throughput multi-well plated instrumentation enable video imaging of cellular and molecular dynamic events from a large number of simultaneous experiments and provide unprecedented opportunities to understand how spatiotemporal dynamic processes work in a cellular/multicellular system [1, 4]. The application of these technologies is becoming a mainstay of the biological sciences worldwide.

We already are at a point where researchers are overwhelmed by myriads of high-quality videos without proper tools for their organization, analysis, and interpretation. This is the main reason why video data are currently underutilized [5]. We believe that the next major advance in imaging of biological samples will come from advancements in the automated analysis of multidimensional images. Having tools that enable processes to be studied rapidly and conveniently over time will, like Hooke's light microscope and Ruska's electron microscope, open up a new world of analysis to biologists and engineers. The analytical methods will enable the study of biological processes in 5D (3D space, time, frequency/wavelength) in large video databases.

1.2 Video Bioinformatics

Genome sequences alone lack spatial and temporal information, and video imaging of specific molecules and their spatiotemporal interactions, using various imaging techniques, are essential to understand how genomes create cells, how cells constitute organisms, and how errant cells cause disease [6]. *The interdisciplinary research field of Video Bioinformatics is defined by Bir Bhanu as the automated processing, analysis, understanding, data mining, visualization, query-based retrieval/storage of biological spatiotemporal events/data and knowledge extracted from dynamic images and microscopic videos.*

The advanced video bioinformatics techniques, fundamental algorithms, and technology will provide quantitative thinking, greater sensitivity, objectivity, and repeatability of life sciences experiments. This will make it possible for massive volumes of video data to be efficiently analyzed, and for fundamental questions in both life sciences and informatics to be answered. The current technology [1, 7–12] to analyze biological videos, which are more complex than structured medical data, is in its infancy. Biological characteristics are not exploited; computer tools are used only for very low-level analysis; and the process is highly human-intensive. One cannot simply use standard image processing, computer vision, and pattern recognition (CVPR) techniques, [3] and expect good results since the domain of biological videos has its own peculiar characteristics and complexities [13–16]. Algorithmic issues in modeling motion, segmentation, shape, tracking, recognition, etc., have not been adequately studied. The varying requirements of users dictate an integrated approach using machine learning rather than handcrafted user-specific solutions to individual problems [17, 18].

Solving the complex problems described above requires life scientists and computer scientists and engineers to work together on innovative approaches. Computer scientists and engineers need greater comprehension of the biological issues, and biologists must understand the information technology and assumptions made in the development of algorithms and their parameters.

1.3 Integrated Life Sciences and Informatics

Conceptually integrated life sciences/informatics research requires us to perform *some* of the following sample tasks:

1. A single moving biological entity (cell, organelle, protein, etc.) needs to be detected, extracted from varying backgrounds and tracked.
2. The dynamics of deformable shape (local/global changes) of a single entity (not in motion) needs to be analyzed and modeled.
3. Entities and their component parts need to recognized and classified.
4. Multiple moving entities need to be tracked and their interaction analyzed and modeled.
5. Multiple moving entities with simultaneous changes in their global/local shape need to be analyzed and modeled.
6. The interactions of component parts of an entity and interaction among multiple entities while in motion need to be analyzed and modeled.
7. Mining of 5D data at various levels of abstractions (e.g., 2D image vs. 1D track) for understanding and modeling of events and detection of anomalous behavior needs to be performed.

The specific computational challenges include algorithmic issues in modeling complex biological motion, segmentation in the presence of complex nonstationary background, elastic registration of frames in the video, complex shape changes, nonlinear movement of biological entities, classification of entities within the cell, recognition in the presence of occlusion, articulation and distortion of shape, adaptation and learning over time, recognition of spatiotemporal events and activities and associated queries and database organization, indexing and search, and computational (space/time) complexity of video processing. A variety of imaging techniques are used to handle spatial resolution from micrometer to millimeter range and temporal resolution from a few seconds to months. The varying requirements of users dictate approaches based on machine learning, rather than handcrafted user-specific solutions to individual problems. As mentioned by Knuth, the noted computer scientist, "Biology easily has 500 years of exciting problems to work on … [19, 20]."

Thus, informatics remains a major challenge. Most of the analysis is done manually which is very time-consuming, and it is quite often very subjective. It is expected that video analysis will become one of the key informatics tasks in modern molecular and cellular biology. The video bioinformatics information related to spatiotemporal dynamics of specific molecules/cells and their interactions in conjunction with genome sequences will provide a deeper understanding from genomes to organisms to diseases.

1.4   Chapters in the Book

Most of the chapters that follow represent work that was completed as part of an NSF funded IGERT program in Video Bioinformatics at the University of California in Riverside [21]. Several of the chapters deal with work that keynote speakers presented at retreats sponsored by this IGERT program (Chaps. 14 and 16). Most other chapters are work done by IGERT PhD Fellows who were selected to participate in this program. The program emphasizes an interdisciplinary approach to data analysis with graduate students from engineering and life sciences being paired to work together as teams. These resulting chapters would likely never have been produced without the cooperation between these two distinct disciplines and demonstrate the power of this type in interdisciplinary cooperation.

The book is divided into six parts. Part 1, which includes the current chapter plus Chap. 2 by Nirmalya Ghosh, provides an introduction to the field of video bioinformatics. Part 2 "Organismal Dynamics: Analyzing Brain Injury and Disease" includes Chaps. 3–6, which deal with the analysis of brain injury in rodent models captured using magnetic resonance imaging (MRI) (Chaps. 3–5) and visualization of cortical brain tissue during seizures in mice using optical coherence tomography (Chap. 6). Part 3, "Dynamics of Stem Cells," presents chapters dealing with segmentation of stem cell colonies and analysis of individual stem cells (Chap. 7), as well as the application of commercially available software to solve problems in video bioinformatics related to stem cell behavior in vitro (Chaps. 8–9). Part 4 deals with "Dynamic Processes in Plant and Fungal Systems." Chapter 10 reviews video bioinformatics problems in plant biology, while Chaps. 11 and 12 present specific examples of problems solved in *Arabidopsis*, and finally Chap. 13 presents a method for analyzing growth and development of a fungus. Part 5 "Dynamics of Intracellular Molecules" presents a high-throughput method for quantifying DNA damage in cells, while Chap. 15 discusses a method using optogenetics to regulate protein (cofilin) transport in cultured cells. Part 6 "Software, Systems and Databases" has four chapters dealing with various software tools and databases that are available for use with video bioinformatics problems (Chaps. 16–19).

Chapter 1 by Bhanu and Talbot provides an overview of the contents of this book dealing with Video Bioinformatics. The term video bioinformatics was introduced by Bir Bhanu [21] to describe the automatic analysis and mining of information from video data collected from nm to mm of spatial resolution and picosecond to months/years of temporal resolution. Because video datasets are usually very large, it is not practical to extract such data manually. Nevertheless, video data are a rich source of new information and provide opportunities for discovery, often not recognizable by the human eye due to the complexity, size, and subtlety of video images. Video bioinformatics software tools not only speedup and automate analysis and interpretation of data, but also avoid biases that may be introduced when such analyses are done manually. Video bioinformatics tools can be applied to any type of video data. In this volume, most examples are of living or biological samples that range from microscopic, such as the movement of a specific protein (cofilin) with small cellular extensions (Chap. 15), to damaged areas of the brain in animal models (Chaps. 3 and 4), to an entire organism such as the analysis of facial expressions in Chap. 18. The methods of video bioinformatics can be applied to animals, humans, plants, and fungi as the following chapters demonstrate or to other types of video data not necessarily covered in this volume.

Chapter 2 by Ghosh introduces the field of video bioinformatics and explains how automated analysis of microscopic images of biological material has progressed. Application of various tools and algorithms from the fields of computer vision, pattern recognition, and machine learning are presented. This chapter discusses state-of-the art tools and their applications to biological problems in image analysis. Research areas that will enable future progress in biological image processing are also presented.

Chapter 3 by Bianchi, Bhanu, and Obenaus addresses problems in imaging traumatic brain damage and the development of automated image analysis tools for extracting information from mild traumatic brain injury images in rodents. Imaging in live animals was done using magnetic resonance imaging (MRI). Both high and low-level contextual features were used to gather information from injured brains. Combining both high and low contextual features provided more accurate segmentation, which can lead to better identification of areas of the brain that have experienced injury and may lead to better methods of treatment for patients with traumatic brain injury.

Chapter 4 by Ghosh, Ashwal, and Obenaus describes methods for the imaging and analysis of neonatal hypoxic ischemia injury (HII), a devastating disease of newborns for which there is limited treatment available. Physicians would be aided by noninvasive imaging information on brain lesions in patients with this disease and such information could improve management of HII. The authors have worked in rodent models and developed a tool called hierarchical region splitting (HRS) that can be used to identify ischemic regions in the brain. This tool is applied to magnetic resonance images (MRI) collected using living animals. The tool provides rapid and robust information on brain lesions and allows dynamic changes in the brain to be followed over time. For example, movement of stem cells into the lesion can be tracked. The method has application in both translational and clinical work. The method can potentially be translated to adult stroke victims.

Chapter 5 by Esfahani, McBride, Shafiei, and Obenaus presents a method that can automatically detect traumatic brain injury (TBI) legions in real time from T2 weighted images collected using magnetic resonance imaging (MRI). The method was developed using rodents but can also be applied to human patients. TBI occurs in numerous Americans each year and better methods of imaging TBI legions are needed. This method agreed well with ground-truth data and did not give any false positive results. It is relatively inexpensive to perform.

Chapter 6 by Eberle et al. applies video bioinformatics analysis to data collected with optical coherence topography (OCT), a label free, minimally invasive imaging technique that can be applied to problems involving visualization of brain activity. The basic principles of OCT are first described, followed by methods for basic preprocessing of OCT images. The authors then present data on the application of this method to the study of seizure progression in vivo in a mouse model. 3D data were analyzed using video bioinformatics tools to provide better visualization of changes in the brain during seizures.

Chapter 7 by Guan, Bhanu, Talbot, and Weng reviews prior work dealing with methods for segmentation and detection of human embryonic stem cells (hESC). Five categories of cells can be recognized when hESC are freshly plated. Three approaches for segmenting hESC are reviewed, and all give better than a 90 % true positive rate with the gradient magnitude distribution method being the most robust and the most rapid to perform. A method is also reviewed for detecting one type of cell in freshly plated hESC populations. The unattached single stem cell can be distinguished from other morphological different types of cells. These tools could be useful in the future in studying cell health and the effects of environmental toxicants on stem cells.

Chapter 8 by Weng, Phandthong, and Talbot uses commercially available software to analyze cultured human embryonic stem cells (hESC). The product, CL-Quant by DRVision, was used to segment time-lapse videos of hESC colonies as they were attaching and spreading on a substrate. Two protocols for segmentation were compared. While the professional version created by DR Vision engineers performed better, the protocol created by an undergraduate with only 1 month of experience was in reasonable agreement with the ground-truth. hESC were evaluated during attachment and spreading in the presence of a ROCK inhibitor or blebbistatin. While spreading occurred faster in the treatment groups, cells appeared to be stressed. The method described could provide a rapid method for evaluating cytotoxic effects of chemicals on cells.

Chapter 9 by Lin, Yip, Phandhong, Davis, and Talbot addresses methods that can be used with commercially available software to extract information from time-lapse videos on cell motility, colony growth, reactive oxygen species production, and neural differentiation. Cells were treated and videos of controls and treated cells were compared using the tools created by the users with CL-Quant software. The software tools described in this chapter could have numerous applications in basic research, toxicological studies, or to monitor quality control in stem culture laboratories.

Chapter 10 by Ung and Raikhel reviews the application of video bioinformatics tools to problems in plant cell biology. The chapter discusses the importance of imaging in understanding plant biology and how imaging methods have evolved. The chapter then focuses on the topics of segmentation for region of interest detection, classifying data using machine learning, and tracking dynamic changes in living plant cells. The chapter concludes with a discussion of the future of plant cell imaging and the role that video bioinformatics can play in improving visualization of cell dynamics.

Chapter 11 by Tambo et al. addresses the specific problem of pollen tube growth in plant cells. Pollen tubes function in the sexual reproduction of plants by carrying sperm to the ovary of a plant where fertilization can occur. Pollen tube growth is a complicated process to model as it depends on many different factors such as interaction of proteins and ions. The chapter begins with a discussion of existing pollen tube growth models. This is followed by sections dealing with a comparison of three existing models and the reasons for developing a simple video-based model. The technical approach used to create the video-based model and the experimental results obtained with this model are presented. The method discussed in this chapter uses only those variables that can be seen or measured in video data to track the growth of the tip of a pollen tube. The model was found experimentally to be robust and capable of studying factors that could affect the growth of pollen tubes.

Chapter 12 by Mkrtchyan et al. deals with the development of an automatic image analysis pipeline for use in studying growth and cell division in the plant *Arabidopsis*. To study cell growth in multilayered multicellular plants, a quantitative high-throughput automated pipeline was developed. The pipeline combines cell image registration, segmentation, tracking, and 3D reconstruction of time-lapse volumetric stacks of confocal microscope images. For each step in the pipeline, the authors first describe the motivation for that step followed by the method for performing the step (e.g., registration) on data collected from the shoot meristem of *Arabidopsis*. They then give examples of results for each step in the pipeline. The automated image pipeline is shown to be capable of handling large volumes of live imaging data and capable of generating statistics on cell growth and cell division, which will be valuable in understanding plant developmental biology and morphogenesis.

Chapter 13 by Cabrera et al. describes the use of video bioinformatics tools to analyze growth and development in a multicellular fungus (*Neurospora crassa*). The chapter begins with a discussion of the complex life cycle of this fungus, which has served for many years as a model organism in biological studies, particularly in genetics. It then describes three assays based on fungal phenotypes and how automated analysis has been possible with video bioinformatics tools. These phenotypes include asexual spore size, hyphal compartment size, and hyphal growth rate. The methods presented allow rapid quantification and can be applied to other fungal gene deletion strains or any organism with similar features.

Chapter 14 by Groesser et al. describes how video bioformatics tools can be used to visualize and quantify DNA damage in cells and the kinetics of DNA repair. This method depends on immunolabeling cycling or stationary cells with antibodies to phosphorylated histone gamma-H2AX (which occurs at the sites of double stranded DNA breaks) and the DNA repair protein 53BP1. Repair kinetics can be quantified by applying video bioinformatics tools, such as segmentation which enables co-localization analysis. BioSig software, which is an imaging bioformatics system, has been adapted for use with this problem. The method described in this chapter enables detection of foci in nuclei and determination of areas where foci for H2AX and 53BP1 co-localize, which would be indicative of DNA repair. The method provides unbiased and rapid handling of large datasets and should be applicable to different cells types.

Chapter 15 by Zahedi, On, and Ethell studies the transport of an actin-severing protein, cofilin, in dendritic spines and the effects of cofilin on synapses. Degeneration of synapses is important in neurological diseases, such as Alzheimer's, and cofilin may play a role in this process. A novel optogenetics approach that generates fluorescent cofilin molecules combined with video bioinformatics analysis of high resolution images was used in conjunction with cultured hippocampal neurons. Cofilin activity was modulated by a photoactivatable probe. This chapter demonstrates the application of video bioinformatics tools to a challenging problem requiring high resolution fluorescent images.

Chapter 16 by Merouane, Narayanaswamy, and Roysam describes an open source toolkit (FARSIGHT) that can be used to perform 5-D cell tracking and linked analytics. Modern optical microscopes have evolved to the point that they can collect multiple types of data over time thereby providing 5-D information (3-D space, time, spectra) that needs to be sorted and analyzed. Realistically, such analyses will need to be done using computer software designed for extracting useful information from large complex datasets. This chapter describes both the algorithms and an open source toolkit for analysis of 5-D microscopy data. The toolkit enables preprocessing, cell segmentation, automated tracking, linked visualization of image-derived measurements, edit-based validation of cell movement tracks, and multivariate pattern analysis tools. The FARSIGHT tool kit could be valuable to any biologist who wishes to achieve rapid experimental and discovery cycles.

Chapter 17 by Thakoor, Cruz, and Bhanu reviews video bioinformatics databases and software. The field of video bioinformatics has been helped by available datasets for testing as well as software that has been developed for image analysis. This review chapter gives valuable information on the resources that are available both in terms of image sets that can be used for analysis or testing and software programs that are available for preforming analyses. The authors provide information on 17 databases useful in video bioinformatics analyses. These databases span a broad range of topics such as the cardiac motion challenge dataset, BITE (the brain images of tumors for evaluation database), the Kahn dynamic Proteomics Database, and the Plant Organelles Database. In addition, the authors provide information on eight software packages that are available for image analysis such as MATLAB, ImageJ, FIJI, and CL-Quant. Each of the databases and software programs are briefly introduced. This chapter will be a valuable resource for anyone entering the area of video bioinformatics or anyone in the field who wants broader knowledge of the resources available.

Chapter 18 by Cruz, Bhanu, and Thakoor deals with the use of video to understand human emotions that underlie facial expressions. The chapter discusses how emotion is projected, in particular in facial expressions, how computers can predict emotion from video data, and some of the difficulties computers have in predicting authentic emotions. Also included is a section on publically available datasets of facial expressions with a description of the characteristic of each example, and explanation of the usefulness of such databases in this type of video bioinformatics approach.

Chapter 19 by Feng, Bhanu, and Heraty addresses the problem of identifying moth species from digital images. Traditionally, classifications have been done manually by examining characteristics of the moth body features and visual characteristics of the wings. With over 160,000 of moths on the planet, manual classification can be a slow and laborious task. This chapter describes a method for automatically identifying moth species based on a probabilistic model that infers Semantically Related Visual (SRV) attributes from low-level visual features of moth images. While developed for moth species identification based on wing attributes, the species identification and retrieval system described in this chapter could be used with other insects.

References

1. (2003) Science. Spec Iss Biol Imag 300:76–102, 4 April 2003
2. (2002) Mathematics in imaging. Spec Iss IEEE Signal Process Mag 19(5), Sept 2002
3. Murphy RF, Meijering E, Danuser G (2005) Guest editorial. IEEE Trans Image Process (Spec Iss Mol Cell Bioimag) 14, Sept 2005. Also see IEEE Signal Process Mag, May 2006
4. Swedlow JR, Goldberg I, Brauner E, Sorger PK (2003) Informatics and quantitative analysis in biological imaging. Science 300:100–102
5. (2012) Nat Methods 9(7):676–682
6. Tsien RY (2003) Imagining imaging's future. Nat Rev Mol Cell Biol 4:SS16–SS21

7. Danckaert A et al (2002) Automated recognition of intracellular organelles in confocal microscope images. Traffic 3:66–73
8. Gerlich D, Mattes J, Elis R (2003) Quantitative motion analysis and visualization of cellular structures. Methods 29:3–13
9. Huang K, Murphy RF (2004) From quantitative microscopy to automated image understanding. J Biomed Opt 9:893–912
10. Marx V (2002) Beautiful bioimaging for the eyes of many beholders. Science, July 2002
11. Parvin B, Yang Q, Fontenay G, Barcellos-Hoff MH (2002) BioSig: an imaging bioinformatic system for studying phenomics. IEEE Comput 65–71, July 2002
12. Sabri S, Richelme F, Pierres A, Benoliel A-M, Bongrand P (1997) Interest of image processing in cell biology and immunology. J Immunol Methods 208:1–27
13. Agarwal PK et al (2002) Algorithmic issues in modeling motion. ACM Comput Surv 34
14. Bray D (2001) Cell movements: from molecules to motility. Garland, New York
15. Lackie JM (1986) Cell movement and cell behavior. Allen & Unwin, Boston
16. Murase M (1992) Dynamics of cellular motility. Wiley, New York
17. Pal SK, Bandyopadhyay S, Ray SS Evolutionary computation in bioinformatics: a review. IEEE Trans Syst Man Cyber Part C 36(5):601–615
18. Fogel GB, Bullinger E, Su S-F, Azuaje F (2008) Special section on machine intelligence approaches to systems biology. IEEE Trans Syst Man Cyber 38(1), Feb 2008
19. Cohen J (2004) Bioinformatics—an introduction for computer scientists. ACM Comput Surv 36:122–158
20. Interview with Knuth. http://www-helix.stanford.edu/people/altman/bioinformatics.html
21. http://www.cris.ucr.edu/IGERT/index.php

Chapter 2
Video Bioinformatics Methods for Analyzing Cell Dynamics: A Survey

Nirmalya Ghosh

Abstract Understanding cellular and subcellular interrelations, spatiotemporal dynamic activities, and complex biological processes from quantitative microscopic video is an emerging field of research. Computational tools from established fields like computer vision, pattern recognition, and machine learning have immensely improved quantification at different stages—from image preprocessing and cell segmentation to cellular feature extraction and selection, classification into different phenotypes, and exploration of hidden content-based patterns in bioimaging databases. This book chapter reviews state of the art in all these stages and directs further research with references from the above-established fields, including key thrust areas like quantitative cell tracking, activity analysis, and cellular video summarization—for enhanced data mining and video bioinformatics.

2.1 Introduction

In the postgenomic era of computational biology, automatic and objective analysis of biomolecular, cellular, and proteomic activities is at the center stage of current bioinformatics research. Microscopes, the prime instrument for observing the cell and molecular world, have treaded a long path of revolution. Widefield microscopy with deconvolution, confocal scanning microscopy, and scanning disk confocal microscopy have facilitated observing cells and their activities and capturing static image and video data for precise and automated analysis, both in 2D and 3D [108, 131]—even closing the gap between live cell imaging and atomic resolution structures in cryo-electron tomography (3–8 nm) [127]. Green fluorescent protein (GFP) markers in the antibody have acted as illuminant in the molecular world to visualize cell activities and brought a new era in cell research [36]. Sometimes

N. Ghosh (✉)
Department of Pediatrics, School of Medicine,
Loma Linda University, Loma Linda, CA, USA
e-mail: nirmalyaghosh11@gmail.com

© Springer International Publishing Switzerland 2015
B. Bhanu and P. Talbot (eds.), *Video Bioinformatics*,
Computational Biology 22, DOI 10.1007/978-3-319-23724-4_2

bright field defocused and/or stereo [131] microscopic imaging are utilized to analyze multiple cells at diverse depth with advantages of low phototoxicity and minimal sample preparation, though lower contrast poses difficulty in segmentation and tracking [80].

The challenge has now shifted from automatic capturing of the slow-varying cell activity in digital media to automated analysis of this vast amount of digital data being stored every day with minimum human interaction [17, 64, 96]. Even an expert cell biologist takes hours to preprocess the microscopic images or videos, analyze numerically the structure of cells, recognize them, recognize the cell activity, and come to a biological conclusion. Automated computational methods are absolute necessities to avoid human fatigue-related errors, to perform intensive data mining beyond human tractability and to make results objective and statistically comparable across international studies [22, 26]. Established techniques in computer vision, pattern recognition, and machine learning fields often come handy to rescue from this tremendous information boom in biology in the recent years [28, 36].

Video bioinformatics is a recently burgeoning field of computational biology that analyzes biological video and image data to automatically detect, quantify, and monitor complex biological phenomena—at molecular, cellular, and tissue levels, internal activities and their interactions, in healthy as well as in injured conditions, and with/without drugs and antibodies injected. A complex end-to-end video bioinformatics procedure generally requires multiple major steps as follows. (1) At first, reduction of computational complexity requires detecting video shots and extracting key frames based on biological activities. This makes established image processing techniques effectively applicable to the static key frames. (2) Images are then enhanced by filtering out noise. (3) Biological regions of interests (ROI) are automatically segmented out and aligned to models if necessary. (4) Different morphological, signal intensity, contrast, shape, and texture features are extracted for different biological objects. (5) Based on their discriminative powers, optimal sets of features are selected to recognize entities. (6) Segmented objects are then classified as different biological entities. (7) Multiple consecutive static images (key frames) are considered again to track entities over space and time and to identify biological activities for an individual entity. (8) Interactions between different entities are then automatically monitored using advanced video data mining techniques. (9) Image and video-based information is then stored in a structured and distributed database for availability and query over the Internet. (10) Machine learning techniques are applied to improve all previous procedures including content-based retrieval.

A big proportion of research is devoted to this marriage of quantitative microscopy, computer vision, and machine learning. A number of research groups have concentrated on processing static images of the cells and classifying them using pattern recognition techniques [11, 17, 20, 26, 31, 75, 83, 94, 106]. Major steps and associated tools that are involved in such complete high-content screening and analysis pipeline have been summarized in recent publications [36, 98, 105, 115, 120]. They derived numerical features from the 2D images and used feature-based classification of the biological molecules. Relatively less effort has been exerted for dynamics of the cells and recognizing the cell activity. Only a small body of research has studied cell dynamics, migration, tracking, bacterial movement, and biological events over the 2D/3D microscopic videos [18, 39, 118, 123, 137, 143] but often lack in automated analysis of such dynamics. This chapter provides reviews of the computational tools for above ten steps that have been already applied in biology or demonstrated potential in mainstream computer vision and pattern recognition (CVPR) field of research for future biological applications. Broad conceptual diagram of a typical video bioinformatics system is summarized in Fig. 2.1. Instead of the mainstream biology, this review chapter is from the perspective of the computational methods applicable in biology.

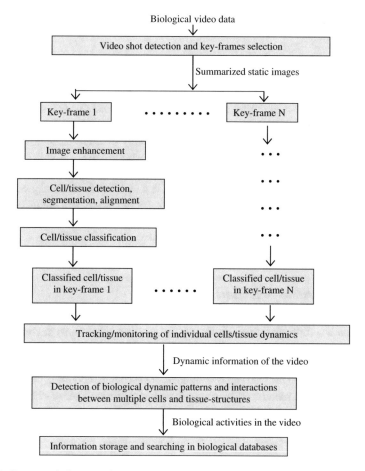

Fig. 2.1 Conceptual diagram of a typical video bioinformatics system

2.2 Salient Video Activity: Shot Detection and Key Frame Extraction

Cell activities are often very slow and corresponding videos often do not contain enough changes in visual information over a long sequence of frames. Hence to reduce computational complexity, images are sometimes captured periodically— i.e., low frames-per-second (fps) video [64, 118, 143] or periodically sampled from a high fps video [31, 119]. These methods are naïve counterpart of key frame selection that might ignore some salient quick and transient cell transformation information.

2.2.1 Shot Detection

Shot detection and key frame selection are two often-used techniques in video processing to reduce computational complexity without losing details, and more contextual in cell activity videos with in general slow dynamics with few quick transients. With low-cost digital storage, taking high-speed (30 fps) cell videos and detecting shots and key frames to trace salient transient cell activities is more practical. Although shot detection is now a relatively matured domain in computer vision, it is surprisingly unused by cell biology community. As cell videos often have fewer types of cells present in the same videos, established shot detection techniques from histograms might work well, e.g., global dissimilarity or temporal changes in different pixel-level features—color components [38], intensity [1, 144], luminance [112] and their distributions and combinations [1, 16], or regional features and likelihood ratios [29, 136] across consecutive video frames.

2.2.2 Key Frame Selection

Key frames are representative frames of a particular video shot, analyzing which one can safely summarize about the frames they represent. A set of key frames are generally selected such that these frames contain enough visual information and its change (dynamics) over the video sequence. The frames acquired periodically or heuristically [31] and analyzed by the cell biologists in the state-of-the-art systems are actually a naïve substitute of these key frames. Key frame selection is also a well-established domain in computer vision. Other than clustering-based techniques [139], most of the keyframing methods attempt to capture the temporal information flow with varying computational complexity—starting from simple first and last frame selection [85], periodic selection [113], after constant amount of change in visual content [19], by minimization of representational error (distortion) in the feature space [47], by iterative positioning of break points (like sub-shots) and key frames (one in each sub-shot) to minimize distortion [61], and by minima in motion feature trend [133]. Unlike these previous methods, sometimes psychoanalytical perception models might be used to automatically decide the number of key frames to be selected depending on change in visual content from the feature trends [40]. From cell video point of view, specifically with morphological transformation (morphogenesis), tracking geometric structures and keyframing based on salient differences [141] might be adopted. Based on complexity and application, similar methods can be envisaged in cellular videos to decide on which frames are to be analyzed to reduce computational burden.

Once the shots and corresponding key frames are decided, cell videos can be analyzed in the same way as single static bioimages as discussed in the following sections. Even for motion-based tracking the cells over the frames, key frames may reduce the computational burden by few orders, specifically for slowly changing cell videos.

2.3 Image Processing and Biological Object Detection

After denoising and preprocessing of static images (from keyframing), cellular and tissue region of interest (ROI) extraction mainly comprises of three stages: detection, segmentation, and alignment (sometimes called "registration"). All these stages are often interrelated, mutually supplementary in nature, and even sometimes inseparable, as depicted in Fig. 2.2. For the simplicity of understanding, they would be dealt separately in following subsections. All these stages directly depend on image features and prior biological models.

Fig. 2.2 Interdependency of different low and midlevel image processing modules in bioinformatics

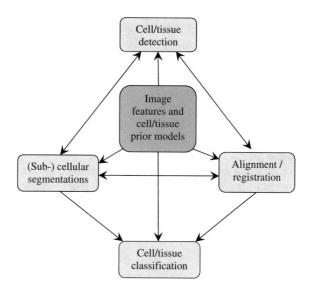

2.3.1 Preprocessing and Noise Reduction

Even after following recommended techniques for sample preparation and data acquisition [48, 108], noise in bioimages is ubiquitous and almost always requires preprocessing and denoising. Though few research works adopt simulated (flat) background without any explicit noise filtering [118], this is an unrealistic assumption for in vivo cell images and videos. During the conversion of patterns of light energy into electrical patterns in the recording device (e.g., CCD camera or photomultiplier tube) random noise is introduced [39]. Specifically because of the high-frequency noise, biological cell shapes loose sharpness and affect segmentation and overall analysis.

For any practical automated analysis of bioimaging data, reduction of random and speckle noises [11, 28, 115, 138] and variations in illumination (e.g., GFP) [123] are the first steps. Usual low-pass filters, besides reducing the high-frequency noise, also reduce the sharpness of the edge and contour features (as they are also high-frequency components of the image). Nonlinear filters (e.g., median filters) often resolve this problem [109, 114]. Sometimes sophisticated anisotropic diffusion filters are used that preserve local characteristics and image qualities [39]. The rational of this method is that image areas containing structure and strong contrast between edges will have a higher variance than areas containing noise only. Hence diffusion algorithms remove noise from an image by modifying the image via partial differential equation. Homogeneous regions are handled by diffusion equation (heat equation) equivalent to Gaussian linear filters with varying kernel size. While anisotropic diffusion filter controls the diffusion process by an "edge-stopping function" that depends on local image features, e.g., magnitude of the edge gradient.

Speckle noise generally comes from small intracell structures that can be reduced by model-based filtering, e.g., modeling cells as ellipse and removing outliers not fitting the model [138]. Sometimes nonlinear least-square-designed FIR filters are used to improve contrast between cell objects and fluid background as they are immersed and then histogram-based dynamic threshold is applied to deal with illumination variation due to fluorescence decay [49]. This contrast improvement might be more effective if some fluorescence decay model is applied [123]. A series of filters often assists in the overall preprocessing—e.g., histogram equalization [109] or auto-density filter increases the contrast, morphological filters (in sequence—dilation, histogram-based intensity threshold and erosion) reduce model-based outliers and finally median filter removes salt-&-pepper (random) noise [31].

In recent reviews [108, 127] of different preprocessing steps, potential methods and pitfalls are provided where starting from selection of particular microscope and acquisition parameters, preprocessing steps like flat field correction and background subtraction, intensity normalization, different Gaussian filtering techniques, and deconvolution strategies are discussed.

2.3.2   Segmentation

After preprocessing, generally explicit segmentation and feature extraction are required for classification. Rare exceptions are the data with no intercell occlusion [31], or where chromosome profiles are extracted using dominant points and variants [104]. Most of current quantitative microscopy data are from cells that are immersed in an in vitro biochemical solutions (beneath a coverslip) and imaged individually [31, 115] or in a nonoverlapping (i.e., without occlusion) situation [118]. Segmentation might be redundant for bioimages in such controlled environment [13]. With the assumption of small roughly uniform background, manual polygonal cropping and dynamic thresholds work well to identify cells [13]. For in vivo data with different cell types and intercell occlusion, these methods are too restrictive and explicit automated segmentation is an absolute necessity. Hence later researchers from Carnegie Mellon University (CMU) have adopted a seeded watershed algorithm for segmentation in 3D microscopic data where seed for each nucleus is created by filtering DNA channel output from the confocal scanning microscope and 93 % segmentation accuracy is reported [49].

An early review paper on interest of image processing in cell biology and immunology [109] proposes three ways of segmenting cells (illuminated by GFP): histogram-based bimodal segmentation, background subtraction, and (heuristic) threshold-based boundary following. Generating outlines of the biological structures, i.e., image segmentation is a challenging task and influences the subsequent analysis [53]. This work on analyzing anatomical tissues (conceptually quite similar to the cell and molecular images) proposes one 2D color image segmentation method. In the grayscale image, segmentation involves distribution of seed points in a microscopic image and generating a Voronoi diagram for these seeds. Gradually, this Voronoi diagram and its associated Delaunay triangulation are modified according to the intensity homogeneity. For the color images, this region-based approach is extended with sequential subdivisions of the bioimage, classifying the subdivisions for foreground (the cell or tissue), or background or both, until each subdivision is uniquely classified. Voronoi statistics (including HSV mean color intensities and their variances) of each subdivision are utilized to classify them. Seed points can be initialized manually or randomly. Then a continuous boundary of the cell or tissue is obtained by fitting splines. Although this procedure is tested for anatomical tissues, like segmenting the lungs, the procedure is generic enough for cell and molecular bioimages and can be extended in 3D using 3D Voronoi diagrams, of course with increased time complexity.

Another review paper [39] addresses a rather innovative way of segmentation in cell images. The method applies multiple levels of thresholds to form a confinement tree that systemizes the knowledge that at what level of threshold, which cells are merged to a single object. Then morphological filtering reconstructs grayscale images in various levels. Thus the method is adaptable to the analysis needs. They also address another edge-based segmentation operating on nonmaximum suppression algorithm and refining the contour by active contours (snakes) with energy function associated with curves. Sometimes morphological operations and regional intensity gradients assist in segmentation. In an application of immunohistochemical stain counting for oesophageal cancer detection [60], the region of interest is first manually cropped, color image is converted to grayscale image, contrast is enhanced by histogram equalization, and morphological TopHat (and other) filtering) is performed for initial segmentation. Then watershed algorithm segment out the nuclei and gradient transform-based edge detection is performed. After two-stage watershed segmentation nuclei are detected.

In pioneering research of Euro-BioImaging group (http://www.eurobioimaging. eu/) in clinical wound-healing video, distinct textural difference between the wound and normal skin is mentioned [67, 77, 107, 118, 147], but for wound segmentation, histogram equalization (to improve contrast), edge detection, and modal threshold are utilized. It is rather surprising that no texture feature is utilized. In another video-based bacterial activity work [118], individual cells are segmented by seed-based region growing algorithm. But seed initialization process is not clear. And in presence of occlusion, which is not considered in this work, region growing procedure may perform poorly. In such cell videos, motion-based segmentation from tracking across frames [140] might help, specifically when the background (however complex it is) does not change too fast. One recent work on automated wound-healing quantification from time-lapsed cell motility video, cascaded SVM-based initial segmentation, and graph cut-based outlier rejection are applied on basic image features [143].

Cell population-based studies (in contrast to study on few cells in an image) sometimes provide more statistical power—specifically for phenotypic changes by drugs, compounds, or RNAi [64]. Centerline, seeded watershed, and level set approaches are common in such applications. Except for such rare cases [50, 64], multicell images are segmented into individual cells before any phenotyping (classification). Segmentation in cell images in presence of speckle noise (intracellular structures, like nucleus, mitochondria, etc.) are dealt systematically by the Lawrence Berkley National Laboratory (LBNL) research group [11] by model-based approach. In multicell images, they model the cells and intracellular structures as ellipses and mathematically demonstrate that, removing the speckle noise and interpolating the cell images accordingly can be done by finding solution to a Laplace equation. They call it "harmonic cut". The cells touching one another are segregated by regularized centroid transform, where normal vectors generated from cell boundaries are clustered to delineate touching cells. This sophisticated method is a generic up to some extent as long as cells can be modeled as ellipses (with smooth quadratic splines). Similar approach has been utilized in model-based detection of cell boundaries and then seeded watershed separation of touching (but non-occluding) cells in the same image [115]. But in many cases, like data used by CMU [13], cells are of irregular shapes. Proper extension of the harmonic cut and regularized centroid transform method for these irregularities is yet to be tested. Recently, "tribes"-based global genetic algorithm is applied to segment cells with partial occlusion by part configuration and learning recurring patterns of specific geometric, topological, and appearance priors in a single type of cell in histology and microscopic images [91]. Different cell shapes (without occlusion) are segmented out from defocused image stack of embryonic kidney cells (HEK 293T), where the best representative slice is first selected by a nonparametric information maximization of a Kolmogorov complexity measure, then active contours are initialized and expanded for level set segmentation [80]. A nice level set and active snake-based multilevel approach segment out core and membrane of cells from uncontrolled background [86]. Seeded watershed algorithm and level set approaches could successfully segment out Drosophila cells and nuclei and then tracked across time-lapsed frames to detect cell divisions and migration with and without drugs [64]. Interested reviewers are encouraged to read CVPR reviews [30] on fusion of different features like color, texture, motion, and shape and unified approach of level set segmentation for potential applications in cell images and videos.

Cellular and subcellular segmentation and colocalization in fluorescence microscopic images are still very relevant research areas [106]. Recently, Fuzzy C-means clustering is found better than baseline hard C-means clustering in segmenting single pap smear cells as well as separating their nuclei and cytoplasms for classification and abnormality detection [23]. In another work, for model-based segmentation of more-confluent (occluded) cell nuclei, predefined patterns in attributed graphs of connected Sobel edge primitives (in different orientations: top, bottom, right, left) are iteratively searched and reassigned as needed to localize nucleus boundaries and then region growing is performed to separate occluded nuclei [4]. Sometimes 2D segmentation results can enhance 3D segmentation from stacks of microscopic images of neuronal nuclei—and also correct some of the 2D under and over segmentation errors by connectivity and centroid clustering [59]. 3D watershed segmentation is the baseline for comparison in this work. In a neuron tracing research, morphological features at multiple levels and in different neuronal parts can successfully segment the entire neuronal cells [76]. Recent review papers [115, 127] critically discuss many such segmentation techniques along with associated advantages and disadvantages.

Texture-based segmentation is one area where future bioimaging research might gain momentum. Few nice reviews [6, 52, 103, 124, 145] summarize well-established texture descriptors that are applied in CVPR applications over decades, including texture-based feature space smoothing that preserves salient edges with supervised [126] or unsupervised methods [33], split-and-merge segmentation by facet models, and region adjacency graphs [72] using multiple resolution [99] texture information measures [100, 101], or region growing segmentation from gradients of textures [45] utilized as inter- and intraclass dissimilarity [130] for random walk [102, 103] or quadtree-based methods [121]—to name a few. Cellular and molecular images have distinct textures for different species and this can immensely enhance segmentation.

Sometimes pixel-, local-, or object-level relations (based in morphology, color, proximity, local texture, shape, and motion) can be represented graphically with objects as nodes and weighted links as strength of interrelations [116]. In such cases graph matching and partitioning methods like normalized graph cut [129] can partition highly connected regions as clusters to segment image objects. Except for few exceptions [4, 143], graph-based segmentation methods are yet to be applied in cell images to their full potential.

2.3.3 Object Alignment

Though not very common, image registration—i.e., aligning the object with a template or model is sometimes required for better feature extraction, quantification, and analysis. This is performed either before the cell segmentation [18, 105] or after it [115]. Although segmentation and registration are dealt separately in most works, they are quite interrelated and mutually cooperative (see Fig. 2.2). For examples, in atlas-based segmentation methods (very common in medical imaging) data model alignment is a prior requirement, while segmented structures assist in landmark-based alignment of test object with the model. Specifically, in shape-based methods segmentation and registration are so similar that a new term "regmentation" is coined in medical image analysis [37]. Due to high variability of cellular and subcellular objects, object alignment is not always possible in a reliable manner and hence not informative for automated analysis. For protein structure alignment—where similar structures and partial resemblance are of importance—optimal paths and distances between atoms are successfully utilized in a graph-matching paradigm [117]. In database search, image registration is required for developing atlas or representative model from similar cellular datasets [96, 98] or for comparing with manually annotated reference images for local features (to overcome variations in sample preparation) before a multireference graph cut [22] or level set [21] does the nuclear segmentation. Registration might help in detecting eccentricity of a test data from the model and thus estimating abnormality for further analysis. Classic image registration algorithms in CVPR [148] or in medical imaging [93] might have immense potential in cellular image analysis [115], specifically when close-loop cooperation between segmentation and registration [37] are adopted in a deformable model approach [44].

2.4   Feature Extraction

A large proportion of the quantitative microscopic analysis research is done with "static" images of the cells and molecules. In cell classification, static features dominate, sometimes due to slow biological processes and sometimes to compromise with the computational burden. Three basic steps in static image analysis are (1) feature extraction, (2) feature selection, and (3) object (cell, biomolecules) classification. These steps are interdependent. Human perception of the cell images provides idea on type of classification strategy expected to perform better. The classifier type influences the selection of features, which in turn guides the image processing strategies. Hence the steps are quite overlapping. This chapter attempts to address these steps individually as far as possible for better understanding.

Image features that are signatures of the object of interest are extracted after the segmented image is processed through different types of image (morphological, textural, intensity-based) operators and filters. Sometimes image processing also covers the occlusion-handling strategy by interpolation or extrapolation of the cells. Image processing (just like corresponding features) can be classified into:

- Morphological (binary silhouette-based)
- Cell regional (color or intensity-based)
- Differential (or contrast edge-based) and Structural (or shape-based)
- Textural (structural periodicity-based)

Relation between these different types of features is summarized in Fig. 2.3. Current section describes the example of these image processing types applied in cell-imaging community, followed by some of the classical CVPR examples to inspire future research.

Fig. 2.3 Different types of extraction, recombination, and selection methods for static image features and interrelations between them

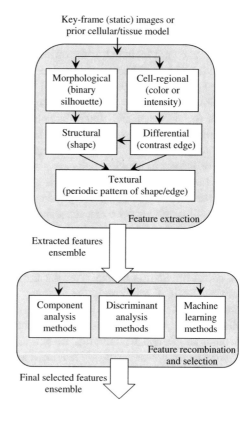

2.4.1 Morphological Features

Morphological image processing techniques basically consider the binary silhouette (obtained from segmentation) of the cells or biological molecules and find different geometric properties [114]. These are probably the lowest level image processing techniques, yet sometimes very useful—especially as different types of cells and biomolecules generally have significantly different outer shapes (morphology).

After the sequence of the human genome is determined, next task is to determine genomic functionality. Proteins encoded by novel human cDNA clones cause morphological changes and/or protein localization at the cellular level which result in various cellular forms [122]. After histogram equalizations, first-order principle component analysis (PCA) of manually segmented sub-images is used as models. They consider 16-bit grayscale images of subcellular compartments like endoplasmic reticulum, Golgi complex, plasma membrane, mitochondrion, nucleolu, peroxisome, etc. Then morphological convolution of the model and original images are done to get the local maxima that are taken as the focal points.

CMU researchers have used an extensive morphological image processing and selected number of features [115]. Extensive list can be found in [82]. Some of the salient ones are: (1) number of subcellular objects per cell, (2) Euler number of the cell (i.e., number of objects minus number of holes per cell), (3) average pixel size, (4) average distance of objects to the center of fluorescence, (5) fraction of fluorescence not included in the objects, (6) fraction of the area of the convex hull not in the object, (7) eccentricity of the hull, (8) average length of the skeletons (or medial axis; found by morphological iterative thinning), (9) average ratio of the skeleton length to the area of the convex hull, (10) average fraction of the object pixels (or fluorescence) within skeleton, and (11) ratio of branch points to the skeleton length. Most of these 2D features are also extended for 3D scanning microscopic data [24, 115]. For images with multiple (same) cells, features like ratio of largest to smallest cells are also considered [50].

In the research of Euro-BioImaging group [118] simple morphological features are extracted for solving "correspondence problem" to track the bacterial cells in the cell motility videos for event detection. The features extracted for each segmented cells include spatial position in the frame (i.e., the centroid of the cell), its area, its length, and width (determined by PCA, along major and minor axes, respectively) and its orientation in the space. These morphological features are used to track and to detect orientation change over the frame sequence for bacterial "tumbling" and other behavioral response to the drugs applied. In a medical tissue diagnostic work from wound-healing video [107], they apply morphological cleaning of the wound area in the image, and compute application-specific morphological features like wound length and wound-area-per-unit-length. From the dynamic variation of the "wound-area-per-unit-length" feature, they decide the healing (or worsening) of the wound with time as drug is applied periodically. LBNL researchers utilize detailed morphometric analysis of TCGA glioblastoma multiforme for tumor categorization from hematoxylin and eosin (H&E) stained tissue where they compute cellularity (density of cells), nuclear size, and morphological differences of nuclei [20].

In recent reviews [25, 105] different morphological operations in cell image analysis—from segmentation to characterization to abnormal cell identification—are nicely summarized where different cell morphological features are discussed including circularity, rectangularity, eccentricity (ratio of minor to major axis length), morphological texture from gray-level co-occurrence matrix: energy, uniformity, entropy, smoothness. These features are also extended to 3D morphology and deformable models [25].

2.4.2   Color and Intensity Features

Intensity value for grayscale cell images and color component values in different color spaces like red-green-blue (RGB), hue-saturation-value (HSV), or other application-specific combinations of them [114] sometimes have the unique region-based features to segment and classify cell and molecular objects in biochemical fluid—especially when salient portions of the cells are illuminated by GFP tags. Color decomposition might also reduce computational load, and might assist thresholding, refinement, and normalization of input image to the base image [20]. Grayscale intensity-based moment of inertia is successfully applied for chromosome slice estimation and profile extraction [104]. According to this work, shape profile is the moment of inertia of the normalized gray value distribution in each slice relative to the tangent to the longitudinal axis at the subdivision point of the slice. Similarly, different grayscale-based features such as brightness, histogram, and amplitude of a region assist in genomic classifications [122].

The prognosis of esophageal cancer patients is related to the portion of MIB-1 positively stained tumor nuclei. An image analysis system is developed on LEICA Image Processing and Analysis System to reduce the subjective, tedious, and inaccurate manual counting of nuclei staining [60]. It can analyze in 15 min. Proliferative activity of tumor is a useful parameter in understanding the behavior of tumor. Correlation between the proliferation activity and overall prognosis has been observed in some tumor. MIB-1 score by immunohistochemical method and stain counting is one affective process. Brown nuclear stain is regarded as cancerous cell and blue nuclear stain as normal cell. Intensity-based classification is performed in RGB space: brown nuclei by red-component-higher-than-blue-one and blue nuclei by the vice versa. Automated systems might suffer from variations in illumination and focusing problems, mainly due to dynamic nature of the protein molecules. Heuristic application-specific filtering [60], fluorescence decay models [123], or illumination-invariant color component-like saturation [53] might overcome such problems.

Haematococcus pluvialis (Chlorophyte) produces carotenoids that are utilized as color pigments and analyzing agents for different degenerative diseases in humans. Haematococcus has two distinct phases in its life cycle: green flagellated motile phase and nonmotile nonflagellated cyst phase formed due to stress conditions. Automated evaluation of red component of imaged cells can give estimate of the carotenoid content without disrupting the cell wall. One work [56] adopts grayscale conversion, histogram equalization, and edge-based segmentation for ROI extraction. Then cell pigment percentage change is detected from hue component by three-layered artificial neural network (ANN) classifier that classifies into two classes: Chlorophyll and Carotenoid, for medical diagnostics.

Another work on semiautomated color segmentation method for anatomical tissue [53], considers mean a variance of color in different voronoi cells dividing the tissues in segmentation and classification of lungs like organs. They have converted RGB tissue images into HSV space and reported that saturation plays important role in distinguishing between biological tissue and cell structures. This coincides with the well-established fact in CVPR that saturation is relatively invariant to the illumination changes, and might be even better than explicit modeling of temporal decay of fluorescence strength (called "leaching effect" of GFP) [123]. Sometimes, to separate out an actual fluorescent tag from noise in low signal-to-noise ratio (SNR) data, cell spots are detected by intensity-based local maxima detection where a "spottiness"-value is computed to characterize the similarity of the intensity signal in the neighborhood of a local maximum with respect to the intensity signal of a theoretical spot [123]. This theoretical spot neighborhood has been modeled using the Gaussian point spread function. Gaussian filtering and interpolation of intensity features have been extensively used in bioimaging [11].

2.4.3 Edge and Shape Features

Edge and shape features are comparatively higher level features than the last two, as they have more uniqueness for object recognition (see Fig. 2.3). Naturally, in cell and biological specimen classification and analysis, edge and shape based features play significant role. Edges are the convolution output of the images from contrast differential operators (number of dimensions same as the data), e.g., Sobel, Roberts, Prewitt, and Canny edge detectors [114]. The edges are connected by boundary-following algorithms to get the contour of the cells/objects in 2D or 3D. These contours are low-level representation of shapes for cell classification.

CMU research group extracts number of edge/shape features from differential operators [82], both in 2D and 3D domain. Salient ones are: (1) fraction of pixels distributed along the edges, (2) measures of magnitude and directional homogeneity of the edges, (3) different Zernike moment features (computed by convolving with Zernike polynomials) to find similarity in shape between the cells and corresponding polynomial. They utilize all these features directly in ANN or other classifier module without trying to develop any shape models. Probably a middle level shape model can improve the classification, as is the case for number of computer vision applications [114].

In an early work [109], shape features are utilized to study human neutrophils exposed to chemotactic stimuli, to describe cell polarization and orientation and to identify chemotactic abnormalities in cells from heavily burnt patients. Information relevant to the mechanisms of adhesive interaction is extracted from the distribution of intercellular distances in cell–cell contact areas. This contact area estimation allows conceptual discrimination between "actual contact" (i.e., with intermembrane distance compatible with molecular interactions) and "apparent contact" (i.e., apparent membrane apposition with intermembrane distance of 50–100 nm). LBNL researchers estimate parameters of the elliptical model of individual cells as shape features [138] and extend their harmonic cut method for iterative tensor voting to refine ill-defined curvilinear structures and for perceptual regrouping of 3D boundaries to separate out touching cells [71]. Sometimes cell shapes are utilized indirectly to classify [31], where the hidden layer of a modular neural network (MNN) might compute the shape features internally, takes into account the shapes of the cell at different scales, and maps directly to different cell classes in the output layer of the MNN. Another work [104] utilizes shape or chromosome boundary and contour curvature to define the singularities (called dominant points) in the chromosome pattern. Longitudinal axes of the chromosome are found by fitting quadratic splines of the distribution of these dominant points and their variants. These axes act as the backbones of the grayscale intensity-based slice determination for extracting chromosome profiles.

In CVPR research shape and boundary-based features are one of the most successful ones for decades [3, 8]. These methods are nicely reviewed in [146] and can be broadly classified under four categories as follows. (1) Scalar boundary transformation techniques: for example, tangents represented as parametric (turning) function of arc lengths; shape centroid methods with polygonal approximation; radial distances [70]; Fourier (frequency) domain features or bending energy of the boundary; circular autoregressive shape models [65]; and central distance of the boundary from fixed length arc placed at different boundary locations. (2) Spatial boundary transformation techniques: for example, multilayered chain code for shapes at different resolution; syntactical coding of strings of primitive shape features; split-&-merge spline approximation with minimum error [27]; hierarchical scale-space representation from multiple-width Gaussian filters [5, 7]; and boundary decomposition by template contour template matching [68]. (3) Scalar global transformation techniques: for example, multiple order 2D or generalized polynomial moments of the silhouette; different shape matrices and vectors from polar raster, concentric circles or maximum shape radius; and granulometries, morphological covariance, geometric correlations, residuals [73]. (4) Spatial global transformation techniques: for example, medial axis, and r-symmetric axis transformations; and shape decomposition based on convex–concave boundary points and fuzzy likelihood. Among these, only a few methods like Fourier shape descriptors [115], spline approximation by iterative refinement of curvilinear structure to separate touching cells [71], elliptical cell shape models [11], Zernike polynomial moments for cellular matching [28], and shape decomposition for neuron tracing [76] are utilized in bioimaging research. Accuracy of these results needs to be evaluated more critically in classification and image retrieval scenarios [3]. Scale-space shape models and features [5, 7, 8] from cellular and subcellular images might improve classification in multicell bioimaging data. One recent work [106] reports a user-friendly freely available software for colocalization in near real time (<1 min for 2D, <5 min for 3D) by segmentation and quantification of subcellular shapes (Squass).

2.4.4 Texture Features

Shape features distinguish objects or cells from what is seen from the outside. Textures are the features of the cells as seen from inside. In cell images, different cells and biomolecules generally have distinct textures (in 2D patch or 3D surfaces) compared to the biochemical fluid (in vivo) or solution (in vitro) they are floating in. Similar is the case for anatomical tissues [107]. These textures are actually periodicity of similar patterns in visual spectrum and are also affected by bio-physical and biochemical properties like viscosity, smoothness, fluorescence absorption, diffuseability, etc. Texture often characterizes the cell or solution when other surrounding conditions remain the same. Hence, cell-image analyzers also apply textures as primary features for cell classification [28, 122] and cell video understanding [118]. Among several texture descriptors utilized by CMU researchers for subcellular localization [28, 82], the key ones are:

- Haralick texture features [6]: These are computed as gray-level co-occurrence matrix (might be extended to color co-occurrence matrix for each components) and then averaged for rotational and translational invariance. Intrinsic statistics including angular second moment, contrast, correlation, sum of squares, inverse difference moment, sum average, sum variance, sum entropy, entropy, difference variance, difference entropy, and information measures are often extracted as features.
- Gabor wavelet texture features: Spatio-intensity periodicity is extracted using Gabor kernel with different scales and orientation. Mean and standard deviation at different abstraction levels are considered as features. Non-orthogonal Gabor wavelets can capture the derivative information of the images.
- Daubechies four wavelet textures features: Cell images are decomposed up to level 10. The average energies of the three high-frequency images at each level are utilized as features. Scales and orientations provide textural fineness and relative arrangements.

Besides above, Low's textures and 15-element feature vector describing symmetries [114] might be utilized in subcellular localization. Recent bioimage analysis research starts to look back on some of the established texture descriptors [52, 124, 145] including Haralick's texture descriptor [6], local binary patterns [87], co-occurrence matrix-based grayscale textural features [122], and learning-based local binary patterns [46] in analysis and classification of 2D-Hela databases and recognition of abnormal smear cells in pap smear medical databases. Multicellular textures are found to be excellent descriptors in monitoring wound-healing and cell-scattering assays in differential interference contrast (DIC) images [143]. There are inherent differences in tissue and cell textures, and many more such research efforts in bioimage analysis are expected in the near future.

2.5   Feature Recombination and Selection

Human vision can recognize different biological cells and their activities in bioimages and videos relatively easily due to complex vision perception experience cycles. But that is neither well understood nor yet implementable in computer programs. Human brain automatically selects best features and their different combinations to analyze the data effortlessly. Automated systems can at the best extract a very large number of low-level image features with the hope that no valuable information is lost, sometimes without knowing the actual usefulness of those features to classify (cells) and recognize (their activities). Cellular and biomolecular images/videos often capture irregular structures and unknown inter-dependent dynamics between them. To analyze them often several features are extracted [82, 115]. But more features mean exponential increase in computational time. This often leads to overlearned complex model that is good only for the seen (training) data. Minimum description length (MDL) principle in machine learning [79] suggests rather simple and generic representation that is valid for unseen (test) data also. Besides relevance in MDL, feature selection also leads to efficient learning of classifiers and data mining in growing complex databases [58]. Feature extraction itself is a dimensional reduction to avoid working with every pixels of the high-resolution bioimage. Selection of salient features based on their distinguishing power is the next step to avoid dimensional explosion [58, 81, 128, 134], sometimes followed by generating better (often complex) features by recombining simple features so that the second-level features fit the application better. Integrating feature from different domains—color, texture, motion, and shape—is gaining importance in recent image informatics [30, 52]. CMU computational bioimaging group is among the very few in bioimaging research to address this major issue in very systemic way [50, 115]. In cell classification domain, classifiers are sometimes limited by underdetermined classification boundaries due to limited number of available cell images in comparison to number of features considered. One solution is feature reduction either by feature recombination or by feature selection utilizing one or combination of algorithms, as summarized in Fig. 2.3 and exemplified as follows.

2.5.1   Component Analysis Methods

Principal Component Analysis (PCA) considers salient eigenvectors of the feature covariance matrix (in general much fewer than number of features) based on corresponding high eigenvalues (i.e., stronger basis vectors). The strongest eigenvectors (above a threshold) often define the linear transformation matrix [122]. This transformation provides weighted linear combinations of the original (sometimes normalized) features in least square error sense to fit the actual feature-set variation—but in a much lower dimensional feature space such that the classes are well separable (with less overlaps) [105].

Nonlinear PCA (NLPCA) is PCA, but with nonlinear transformation and combinations of the original features. From infinite possibilities to get such nonlinear transformations, one way is to learn it from a symmetric neural network [14, 35]. At the trained state, the first layer of the ANN structure converts the input original features to a linear combination and the second layer transforms them in a nonlinear way. Then they are inversely transformed back (first nonlinearly, and then linearly) to the output features, which are identical (or very close) to the input original features. Second layer outputs are the NLPCA-recombined features.

Kernel PCA (KPCA) adopts a nonlinear kernel function—like polynomial function, multilayer perceptron (MLP), radial basis function (RBF) or any other nonlinear function that first transforms the original feature space to a very high-dimensional feature space. Thus in a way, KPCA is feature extractor as well. Then linear PCA reduces the huge ensemble of features to very few recombined features compared to the original input features.

Independent Component Analysis (ICA) makes use of the fact that less the dependency among individual features for the acquired dataset, more mutually cooperative and precise the feature set is to describe the (biological) features of the (cell) image [105]. Criteria such as non-Gaussian nature are utilized to maximize the independence among recombined features (sometimes with both linear and nonlinear transformations) to cover larger area in potentially infinite dimensional feature space.

2.5.2   Discriminant Analysis Methods

Classification or Decision Trees (DT) is formed with individual original feature where the effective classification power of each feature is measured by entropy-based information gain and penalized for too much fragmenting of the data by a split information feature as defined by C4.5 algorithm [79]. Features with higher information gain separates different classes better and hence might be selected for classification.

Fractal Dimensionality Reduction (FDR) works on the principle that few features are often redundant because same information is shared by multiple features, while few other features might be intrinsic because that data points are cohesive and better classified in the corresponding space. The fractal dimensionality of the data set, often represented by correlation fractal dimension, describes self-similarity of the data points and is a good approximation of the intrinsic dimensionality of the data. Correlation-based fractal dimensionality of the whole data set is computed first. Partial fractal dimensionality of a feature is measured in the same way, but without using that particular feature. Feature leading to minimum decrease in correlation is considered noise and hence not selected for further consideration. Thus iterative backward elimination is continued to reduce number of features.

Linear Discriminant Analysis (LDA) selects those features that separate the classes best (with least classification errors) with linear class boundaries. The criterion to be minimized is directly proportional to intraclass variations and inversely proportional to interclass distances between means [35]. Thus minimizing this measure class cohesiveness and as well as interclass separation can be increased. Feature set that minimizes this criterion is selected [23]. But the number of possible feature sets explodes with the feature dimensions.

Stepwise Discriminant Analysis (SDA) applies a (split-&-merge like) greedy search approach to solve computational explosion in LDA. It adopts same criterion as in LDA for the present feature set and computes F-statistics for each left-out feature to enter into and for each currently selected feature to exit from the current set. The feature with highest F-to-enter is added and the feature with lowest F-to-exit is eliminated in turn based on F-statistics computed in between. Thus forward selection and backward elimination are iterated till the criterion value stabilizes [49].

2.5.3 Evolutionary Learning Methods

Genetic Algorithm (GA) attempts to avoid the usual problem of entrapment in local minima in the feature-dimensional search space inflicting the greedy search algorithm of SDA. GA follows "survival of the fittest" rule from evolution theory and utilizes "mutational" randomness to come out of the local minima in the search of the global minima. It considers a string of bits (1: for feature being selected and 0: for feature being left out) as a species. Systematic variations are applied by "crossovers" and randomness by "mutations" among the current "chromosomes" to change the initial populations toward more fitted populations. Defining proper fitness function is critical for evaluation of intermediate populations and individual species. For feature selection, classification error often defines the fitness function such that reduced error means better feature sets. To bias the selection toward minimum possible sets of optimal feature, MDL constraint sometimes works in parallel with the classification errors [66, 79].

2.5.4 Performance Analysis and Future Scope

CMU work [115] in recognition of proteomic subcellular location patterns in cellular images reveals that understanding protein functionalities is facilitated by localization of subcellular compartments as they create unique biochemical environment for protein folding and other functionalities [84, 97]. To classify these patterns with less number of available static images, they start with host of features (discussed earlier) and then select features with evaluation by above-mentioned strategies for all ten major subcellular patterns in HeLa cells. Their results reveal that in this specific application, feature selection procedures (DT, FDR, LDA, SDA, and GA) perform better than feature recombination procedures (PCA, NLPCA, KPCA, and ICA). SDA performs the best with reduction of feature dimensionality by 0.46 factor while increasing the classification accuracy by 2.2 % [50]. GA-based method is the close second with reduction in dimensionality by 0.51 factor while increasing the accuracy by 2.3 %.

One possible future research direction for GA-based feature selection might be cascading a classifier like ANN to evaluate performance of the current set of features. ANN output is the value of the fitness function that is fed back to GA for crossover and mutation decisions to reach the fittest set of features in GA output. For feature recombination, novel ideas of synthetic feature generation adopting genetic programming (GP) methods might be successful as in CVPR applications [66]. In GP method, sequential image operators are represented by tree-like structure and crossover or mutation of branches (as in GA) leads to generation of novel synthetic (recombined) features, that might not make sense ordinarily but might define the best feature to define class boundaries in a lower dimensional feature space. CMU work [50] underscores the need of proper feature selection procedures before the classification stage. One recent review [120] discusses different applications of machine-learning techniques in cell biology—from preprocessing, detection, feature extraction, feature selection, supervised and unsupervised classification, performance optimization, and availability of such software packages for cell biologists.

2.6 Cell Classification

In static image bioinformatics, cellular or subcellular recognition or classification is often the ultimate goal. As infinite structural variations among cells and intercellular organelles and molecules are possible, high-throughput automated recognition of biological structures and distributions requires both robust image feature sets and accurate classifiers. The dynamic characteristics of the cells and biochemical activities make classification even more difficult. As an example, the most typical Golgi images are characterized by compact structure, while prior to mitotic cell division, Golgi complex undergoes fragmentation to reunite at the later stage in two offspring cells. This dynamic fragmentation in Golgi complex gives problem in simple morphology-based structural classifications even for manual detections, while computed texture descriptors (imperceptible to human eyes) recognize them much better [90]. Similarly, in human protein atlas, manual annotations are corrected by automated classification in support vector machine (SVM) classifier followed by hierarchical clustering [63]. Sometimes, without explicit segmentation, morphologically preprocessed images themselves are fed to modular neural network [31]. Modularity of the hidden layer considers the image at different scales and for different regions. As the bioimage data used in this work have no cell-to-cell occlusion, the different regions of the images generally contain individual cells (i.e., segmentation is implicit) and provide good classification results.

CMU group applies several supervised and unsupervised classifiers to recognize subcellular localizations from multiple biological image sets [50, 115]. They list links to several proteomic databases that are freely available for comparison, but notify that unified framework for all expressed proteins in different cell types under many biological conditions is still a big challenge [90]. Their work reports recognition accuracy of 95 % for 2D and 98 % for 3D HeLa cell images. Comparison of results for each image from these classifiers permits estimation of the lower bound classification error rate for each subcellular pattern, which they interprets as to reflect the fraction of cells whose patterns are distorted by mitosis, cell death, or acquisition errors. They claim that sometimes automatic classification can outperform human visual classification. For easily confused endomembrane compartments (endoplasmic reticulum, Golgi, endsomes, lysosomes) pattern classification is improved by 5–15 % over the human classification accuracy [49]. Specifically, cell and organelle characteristics like Gabbor texture features and Daubechies −4 wavelet features cannot be measured visually for manual classification and that makes the difference in favor of statistical pattern classification strategies. They distinguish ten major eukaryotic subcellular location patterns in 2D microscopic images and eleven in 3D images, including few that are not discernible in human eyes.

Some of the image-based cell molecular applications utilize simple histogram-based thresholds or seed-growing segmentation for classification. For the automated analysis of epithelial wound-healing process from time-lapsed image sets [107], simple region-based segmentation and seed-growing technique are used as the classification between lacerated wound and unaffected skin—although seed initialized method is not clear. Similar technique is also applied segmenting and classifying bacteria [118], tracking them individually, estimating spatiotemporal tracks and recognizing biological activity with and without application of drugs. When there are similar cellular or subcellular structures in a distinctly different background (in vitro solutions), segmentation itself is type of pixel-based classification, where geometrical models and intensity clustering are adopted in several works [11]. Sometimes location-based region adjacency graphs could distinguish lumenal epithelial cells, stromal cells and nuclei from in vitro subcellular images [20, 22] or sparse features are learnt to classify tumors in histopathology [89]. When several different proteins are present for each location class, local features information from protein as well as reference marker images (acquired in parallel) might be useful to correct classification [28]. In a recent work [23] on classification and abnormality detection in Pap smear cells, multiclass datasets (4 and 7 classes) are merged for 2-class problems—normal and abnormal—to compare performance. Five classifiers are tested—Bayesian classifier, LDA, K-nearest neighbor (KNN), ANN, and SVM—where ANN performed the best—with >95 % accuracy for multiclass and >97 % for 2-class problems.

Major pattern classification methodologies used by different bioimaging groups are depicted in conceptual diagram in Fig. 2.4 and briefly addressed below with pointers to references for necessary details. Interested readers might review recent survey papers on complete cell analysis systems that summarize different unsupervised, semi-supervised and supervised machine learning-based classifiers [115, 120].

2.6.1 Artificial Neural Network (ANN)

This are layered directed acyclic graph (DAG) imitating biological neural networks to transform input signals or features to output activations through linear or complex nonlinear functions. Complex nonlinear multi-input-multi-output (MIMO) mapping functions are learnt from a training dataset by supervised learning rules like back-error propagation, reinforcement learning, competitive learning, etc. [14]. This is generally applied when the mapping function might be considered as a black box and no further structural analysis is warranted. There are number of variants of

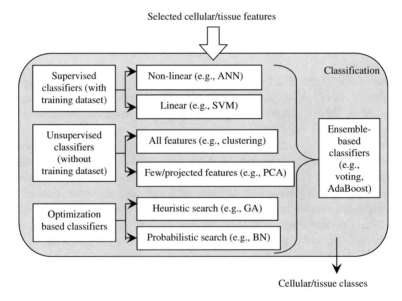

Fig. 2.4 Major pattern recognition techniques utilized in cellular and tissue classification in biological datasets

ANN structures [12], although fully connected multilayered-perceptron (MLP) with sigmoid activation function and back-error propagation learning rule is the most common and successful [13]. In the feature-based supervised classification of cellular images, input features and correct output class labels (often visually classified) are utilized to tune the neuronal connection weights (only changeable parameters in ANN) over the training cycles and once trained can classify unseen (test) data [23, 25, 49, 56].

One work [31] adopts a modular neural network (MNN) trained with sets of confocal sections through cell lines fluorescently stained for markers of key intracellular structures. MNN is developed as three 2D layers of input. In the modular structure, MNN input layer obtains monocular pixel-level intensity values, hidden layer considers different sections of the cellular image at different resolutions/scale to capture the overlap and the structural relations and the output layer produces the classes like mitotic nucleus, nucleus, Golgi, etc. Training is done with standard back-error propagation with 67 % of the randomly sampled data. Key feature of MNN is capability to capture structure of the organelles in 2D.

2.6.2 Support Vector Machines (SVM)

Unlike ANN, which is a nonlinear classifier, SVM generally is a linear classifier that finds a hyperplane between the two classes that maximizes the distance between the plane itself and the data points from different classes closest to the hyperplane [125]. It is an example of margin-based classifier in pattern recognition. The plane is supported by multiple pairs of closest data points from different biological classes, where the plane is at equal distance from both the points in a particular pair. Actual feature dimension is mapped nonlinearly to a very large dimensional space with the hope that class overlaps can be nullified or reduced for better classifications with less error. General two-class SVMs are extended to multiclass applications by max-win, pairwise, and classifier DAG. SVMs can adopt different kernel functions for mapping different low-dimension to high-dimension, like linear, radial basis functions (RBF), exponential RBFs, polynomials, etc. One way to decide which one to select is to start with complex kernels and gradually reduce order of complexity and compare performance to stop at an optimum level for a particular application. In CMU work, SVMs with exponential RBF perform the best for image-based subcellular location classification [50, 63, 115]. SVM classifiers are very successful in several bioimaging applications [23, 25].

2.6.3 Ensemble Methods

To avoid the general problem of entrapment in local minima of the error surface during training (as in ANN) [35, 125], one way of robust classification is considering multiple classifiers with different structures and parameters and fusing the results from these individual classifiers to get the final result. This is the basis of ensemble methods with many variants based on the technique of fusing the base classifiers [12]. AdaBoost learning starts with weak classifiers and gradually increases weights of those classifiers that have made wrong classification in last iteration. Thus more stress is exerted to learn to classify confusing examples. On the other hand, Bagging method tries to increase the training dataset by resampling with replacement and to learn base classifiers with different input sets. Results from them are combined to get the final result. Mixture of experts is another ensemble technique that follows divide-and-conquer strategy so that training set is partitioned by some similarity, classified by different base classifiers for different partitions (like, gating network) and then combined (e.g., by local expert network). Majority or consensus voting is one simple ensemble method that, married with Kaplan Meier test, can effectively perform tumor subtyping [20].

2.6.4 *Hierarchical Clustering*

When subcellular images cannot be classified by visual inspections (like in Golgi compartments, due to mitotic cell division), supervised learning is not possible. Clustering is an unsupervised learning technique to know the classes from unlabeled training data from proximity in the feature space. After proper normalization of different features (to give proper weights on different dimensions), proximity can be measured with any standard distance metric [35], like Euclidean distance or Mahalanobis distance using feature covariance matrices, etc. Individual classes can be formed with proximity and multiple discriminant analysis (MDA) of the clusters in feature space. K-means spectral clustering [25] is one such variant. CMU researchers adopt bottom-up approach to learn a hierarchical subcellular location tree (SLT) from clustering, where at first every organelle structure is taken as individual classes and then classes are iteratively merged by similarity (proximity and MDA) in layers to form a classification tree [115]. This type of hierarchical classification tree or SLT is a very high-level tool for cell molecular biology and can be applied in other medical diagnostic systems as well. A SLT is automatically developed from cellular and subcellular image sets [90, 115] where classes are merged correctly as expected from biological knowledge (like first to merge were giantin and gpp130 as both are the Golgi proteins). In protein regulatory networks, graphical connectivity-based hierarchical clustering is applied to classify cell lines and data mine parts of them to categorize whether it is "living" or "dying" [95]. The biggest problem with unsupervised learning is visual or manual validation of the classification results is not possible as there is no labeled training set.

2.6.5   PCA Subspace-Based Classifiers

Classifying cellular forms of proteins encoded by human cDNA clones is a primary step toward understanding the biological role of proteins and their coding genes. Classifier surface is successfully estimated in PCA subspace to classify protein structures with a novel framework I-GENFACE for protein localization [122]. Morphological, geometrical, and statistical features, such as brightness of a region of pixels, the object boundary, and co-occurrence matrix-based grayscale textural features, spot and line features, histogram, and amplitude features, etc. are extracted semiautomatically. Distance-based metric in PCA subspace is adopted to classify the protein forms and then the corresponding images. Classification accuracy achieved is approximately 90 % for seven subcellular classes.

2.6.6   Performance Comparison

Beside standard performance evaluation tools in CVPR [12, 35], sometimes application-specific criteria are defined in cell imaging [49].

(1) Complexity of the decision boundaries that a classifier can generate For a pair of Golgi proteins, giantin, and gpp130 (ones difficult to classify visually) and two most informative features derived by SDA-based feature selection (namely, fraction of fluorescence not in any object and convex hull eccentricity) CMU research illustrates the complexity on the 2D scatter plot. This is one way to check the complexity of the classifier needed, like the order of the activation function in ANN, or polynomial order in kernel-based SVM, etc. Although complex classifiers sometimes classify small dataset (with less variation) utilizing complex features, according to minimum description length (MDL) principle [79] these are overfitted classifiers as they loose generality for the dynamic organic environment.

(2) Dependence of the classifier performance on the size of the training set This is the capability to learn from limited training set and insensitiveness to the presence of outliers in the data. In cell molecular biology, complexity of the dynamic environment might demand multiparameter classifier that in turn needs larger training sets (to fit the parameters iteratively) which is not always available. CMU work shows [50] that with more training data classification accuracy improves, as expected. Even without access of the complete dataset, probabilistic active learning could model and discover biological response in gene datasets [88]. In this work, greedy merge structure learning iteratively combines distributions with the assumption that same conditions affect the unseen variants similarly. Outliers generally affect in incremental learning modes of different classifiers, like ANN and reduce accuracy.

(3) Sensitivity of performance to the presence of uninformative features All the features may not contribute cooperatively toward classification. CMU research claims that ANN perform better for 3D cell images than 2D images, which is

somewhat unexpected and underlines the importance of feature selection. They adopt SDA feature ranking based on information content and gradually add features according to high-to-low ranking to compare how the classifiers behave [50, 115]. This is a classical pattern recognition method to check how many features are adequate to work with a particular classifier applied to a particular task [35]. They also conclude that the ability of a classifier to adapt to more noisy features depends on the feature space itself.

Above indices are very general for any applications. CMU work also evaluates classifiers based on statistical paired t-test. They conclude that SVM with exponential RBF kernel performs most consistently for different subcellular location feature sets [50].

2.6.7   Other Methods and Future Scope

Evolutionary computation a key machine learning paradigm not yet utilized to its full potential in bioimaging. Methods like genetic algorithms (GA), genetic programming (GP) (see Sect. 2.5) and their variant like colony optimization, particle swarm optimization, "tribe"-based global GA, etc. fall in this class. Only few recent works report successful usage of such techniques for segmentation [91], feature selection, and classification [58]. Bayesian learning [57] is another area where very limited cell classification research is so far invested [23] but might lead to success—specifically as cellular localizations can easily be represented as cause-and-effect relations with the surrounding biological processes and injected drugs and antigens.

2.7   Dynamics and Tracking: Cell Activity Recognition

Cellular processes are heterogenous and complex, yet very organized. For understanding molecular mechanisms at the systems level, like cell migration and signal transduction [137], complex spatiotemporal behaviors of cellular processes are needed to be datamined with objective computational tools. Cell migration, besides motion, involves shape changes (morphogenesis) and interactions at different levels, from single cell flagella-driven movement to bacterial swarming to collective stem cell migration toward chemoattractants. Estimated motion leads to preferred migratory paths as well as related shape deformations. Sometimes correlation between signaling events to spatial organization of the biological specimens might enhance understanding biological processes [105]. One recent work [18] proposes a complete pipeline of computational methods for analyzing cell migrations and dynamic events, starting from acquisition, registration, segmentation, and classification, and finally addressing cell tracking, event analysis, and interpretation. Changes in shape and topology are tracked and motion fields are computed. Another application oriented review [105] of developments and challenges in automated analysis summarizes few examples of intracellular dynamics, cell tracking, and cellular events as follows: (1) Estimation and control of cell cycle state and its rate of change directly will link to cancer and DNA damage. (2) Very little is known on intricate dendrite branching pattern unique for each neuronal class. Tracking of dendrite arbors in 3D microscopy might help estimating dynamic relationship between dendrite growth and synaptogenesis. (3) Embryonic heart development and embryogenesis of Zebra fish could be monitored to enhance understanding and quantifying structural phenotypes in tissue.

The challenge of the postgenomic era is functional genomics, i.e., understanding how the genome is expressed to produce myriad cell phenotypes [11, 88, 94]. To utilize genomic information to understand the biology of complex organisms, one must understand the dynamics of phenotype generation and maintenance. Also cell signaling and extracellular microenvironment have a profound impact on cell phenotype. These interactions are the fundamental prerequisites to control cell cycles, DNA replication, transcription, metabolism, and signal transduction. All the biological events in the list require some kind of particle tracking and then classification of the dynamics. Signal transduction is believed to be performed by protein molecules passing across the cell membrane, carrying some electrochemical massage to the target cell or organelle, where it initiates some biochemical event [119]. Hence tracing signals and analyzing their implications also require tracking over image sequences. Biochemical activities in the molecular (e.g., protein, DNA, RNA, genes etc.) and atomic levels (e.g., protein folding leads restructured form of the proteins followed by higher level activities like mitotic fragmentation) in intracellular compartments (e.g., mitochondria, Golgi bodies, nuclei, cytoplasm etc.) and intercellular signaling are one of the prime signs of life [97]. Higher level activities in living organism trace back to these molecular level biochemical activities [36].

One crucial point in bioinformatics is that the biochemical processes are often very slow, while few transient processes are very fast—e.g., red blood corpuscles (RBC) are dying in hundreds every minute, and new RBCs replace them to keep the equilibrium. After effective shot detection and key frame selection depending on rate of change of information content of the cellular video (see Sect. 2.2), cell tracking, and biological activity analysis can be efficiently performed only with those key frames (see conceptual diagram in Fig. 2.1).

Like RBC lifecycle, many other cellular events are transient in nature—including plant cell activities. In plant physiology research, cell dynamics analysis plays a significant role—from monitoring cell morphogenesis [62] to dynamic gene activity [78] for developmental processes. In one in vivo study of CA^{2+} in the pollen grain and papilla during pollination in Arabidopsis in fluorescence and ratiometric imaging, yellow cameleon protein indicator is utilized to detect change in CA^{2+} dynamics over different phases of pollination [54]. Unfortunately, growth rate of pollen tube is measured manually by the rulers that, unlike in computational methods, reduces statistical reliability. Similarly, interactions between B and T cells are essential for most antibody responses, but the dynamics of these interactions are poorly understood [92]. By two-photon microscopy of intact lymph nodes, it is

demonstrated that upon exposure to antigen, B cells migrate with directional preference toward the B-zone-T-zone boundary in a CCR7-dependent manner. There are
salient variations in velocity, speed, and duration of activity based on antigen doses.
These findings provide evidence of lymphocyte chemotaxis in vivo, and can define
similar dynamics associated with T cell-dependent antibody responses. Development
of many vertebrate tissues involves long-range cell migrations that are often quantified from time-lapsed images and few samples of data. One work [132] utilizes
two-photon laser scanning microscopy and quantitative analysis of four-dimensional
cell migration data to investigate the movement of thymocytes through the cortex in
real time. This work tracks the thymocytes over multiple frames of cell video, forms
time-stamped spatiotemporal trajectories to classify into two classes of motility rates
(higher and lower), and concludes that displacement from origin varies differently for
these motility rates (lower motility follows linear rule, while higher ones follow
quadratic rule). And these two distinct migratory behaviors within wild-type cortical
thymocytes are analyzed for further higher level biological decisions. Cell activities
like cell division, cell migration, protein signaling, and protein folding in biological
videos should be computationally analyzed and classified into spatiotemporal processes to understand the dynamics behind them.

Euro-BioImaging consortium is one of the very few groups actually analyzing
dynamics in cell videos. They analyze bacterial motility videos taken under light
microscopy and in vitro solutions [118]. They record the trajectories of
free-swimming bacteria of the species Rhodobacter spheroids under a variety of
incubation conditions, and estimated trajectories of the rotations of these bacteria
tethered to glass coverslips using an antiflagellin antibody. The rapid rotations of
helical flagella by proton-powered molecular rotary motors (embedded in the
bacterial membrane) cause free-swimming bacteria to swim forward in curved
trajectories. Brief reversal of the direction of rotation of the motors in a single
bacterium induces its flagellar bundle to fly apart, causing the cells to undergo a
"tumble", leading to randomizations of the new direction of motion upon
resumption of the normal rotational direction. Bacteria swimming up a concentration gradient of a chemoattractant (e.g., glucose) tumble less frequently than
bacteria entering a hostile (e.g., acidic) environment. Euro-BioImaging group
studies these bacterial responses to environmental stimuli by considering spatiotemporal trajectories of individual bacteria and the times and positions of bacterial events, such as tumbles and reverses. It would be impractical and intractable
to undertake such detailed analysis and annotations of events manually.
Image/video processing techniques are indispensable for such analysis. Hence they
track multiple bacterial motions, form spatiotemporal traces (after smoothing and
merging of tracks as in "boundary-tracking" algorithm in CVPR), and detect cell
biological states like "swimming forward" (flagella rotating counterclockwise),
tumbling (flagella rotating clockwise), and stopped. These state transitions are then
mapped to different biochemical ambience and corresponding responses in terms of
bacterial speed, rotational frequency, etc.

Automated monitoring of wound-healing assay in time-lapsed microscopic data
is another application where cell motility dynamics might help clinically [118, 143].

Euro-BioImaging group performs textural segmentation and quantify healing by rate of change of dimensional spread with or without drug [118]. In a recent sophisticated work on wound-healing and cell-scattering data, cascade of SVMs performs initial classification of local patches, and then graph cut corrects and reclassifies [143]. Acceleration effect of Hepatocyte growth factor/scatter factor (HGF/SF) is utilized for monitoring. Few recent works in cellular biology point out important future application of CVPR strategies to understand cellular events. In one system pharmacological application [64], cell population is tracked by morphology, proximity, and specifically the motion fields obtained from particle filters and interacting multiple model filters. Morphological phenotypes help identification while SVM, factor analysis, and Gaussian mixture models classify the profiling types. All the above examples demonstrate how CVPR strategies can enhance the cellular activity analysis, and how understanding several other similar biological processes, upon computational analysis and exploratory data mining, can enrich our higher level knowledge (see Fig. 2.1). Different established object tracking [140] and structure from motion algorithms [12] could be adopted to analyze these applications.

2.8   Bioimaging Databases and Content-Based Retrieval

Vitality of living cells and their dynamic behaviors separate them from innate rigid objects [119]. Innate objects generally have same shapes (e.g., rigid objects like a car can change its 2D projection due to motion and viewpoint change, but always have same 3D structure [43]) or shape changes in few discrete predictable ways (e.g., flexible objects like a military fighter jet changes its 3D structure due to different missile attachments, wing positions, and other artillery manipulations; but definitely in a few discrete and predictable ways [32]). In case of living cells and their intracellular molecules, environmental effects are more complex and often not yet understood and cause the cell shapes to change in very unpredictable ways [17, 115]. Like a motile cilium changes its shape unpredictably to extend one part (like a leg), get hold (like a hand) and then shifts the body organisms toward that direction. Its shapes have sharp differences, yet it is a cilium [119]. Time-lapsed imagery demonstrates wide variations that are to be stored in a dynamic database for future search, research, and analysis. Additionally proper content-based image retrieval schemes are to be adopted so that mere shapes cannot misguide the retrieval [26, 88]. Even a single type of cell might have high variability as different biologists study it at different conditions [22]. Distributed databases and web-based querying facilities increase cooperative efforts of experts from different parts of the world [77]. But at the same time this underscores demanding requirements of necessary mapping between different nomenclatures and formats [147], content-based indexing strategies [26], and machine learning methods for improved retrieval [36, 96, 98, 120]. Besides phenotyping and understanding dynamics,

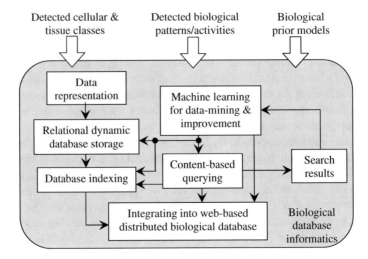

Fig. 2.5 Bioinformatics modules in large-scale biological dynamic databases that are often distributed across the globe and datamined by complex content-based queries over the Internet

database might be useful for building atlas for model organisms or for 3D reconstruction of brain wiring from international data with wide variability [96].

Feature-based image representation, dynamic database maintenance, content-based indexing, content-based image retrieval (CBIR), and learning-based improvement of retrieval performance are relatively matured subfields in CVPR. Interrelations between these broad CVPR modules in bioinformatics context are summarized in Fig. 2.5. In cell biological image processing, only few works are reported in this direction, mostly for static images [22, 26, 88, 98] and rarely for cell dynamics in videos [77, 147]. These efforts are summarized below with relevant pointers to future research directions.

2.8.1 Data Representation

The first step of any database design is the determination of the scope of the data to be stored, i.e., together with their complete description, taking into account the targeted group of users and the applications of interest. To determine these generally biologists and microscopy experts develop a list of biological descriptors for data representation and database querying. These descriptors guide modules down the line including image processing, feature extraction, classification, database management, and structural query systems.

Organisms express their genomes in a cell-specific manner, resulting in a variety of cellular phenotypes or phenomes [22]. Mapping cell phenomes under a variety of experimental conditions is necessary in order to understand the responses of organisms to stimuli. Representing such data requires an integrated view of experimental and informatics protocols. This is more critical when experimental procedure varies [22] or even nomenclature differs between research groups [147]. BioSig system developed in LBNL [20, 94] adopts a hierarchical data model to capture experimental variables and map them to light microscopic image collections and their computed representations (features) at different levels—sample tissues, cells, and organelles. At each layer, information content is represented with an attributed graph of cellular morphology, protein localization, and cellular organization in tissue or cell culture.

There are two kinds of information associated with visual objects (image or video): information about the object, called its metadata, and information contained within the object, called visual features [34, 67]. Metadata (such as the name of a protein) is alphanumeric and generally expressible as a schema of a relational or object-oriented database [147]. Visual features, in contrast, are mathematical properties of the image derived by computational tools from image processing, CVPR, or geometric routines discussed in earlier sections [11, 105, 115]. A database system that allows a user to search for objects based on contents quantified by above-mentioned features is said to support content-based image retrieval (CBIR).

Euro-BioImaging group [34, 147] utilizes following (quite generic) data representation covering wide information variability including the rare support to cell video data.

- General information: administration and organizations: submitter's name, title, funding, references, and contacts.
- General information: microscopic data: metadata on location, format, and size, channel and axes information (coordinate system), annotated figures.
- Biological data: details of the biological specimens (taxonomic information and parameters that depend on the type of specimen) and observable biological features.
- Experimental details: sample preparation: experiment, preparation steps, their biochemical and physical parameters, instruments used (e.g., for microinjection). Free-text comments for nonstandard information can be provided as well.
- Experimental details: data acquisition and instrumentation: for reproducibility, microscopic settings, and image-recording schemes are stored.
- Experimental details: image processing: ranging from simple enhancement to complex 3D reconstruction algorithmic information.

The need of unified data representation comes from the diversity of experimental procedures followed by biological researchers around the globe [22], like different application-specific microscopic systems are used: all kinds of light, electron and scanning probe microscopy with different resolutions. The biological targets are also diverse, ranging from entire organisms as observed by developmental biologists to the macromolecules studied by structural biologists [36, 98]. Datasets are also of quite different sizes ranging from less than 1 MB for many electron microscopic datasets to hundreds of MB for scanning microscopic videos in cellular dynamics contexts [67]. Dimensionality of the datasets also differs a lot [108].

Atomic force microscopy (AFM) images are two-dimensional. Three-dimensional density distributions are reconstructed from 2D electron microscopy data. Video microscopy generates 3D files with two spatial axes and one time axis. And confocal light microscopes can even record 3D datasets as a function of time. In addition complementary information may be stored in several channels in multilabeling fluorescence microscopy [28]. Sometimes infrared high-spectral frequencies are utilized as an additional dimension for lung cancer tissue diagnosis [2]. Dynamic database should have flexibility to handle such multidimensional data. Even recent reviews on spatiotemporal dynamics of cellular processes inform that representation of behavioral knowledge in biological database is still a great challenge [137]. For distributed and web-based databases that is being accessed by hundreds of researchers around the globe with diverse datasets, unified data representation needs number of seemingly trivial information to be stored and incorporated in the relational database model [77].

2.8.2 Database and Indexing

Database indexing is an established field with enormous success in text-based systems. Success of these systems stands upon the general user-independent definitions or meanings of text-based database elements. Images and videos often contain richer information than textual explanations and informatics researchers work on content-based indexing and querying the image/video databases [10, 55]. There are also specialized databases like those with neuronal morphology [76]. Indexing and CBIR in relational database management systems (RDBMS) with multidimensional biological and cellular images are very challenging [26].

In biological databases, querying should be based on implicit information content, rather than by their textual annotations only. "Query-by-content" generally makes reference to those data modeling techniques in which user-defined functions aim at "understanding" the informational content of the datasets (at least to some extent) from the quantified descriptors (features). Euro-BioImaging consortium of multiple European nations is engaged in a pioneering effort (http://www.eurobioimaging.eu/) of developing one such web-based distributed database prototype [34, 77]. There are number of similar web-based biological databases, not necessarily image databases, like databases of sequences of nucleic acids (GenBank and EMBL Data Library) and those for protein sequences (SWISS-PROT and PIR). A digital neuronal database can help neuromorphological pattern analysis and brain atlas modeling (a significant paradigm shift) from advanced microscopy and image processing. A recent critical review discusses challenges in such applications—including dynamics, machine learning, and associated computations [98].

The complexity of stored information sometimes requires a unique database modeling tool, like Infomodeler used by the Euro-BioImaging group [67] to form a entity–relation (E-R) diagram with biological entities, their attributes, and relationships (like generic: "is related to", aggregations: "is composed of", and inheritance:

"is a" relations). Infomodeler provides database design in two abstraction levels. First, it allows an object-oriented approach (object role modeling; ORM) and second, it allows design in a logical model (E-R diagram). They also mention a denormalization step where redundancy is introduced to improve database performance. Among many entities already entered and several attributes defined among them, not all of them are relevant for a particular submission, since some of them depend on the microscopy technique or on the specimen [77]. As an example a commercial microscope can have number of optional laser beam, objective lens, and filter settings, only few of which are available at a particular location, and still fewer selective ones are actually used for a particular biological experiment. Hence to reduce the burden on the submitter, inheritance-based schemes are utilized just to specify the personal settings, and then the submission database fills out the rest with the default values (if entered earlier).

Euro-BioImaging database provides pointers and links to relevant databases at appropriate place, like SWISS-PROT protein database, EMBL Nucleotide database, protein data bank (PDB), etc. Their database comprises of three primary interfaces: submission interface, query interface, and visualization interface and two database modules: submission database and production database (they are independent to ensure security) [67]. Submission interface is the most complex one as, beside handling queries and incorporating results for visualization, it should also normalize incomplete datasets (by itself or by forcing the user to provide mandatory information) and interact integratively with the database in the background. The data is temporarily stored in submission database. Database curator modules then review the input, complete the unfilled format if necessary and migrate the data to production database. Query interface converts the submission into a structural query language (SQL) code with logical operations. Visualization interface handles the results from the SQL code converting to user-understandable forms to incorporate with the display page.

The backbone of any database system is the RDBMS. Due to high complexity of the cell video data in biological dynamics, and due to the semantic level queries preferred by the experts [26, 77], biological databases require Object-Relational Database Management System (ORDBMS), as it supports complex relationships between the biological entities [67, 147]. The complexity of the queries demands extension of the standard SQL for 3D data handling, named SQL-3. Queries are often needed to be modified for unified framework before actual database search.

2.8.3 Content-Based Querying

In contrast to other databases, the term "query-by-content" (QBC) is seldom used in the context of biological databases. However, some of the functionality implied by this term is in common usage in biological databases [34, 67, 77]. When a new gene sequence is searched for similar sequences in GenBank without using textual annotations, algorithms like Fast-All (FASTA) will provide a rank-ordered similar gene sequence list. Besides textual descriptions, structural databases (e.g., PDB) store thousands of atomic resolution structures of proteins and nucleic acids with a list of coordinates of the atoms. In such databases, alongside keywords, queries might contain organisms, resolution, etc. as structural information and hence considered QBC. Searching 3D structural similarity could help discovering novel biologically active molecules and investigating the relationship between proteins' structures and their functions. Web-based QBC system by Euro-BioImaging group [34, 147] is one such protocol which searches for similar 3D structures of the macromolecules where similarity is measured in terms of features like 3D bounding size, multiscale shapes, channels of low density areas, internal cavity, and geometric symmetry. First two features are generic ones, while others are application-specific. Last type of features, although constrains the applicability and query space, makes the search space more dense with potential match and increases precision and accuracy. These are more relevant for database querying in terms of features like run lengths, velocities, and frequencies, and events like durations and patterns of bacterial tumbles, and correlated bacterial stops and reversals with changes in environmental conditions [118].

One of the most challenging issues is to choose an effective measure of structural resemblance (i.e., biological similarity) between two biological objects [26, 67, 77]. To align a pair of proteins, inter-atom distances of a 3D structure are often represented as 2D matrices and found useful for comparison since similar 3D structures have similar inter-residue distances. So, the problem of matching two proteins structures boils down to graph-matching problem where fundamental graph-partitioning and graph-matching methods [116, 129] can be applied to partition the proteins into smaller subgroups by forming hierarchical structural relations and quantifying matching percentages. One similar hierarchical graph cut method represents eXtended Markup Language (XML) data of complex protein regulatory networks as connected graphs, decomposes it spectrally into cohesive subnets at different abstraction levels and then data-mines for hidden cell motifs and cancerous activities [95]. Another graph-based similarity measure [117] applies combinatorial extension (CE) of the optimal path to find an optimal 3D alignment of two polypeptide chains and utilizes characteristics of local geometry (defined by vectors between C-alpha positions) of the compared peptides. In this web-based system, users submit complete or partial polypeptide chains in PDB format. Then, statistical results are returned along with the aligned sequence resulting from the structure alignment. Similar protocols are adopted for 3D searching in databases of small molecules to facilitate drug and pesticide discovery [84].

There are two different styles for providing examples or queries [10, 12]: (1) pictorial example (Virage Image Engine and NETRA system) and (2) feature value (like color, region area, texture, etc.) and expected percentage similarity as example (QBIC engine from IBM). In the first style of querying, features are first computed for the query example and the target images in the database and then it boils down to the second method. Distance metric is defined as a monotonically increasing function (e.g., weighted Euclidean measure) of these features to give a unique value and this metric should satisfy axioms of validity [79]. Generally CBIR focuses more on 2D images, less on videos [118] and still lesser for 3D images. In a content-based 3D neuroradiologic image retrieval system [69], a multimedia database contains a number of multimodal images—namely magnetic resonance and computer tomography (MR/CT) images as well as patient information (patient's age, sex, symptom, etc.). With proper CBIR tool, such a system could help medical doctors to confirm diagnoses, as well as for exploring possible treatments by comparing the image with those stored in the medical knowledge databank.

2.8.4 Learning in Content-Based Image Retrieval

When a content-based retrieval system is applied to any specific domain it needs to answer two pivotal questions discussed earlier in details: (1) feature selection: of the extended list of features discussed in earlier sections, which computable features are sufficient to describe all images in the domain and (2) classification: what mathematical function should be used to find a measure of similarity between two objects. The second one poses more problems due to subjectivity of perceptual similarity among the observers. Two cells in two biological images can be decided as "similar" by one biologist due to their partwise structural similarity (e.g., they consists of a central cell body and cilia projections with the same pattern), while another biologist may classify them as different due to their functionalities. This dynamic nature of similarity measure makes CBIR more challenging. Machine learning strategies based on relevance feedback [12, 110] might help in such cases where the similarity measure (or even weights for combining different features) could be learned from user feedback (interactive manual inputs) regarding relevance of the result. This method learns the user query, structures the query and the search space to take into consideration more potential matches and incrementally improves the retrieval results over multiple iterations for the same user (see Fig. 2.5). Moreover, this method is extended for short-term learning from a single user and long-term learning from multiple users using the system several times to improve the overall retrieval performance [10]. Query-by-content in biological databases is yet to adopt this type of practical learning strategies.

2.8.5 Distributed Databases and Web-Based Querying

Distributed computation and web-based dissemination strategies are required [77] because of several necessary qualities of large dynamic databases: (1) flexibility to enter new data and update RDBMS continuously and incrementally with streaming data from users around the world; (2) integration of apparently diverse frameworks to analyze data and results including cross-database search in a seamless unified way; (3) fault tolerance of multiserver systems for computationally expensive database manipulations that can be split into parallel and multithreaded modules. More is the complexity and abstractness of the data (like images, videos, cell activities), more is such requirements. For emerging field like bioinformatics, where hundreds of research groups are working globally on similar (but not exactly the same) biochemical processes, distributed database systems, web-based querying and platform-independent visualization tools are absolute necessities [77, 96]. These will give the researcher facilities to enrich the database, share their results with international community, statistically compare their results with existing knowledge and cooperatively work toward better results [36, 105].

A very nice overview of web database operation, implementation, scalability, interoperability, and future directions are discussed by the Euro-BioImaging group [77], including methods to cope up with mapping different names used for the same entities, integrating the data diversity, and updating the web database incrementally while the storage location, experimentation and associated parameters are continuously changing [147]. This consortium does a pioneering research in developing a web-based online biological database system. It describes the ongoing research on developing the dynamic database on an Informix Dynamic Server with Universal Data Option [67, 77]. This object-relational system allows handling complex data using features such as collection types, inheritance, and user-defined data types. Informix databases are used to provide additional functionality: the Web Integration Option enables World Wide Web (WWW) access to the database; the Video Foundation Blade handles video functionality. WWW facility provides the necessary structure for worldwide collaboration and information sharing and dissemination [34, 147]. Future scopes lie in incorporating new microscopy techniques, customizing WWW visualization interface that depends on user profile, and tighter interaction with collaborating databases [147]. Current biomolecular databases [20, 26] basically follow similar RDBMS structures, just from different providers.

Database over the web has to bear extra burden of providing simple interfaces for the (sometimes computer-naïve) biologists and at the same time ensure security and integrity of the distributed and dynamic database from intrusion and misleading submissions. Hence it is better to separate out submission module and actual database by a buffer database. The standard approach to connect with a database involves calling a CGI application (a program running on the web server) through calls from flat files containing HTML text [67, 94]. The alternative approach involves making a direct call to a database program with the page names and all relevant parameters. In both cases, SQL code is incorporated into standard HTML code. When the WWW browser requests the HTML file, the SQL code segment is extracted, passed to the RDBMS, and interpreted [147]. The result is formatted in standard HTML for visualization. Web pages are created dynamically, i.e., some template formats are modified based on user needs to create a specific web page. It also reduces the development time. Other domain-specific creation is stressed with user-defined web tags. Importantly, all the semantic web standards can still be combined and this generic framework can be extended to many other web database applications (e.g., medicine, arts) [77]. BioSig system developed in LBNL [94] also makes their computational biology framework distributed and platform-independent using eXtended Markup Language (XML) protocol generated by biological experiment and handling those to reach bioinformatics decisions [20].

2.9   Future Research Directions

Many future research scopes are already discussed under individual sections followed by relevant pointers toward related works from CVPR and other research fields. To summarize, cell video analysis can be immensely enhanced by future research areas including: (1) cell video summarization methods [40], (2) texture [52] and graph cut-based segmentation [116], (3) close-loop cooperation between segmentation and object alignment with deformable models [37, 44], (4) synthesizing combinatorial features by genetic programming [66], (5) evolutionary learning in feature selection and classification [91, 120], (6) utilizing hyperspectral image features beyond currently applied frequency ranges [2, 9], (7) application of Bayesian classifiers [57], (8) improvement of performance even from small training dataset often encountered in bioimaging [10, 84, 88], (9) motion-based segmentation [15, 111, 135, 142, 143] and tracking [140] in cell videos, and (10) continuous learning by relevance feedback to improve database retrieval [10].

Learning enumerable phenotypes and distinguishing them requires parametric models that can capture cell and nuclear shapes as well as nonparametric models to capture complex shapes and relationships between them. Such generative models could be learned from static datasets [17]. Interesting future direction will be making those generative models dynamic—to capture temporal evolutions—possibly by dynamic Bayesian networks [57]. But unified framework of generative models to handle behavior of cells from diverse pedigree is still a very challenging task [84]—as model topology itself need to change within and across time. Morphogenesis and cell fragmentation complicate the Bayesian graphical analysis even more. Expectation maximization (EM) learning [35] cannot handle such flexibility to learn widely varying cell shapes, protein distributions within organelles and subcellular location patterns [97]. Recently an evolvable Bayesian graph has been proposed in incremental 3D model building application [43]. This generic probabilistic graphical model has flexibility [41] to represent unpredictable structural changes of the same cells, replication, and effects of drug or antigen applications over time. This framework also has potential [42] of modeling cellular behavior caused by different biochemical environments, analyzing interrelations among neighboring organelles in uncontrolled unpredictable environment, and even handling content-based video querying in complex database search engines.

2.10   Conclusions

Automated analysis of microscopic bioimages is making significant progresses with application of established tools and algorithms from computer vision, pattern recognition, and machine learning. Quantification and exploration of dynamic activities and evolving interrelations between them from cellular and biomolecular videos are the current key frontiers [36, 98, 105, 120, 137]. More cohesive and cooperative merger between computational and biological sciences [51] are expected to overcome these challenges toward achieving better understanding of the hidden patterns in living universe [74].

Acknowledgments The author would like to acknowledge Department of Pediatrics at Loma Linda University (LLU) School of Medicine, Loma Linda, CA, USA and Center for Research in Intelligent Systems at University of California Riverside (UCR) Department of Electrical Engineering, Riverside, CA, USA for supporting his research over the years—especially Dr. Stephen Ashwal (LLU) and Dr. Bir Bhanu (UCR) for encouraging exploratory research.

References

1. Aigrain P, Joly O (1994) The automatic real-time analysis of film editing and transition effects and its applications. Comput Graph 18(1):93–103
2. Akbari H, Uto K, Kosugi Y, Kojima K, Tanaka N (2011) Cancer detection using infrared hyperspectral imaging. Cancer Sci 102(4):852–857
3. Amanatiadis A, Kaburlasos VG, Gasterator A, Papadakis SE (2011) Evaluation of shape descriptors for shape-based image retrieval. Inst Eng Tech (IET) Image Proc 5(5):493–499
4. Arslan S, Ersahin T, Cetin-Atalay R, Gunduz-Demir C (2013) Attributed relational graphs for cell nucleus segmentation in fluorescence microscopy images. IEEE Trans Med Imaging 32(6):1121–1131
5. Asada H, Brady M (1986) The curvature primal sketch. IEEE Trans Pattern Anal Mach Intell 8:2–14
6. Attig A, Perner P (2011). A comparison between Haralick's texture descriptors and the texture descriptors based on random sets for biological images. In: Perner P (ed) Machine learning and data mining in pattern recognition, vol 6871. Springer. LNAI, pp 524–538
7. Babaud J, Witkin A, Baudin M, Duda R (1986) Uniqueness of the Gaussian kernel for scale-space filtering. IEEE Trans Pattern Anal Mach Intell 8:26–33
8. Berrada F, Aboutajdine D, Ouatik SE, Lachkar A (2011) Review of 2D shape descriptors based on the curvature scale space approach. Proc Int Conf Multimedia Comput Syst (ICMCS) 1–6
9. Bhanu B, Pavlidis I (eds) (2005) Computer vision beyond the visible spectrum. Springer
10. Bhanu B, Chang K, Dong A (2008) Long term cross-session relevance feedback using virtual features. IEEE Trans Knowl Data Eng 20(3):352–368
11. Bilgin CC, Kim S, Leung E, Chang H, Parvin B (2013) Integrated profiling of three dimensional cell culture models and 3D microscopy. Bioinformatics 29(23):3087–3093
12. Bishop CM (2006) Pattern recognition and machine learning, 6th edn. Springer, New York
13. Boland MV, Murphy RF (2001) A neural network classifier capable of recognizing the patterns of all major subcellular structures in fluorescence microscope images of HeLa cells. Bioinformatics 17(12):1213–1223

14. Bose NK, Liang P (1996) Neural networks fundamentals with graphs, algorithms, and applications. McGraw-Hill Publications
15. Bradski GR, Davis JW (2002) Motion segmentation and pose recognition with motion history gradients. Mach Vis Appl 13:174–184
16. Brunelli R, Mich O, Modena CM (1999) A survey on the automatic indexing of the video data. J Vis Commun Image Represantation 10:78–112
17. Buck TE, Li J, Rohde GK, Murphy RF (2012) Toward the virtual cell: Automated approached to building models of subcellular organization "learned" from microscopy images. BioEssays 34(9):791–799
18. Castaneda V, Cerda M, Santibanez F, Jara J, Pulgar E, Palma K et al (2014) Computational methods for analysis of dynamic events in cell migration. Curr Mol Med 14(2):291–307
19. Chang HS, Sull S, Lee SU (1999) Efficient video indexing scheme for content-based retrieval. IEEE Trans Circ Syst Video Technol 9(8):1269–1279
20. Chang H, Fontenay GV, Han J, Cong G, Baehner FL, Gray JW et al (2011) Morphometric analysis of TCGA glioblastoma multiforme. BMC Bioinf 12:484
21. Chang H, Han J, Spellman PT, Parvin B (2012) Multireference level set for the characterization of nuclear morphology in glioblastoma multiforme. IEEE Trans Biomed Eng 59(12):3460–3467
22. Chang H, Han J, Borowsky A, Loss L, Gray JW, Spellman PT, Parvin B (2013) Invariant delineation of nuclear architecture in glioblastoma multiforme for clinical and molecular association. IEEE Trans Med Imaging 32(4):670–682
23. Chankong T, Theera-Umpon Auephanwiriyakul S (2014) Automatic cervical cell segmentation and classification in Pap smears. Comput Methods Programs Biomed 113 (2):539–556
24. Chen X, Murphy RF (2004) Robust classification of subcellular location patterns in high resolution 3D fluorescence microscope images. Proc Int Conf IEEE Eng Med Biol Soc 1632–1635
25. Chen S, Zhao M, Wu G, Yao C, Zhang J (2012) Recent advances in morphological cell image analysis. Comput Math Methods Med 10:101536
26. Cho BH, Cao-Berg I, Bakal JA, Murphy RF (2012) OMERO.searcher: content-based image search for microscope images. Nat Methods 9(7):633–634
27. Chung P, Tsai C, Chen F, Sun Y (1994) Polygonal approximation using a competitive Hopfield neural network. Pattern Recogn 27:1505–1512
28. Coelho LP, Kangas JD, Naik AW, Osuna-Highley E, Glory-Afshar E, Fuhrman M et al (2013) Determining the subcellular location of new proteins from microscope images using local features. Bioinformatics 29(18):2343–2349
29. Corridoni JM, del Bimbo A (1995) Film semantic analysis. Proc Int Conf Comput Archit Mach Percept (CAMP) 201–209
30. Cremers D, Rousson M, Deriche R (2007) A review of statistical approaches to level set segmentation: Integrating color, texture, motion and shape. Int J Comput Vis 72(2):195–215
31. Danckaert A, Gonzalez-Couto E, Bollondi L, Thompson N, Hayes B (2002) Automated recognition of intracellular organelles in confocal microscope images. Traffic 3:66–73
32. Das S, Bhanu B (1998) A system for model-based object recognition in perspective aerial images. Pattern Recogn 31(4):465–491
33. Davis LS, Mitiche A (1982) Mites: a model-driven, iterative texture segmentation algorithm. Comput Graph Image Process 19:95–110
34. de Alarcon PA, Gupta A, Carazo JM (1999) A framework for querying a database for structural information on 3D images of macromolecules: a web-based query-by-content prototype on the BioImage macromolecular server. J Struct Biol 125:112–122
35. Duda RO, Hart PE, Stork DG (2001) Pattern classification (2nd edn). Wiley
36. Eliceiri KW, Berthold MR, Goldberg IG, Ibanez L, Manjunath BS, Maryann ME et al (2012) Biological imaging software tools. Nat Methods 9(7):697–710
37. Erdt M, Steger S, Sakas G (2012) Regmentation: a new view of image segmentation and registration. J Radiat Oncol Inf 4(1):1–23

38. Gargi U, Oswald S, Kosiba DA, Devadiga S, Kasturi R (1995) Evaluation of video sequence indexing and hierarchical video indexing. Proc SPIE Conf Storage Retrieval Image Video Databases III 2420:144
39. Gerlich D, Mattes J, Elis R (2003) Quantitative motion analysis and visualization of cellular structures. Methods 29:3–13
40. Ghosh N, Bhanu B (2006) A psychological adaptive model for video analysis. Proc IEEE Int Conf Pattern Recogn (ICPR) 4:346–349
41. Ghosh N, Bhanu B (2008). How current BNs fail to represent evolvable pattern recognition problems and a proposed solution. *Proc. IEEE Intl. Conf. on Pattern Recognition (ICPR)*. Pg 3618–3621
42. Ghosh N, Bhanu B, Denina G (2009) Continuous evolvable Bayesian nets for human action analysis in videos. Proc ACM/IEEE Int Conf Distrib Smart Cameras (ICDSC) 194–201
43. Ghosh N, Bhanu B (2014) Evolving Bayesian graph for 3D vehicle model building from video. IEEE Trans Intell Transp Syst 15(2):563–578
44. Glocker B, Sotiras A, Komodakis N, Paragios N (2011) Deformable medical image registration: setting the state of the art with discrete methods. Annu Rev Biomed Eng 13:219–244
45. Grinker S (1980) Edge based segmentation and texture separation. Proc IEEE Int Conf Pattern Recogn 554–557
46. Guo Y, Zhao G, Peitikainen M (2012) Discriminative features for texture description. Pattern Recogn 45(10):3834–3843
47. Hanjalic A, Lagendijk RL, Biemond J (1997) A new method for key frame based video content representation. Ser Softw Eng Knowl Eng 8:97–110
48. Hu Y, Murphy RF (2004) Automated interpretation of subcellular patterns from immunofluorescnece microscopy. J Immunol Methods 290:93–105
49. Huang K, Murphy RF (2004) From quantitative microscopy to automated image understanding. J Biomed Optics 9(5):893–912
50. Huang K, Murphy RF (2004b) Automated classification of subcellular patterns in multicell images without segmentation into single cells. Proc IEEE Int Symp Biomed Imaging (ISBI) 1139–1142
51. IEEE TIP Special Issue (2005) IEEE Trans Image Process. Spec Issue Mol Cell Bioimaging 14(9):1233–1410
52. Ilea DE, Whelan PF (2011) Image segmentation based on the integration of color-texture descriptors—a review. Pattern Recogn 44(10–11):2479-2501
53. Imelinska C, Downes MS, Yuan W (2000) Semi-automated color segmentation of anatomical tissue. Comput Med Imaging Graph 24:173–180
54. Iwano M, Shiba H, Miwa T, Che FS, Takayama S, Nagai T et al (2004) Ca^{2+} dynamics in a pollen grain and papilla cell during pollination of Arabidopsis. Plant Physiol 136(3):3562–3571
55. Kafai M, Eshghi K, Bhanu B (2014) Discrete cosine transform based locality-sensitive hashes for retrieval. IEEE Trans Multimedia (In press)
56. Kamath SB, Chidambar S, Brinda BR, Kumar MA, Sarada R, Ravishankar GA (2005) Digital image processing – an alternative tool for monitoring of pigment levels in cultured cells with special reference to green alga Haematococcus pluvialis. Biosens Bioelectron 21 (5):768–773
57. Korb KB, Nicholson AE (2011) Bayesian artificial intelligence (2nd Edn). Chapman & Hall/CRC Publication
58. La-Iglesia BD (2013) Evolutionary computation for feature selection in classification problems. Data Min Knowl Disc 3(6):381–407
59. LaTorre A, Alonos-Nanclares L, Muelas S, Pena JM, DeFelipe J (2013) 3D segmentations of neuronal nuclei from confocal microscope image stacks. Frontiers Neuroanat 7:49
60. Law AKW, Lam KY, Lam FK, Wong TKW, Poon JLS, Chan HY (2003) Image analysis system for assessment of immunohistochemically stained marker (MIB-1) in oesophageal squamous cell carcinoma. Comput Methods Programs Biomed 70(1):37–45

61. Lee HC, Kim SD (2003) Iterative key frame selection in the rate-constraint environment. Sig Process Image Commun 18(1):1–15
62. Li Y, Sorefan K, Hemmann G, Bevan MW (2004) Arabidopsis NAP and PIR regulated Actin-based cell morphogenesis and multiple development processes. Plant Physiol 136 (3):3616–3627
63. Li J, Newberg JY, Uhlen M, Lundberg E, Murphy RF (2012) Automated analysis and reannotation pf subcellular locations in confocal images from the human protein atlas. PLoS ONE 7(11):e50514
64. Li F, Yin Z, Jin G, Zhao H, Wong STC (2013) Bioimage informatics for systems pharmacology. PLoS Comput Biol 9(4):e1003043
65. Lin Y, Dou J, Wang H (1992) Contour shape description based on an arch height function. Pattern Recogn 25:17–23
66. Lin Y, Bhanu B (2005) Object detection via feature synthesis using MDL-based genetic programming. IEEE Trans Syst, Man Cybern, Part B 35(3):538–547
67. Lindek S, Fritsch R, Machtynger J, de Alarcon PA, Chagoyen M (1999) Design and realization of an on-line database for multidimensional microscopic images of biological specimens. J Struct Biol 125(2):103–111
68. Liu H, Srinath M (1990) Partial shape classification using contour matching in distance transformation. IEEE Trans Pattern Anal Mach Intell 12(11):1072–1079
69. Liu Y, Dellaert F (1998) A classification based similarity metric for 3D image retrieval. Proc IEEE Int Conf Comput Vis Pattern Recogn (CVPR) 800–805
70. Loncaric S (1998) A survey of shape analysis techniques. Pattern Recogn 31(8):983–1001
71. Loss L, Bebis G, Parvin B (2011) Iterative tensor voting for perceptual grouping of ill-defined curvilinear structures. IEEE Trans Med Imaging 30(8):1503–1513
72. Lumia R, Haralick RM, Zuniga O, Shapiro L, Pong TC, Wang FP (1983) Texture analysis of aerial photographs. Pattern Recogn 16(1):39–46
73. Maragos P (1989) Pattern spectrum and multiscale shape representation. IEEE Trans Pattern Anal Mach Intell 11(7):701–716
74. Marx V (2002) Beautiful bioimaging for the eyes of many beholders. Science 297:39–40
75. Meijering E, Niessen W, Weickert J, Viergever M (2002) Diffusion-enhanced visualization and quantification of vascular anomalies in three-dimensional rotational angiography: results of an in-vitro evaluation. Med Image Anal 6(3):215–233
76. Meijering E (2010) Neuron tracing in perspective. Cytometry Part A 77A(7):693–704
77. Miles A, Zhao J, Klyne G, White-Cooper H, Shotton D (2010) OpenFlyData: An exemplar data web integrating gene expression data on the fruit fly Drosophila melanogaster. J Biomed Inf 43(5):752–761
78. Mirabella R, Franken C, van der Krogt GNM, Bisseling T, Geurts R (2004) Use of the fluorescent timer DsRED-E5 as reporter to monitor dynamics of gene activity in plants. Plant Physiol 135(4):1879–1887
79. Mitchell T (1997) Machine learning. McGraw Hill Publication
80. Mohamadlou H, Shope JC, Flann NS (2014) Maximizing kolmogorov complexity for accurate and robust bright field cell segmentation. BMC Bioinf 15(1):32
81. Molina LC, Belanche L, Nebot A (2002) Feature selection algorithms: a survey and experimental evaluation. Proc IEEE Int Conf Data Min (ICDM) 306–313
82. Murphy RF, Velliste M, Porreca G (2003) Robust numerical features for description and classification of the subcellular location patterns in fluorescence microscope images. J VLSI Sign Process Syst Sign, Image Video Technol 35(3):311–321
83. Murphy RF (2011) An active role for machine learning in drug development. Nat Chem Biol 7(6):327–330
84. Murphy RF (2012) Cell Organizer: Image-derived models of subcellular organization and protein distribution. Methods Cell Biol 110:179–193
85. Nagasaka A, Tanaka Y (1991) Automatic video indexing and full-video search for object appearances. Proc Working Conf Vis Database Syst 119–133

86. Nam D, Mantell J, Bull D, Verkade P, Achim A (2014) A novel framework for segmentation of secretory granules in electron microscopy. Med Image Anal 2014(18):411–424
87. Nanni L, Lumini A, Brahnam S (2010) Local binary patterns variants as texture descriptors for medical image analysis. Artif Intell Med 49(2):117–125
88. Naik AW, Kangsas JD, Langmead CJ, Murphy RF (2013) Efficient modeling and active learning discovery of biological responses. PLoS ONE 8(12):e83996
89. Nayak N, Chang H, Borowsky A, Spellman P, Parvin B (2013) Classification of tumor histopathology via sparse feature learning. Proc IEEE Int Symp Biomed Imaging 410–413
90. Newberg J, Hua J, Murphy RF (2009) Location proteomics: systematic determination of protein subcellular location. Methods Mol Biol 500:313–332
91. Nosrati MS, Hamarneh G (2013) Segmentation of cells with partial occlusion and part configuration constraint using evolutionary computation. Med Image Comput Assist Interv 16(Pt 1):461–468
92. Okada T, Miller MJ, Parker I, Krummel MF, Neighbors M, Sb Hartley et al (2005) Antigen-engaged B cells undergo chemotaxis toward the T zone and form motile conjugates with helper T cells. PLoS Biol 3(6):e150
93. Oliveira FP, Tavares JM (2014) Medical image registration: a review. Comput Methods Biomech Biomed Eng 17(2):73–93
94. Parvin B, Fontenay G, Yang Q, Barcellos-Hoff MH (2003) BioSig: an imaging bioinformatics system for phenotypic analysis. IEEE Trans Syst, Man, Cybernatics Part B 33(5):814–824
95. Parvin B, Ghosh N, Heiser L, Knapp M, Talcott C, Laderoute K, et al (2007) Spectral decomposition of signaling networks. Proc IEEE Symp Comput Intell Bioinf Comput Biol (CIBCB) 76–81
96. Peng H (2008) Bioimage informatics: a new area of engineering biology. Bioinformatics 24 (17):1827–1836
97. Peng T, Murphy RF (2011) Image-derived, three-dimensional generative models of cellular organization. Cytometry Part A 79A(5):383–391
98. Peng H, Roysam B, Ascoli GA (2013) Automated image computing reshapes computational neuroscience. BMC Bioinf 14:293
99. Pietimainen M, Rosenfeld A (1981) Image segmentation by texture using pyramid node linking. IEEE Trans Syst Man Cybern 11(12):822–825
100. Raafat HM, Wong AKC (1980) Texture information directed algorithm for biological image segmentation and classification. Proc Int Conf Cybern Soc 1003–1008
101. Raafat HM, Wong AKC (1986) Texture based image segmentation. Proc IEEE Int Conf Comput Vis Pattern Recogn (CVPR) 469–475
102. Reed TR, Wechsler H, Werman M (1990) Texture segmentation using a diffusion region growing technique. Pattern Recogn 23(9):953–960
103. Reed TR, Dubuf JM-H (1993) A review of recent texture segmentation and feature extraction techniques. Comput Vis Graph Image Process: Image Underst 57(3):359–372
104. Ritter G, Schreib G (2001) Using dominant points and variants for profile extraction from chromosomes. Pattern Recogn 34(4):923–938
105. Rittscher J (2010) Characterization of biological processes through automated image analysis. Annu Rev Biomed Eng 12:315–344
106. Rizk A, Paul G, Incardona P, Buqarski M, Mansouri M, Neimann A et al (2014) Segmentation and quantification of subcellular structures in fluorescence microscopy imaged using Squassh. Nat Protoc 9(3):586–596
107. Rodriguez A, Shotton DM, Trelles O, Guil N (2000) Automatic feature extraction in wound healing videos. In: Proceedings of 6th RIAO conference on content-based multimedia information access. Paris
108. Ronneberger O, Baddeley D, Scheipi F, Verveer PJ, Burkhardt H, Cremer C et al (2008) Spatial quantitative analysis of fluorescently labeled nuclear structures: problems, methods, pitfalls. Chromosom Res 16(3):523–562

109. Sabri S, Richelme F, Pierres A, Benoliel AM, Bongrand P (1997) Interest of image processing in cell biology and immunology. J Immunol Methods 208(1):1–27
110. Salton G, Buckley C (1990) Improving retrieval performance by relevance feedback. J Am Soc Inf Sci 41(4):288–297
111. Schoenenmann T, Cremers D (2006) Near real-time motion segmentation using graph cuts. In: Franke K et al (Eds) Pattern recognition. Springer, LNCS 4174, Berlin, pp 455–464
112. Sethi IK, Patel N (1995) A statistical approach to scene change detection. Proc SPIE Symp Storage Retrieval Image Video Databases III 2420:329–338
113. Shahrary B, Gibbon DC (1995) Automatic generation of pictorial transcript of video programs. Proc SPIE Digital Video Compression: Algorithms Technol 2420:512–518
114. Shapiro LG, Stockman GC (2001) Computer vision. Prentice Hall Publications
115. Shariff A, Kangas J, Coelho LP, Quinn S, Murphy RF (2010) Automated image analysis for high-content screening and analysis. J Biomol Screen 15(7):726–734
116. Shi J, Malik J (2000) Normalized cuts and image segmentation. IEEE Trans Pattern Anal Mach Intell 22(8):888–905
117. Shindyalov IN, Bourne PE (1998) Protein structure alignment by incremental combinatorial extension (CE) of the optimal path. Protein Eng 11(9):739–747
118. Shotton DM, Rodriguez A, Guil N, Trelles O (2000) Object tracking and event recognition in biological microscopy videos. Proc IEEE Int Conf Pattern Recogn (ICPR) 4:226–229
119. Soll DR, Wessels D (eds) (1998) Motion Analysis of Living Cells. Wiley-Liss Publication, New York
120. Sommer C, Gerlich DW (2013) Machine learning in cell biology – teaching computers to recognize phenotypes. J Cell Sci 126(24):5529–5539
121. Spann M, Wilson R (1985) A quad-tree approach to image segmentation which combines statistical and spatial information. Pattern Recogn 18(3):257–269
122. Tachino RM, Kabuyama N, Gotoh T, Kagei S, Naruse M, Kisu Y et al (2003) High-throughput classification of images of cells transfected with cDNA clones. CR Biol 326(10):993–1001
123. Thomann D, Dorn J, Sorger PK, Danuser G (2003) Automatic fluorescence tag localization II: improvement in super-resolution by relative tracking. J Microsc 211(3):230–248
124. Tuceryan M, Jain AK (1998) Texture analysis. In: Chen CH, Pau LF, Wang PSP (eds) The handbook of pattern recognition and computer vision (2nd edn). World Scientific Publishing Co Pte Ltd 207–248
125. Vapnik VN (1998) Statistical learning theory. Wiley, New York
126. Verbeek PW, DeJong DJ (1984) Edge preserving texture analysis. Proc IEEE Int Conf Pattern Recogn (ICPR) 1030–1032
127. Volkmann N (2010) Methods for segmentation and interpretation of electron tomographic reconstructions. Methods Enzymol 483:31–46
128. Wang G, Song Q, Sun H, Zhang X, Xu B, Zhou Y (2003) A feature subset selection algorithm automatic recommendation method. J Artif Intell Res (JAIR) 47:1–34
129. Weiss Y (1999) Segmentation using eigenvectors: a unifying view. Proc IEEE Int Conf Comput Vis (ICCV) 2:975–982
130. Wermser D (1984) Unsupervised segmentation by use of a texture gradient. Proc IEEE Int Conf Pattern Recogn (ICPR) 1114–1116
131. Willis B, Turner JN, Collins DN, Roysam B, Holmes TJ (1993) Developments in three-dimensional stereo brightfield microscopy. Microsc Res Tech 24(5):437–451
132. Witt CM, Roychaudhuri S, Schaefer B, Chakraborty AK, Robey EA (2005) Directed migration of positively selected thymocytes visualized in real time. PLoS Biol 3(6):e160
133. Wolf W (1996) Key frame selection by motion analysis. Proc IEEE Int Conf Acoust, Speech, Sign Process (ICASSP) 2:1228–1231
134. Wolf L, Shashua A (2005) Feature selection for unsupervised and supervised interface: the emergence of sparsity in a weight-based approach. J Mach Learn Res 6:1855–1887
135. Xiao J, Shah M (2005) Motion layer extraction in the presence of occlusion using graph cut. IEEE Trans Pattern Anal Mach Intell 27(10):1644–1659

136. Xiong W, Li CM, Ma RH (1997) Automatic video data structuring through shot partitioning and key-frame computing. Mach Vis Appl 10(2):51–65
137. Yang G (2013) Bioimage informatics for understanding spatiotemporal dynamics of cellular processes. Wiley Interdisc Rev: Syst Biol Med 5(3):367–380
138. Yang Q, Parvin B (2003) Harmonic cut and regularized centroid transform for localization of subcellular structures. IEEE Trans Biomed Eng 50(4):469–475
139. Yeung MM, Liu B (1995) Efficient matching and clustering of video shots. Proc IEEE Int Conf Image Process (ICIP) 1:338–342
140. Yilmaz A, Javed O, Shah M (2006) Object tracking: A survey. ACM Comput Surv 38(4):13
141. Zabih R, Miller J, Mai K (1999) A feature-based algorithm for detecting and classifying production effects. Multimedia Syst 7(2):119–128
142. Zappella L, Llado X, Salvi J (2008) Motion segmentation: a review. Proc Conf Artif Intell Res Dev 398–407
143. Zaritsky A, Natan S, Horev J, Hecht I, Wolf L, Ben-Jacob E, Tsarfaty I (2011) Cell motility dynamics: a novel segmentation algorithm to quantify multi-cellular bright field microscopy images. PLoS ONE 6(11):e27593
144. Zhang HJ, Kankanhalli A, Smoliar SW (1993) Automatic partitioning of full-motion video. Multimedia Syst 1(1):10–28
145. Zhang J, Tan T (2002) Brief review of invariant texture analysis methods. Pattern Recogn 35(3):735–747
146. Zhang D, Lu G (2004) Review of shape representation and description techniques. Pattern Recogn 37(1):1–19
147. Zhao J, Miles A, Klyne G, Shotton D (2009) Linked data and provenance in biological data webs. Briefings Bioinf 10(2):139–152
148. Zitova B, Flusser J (2003) Image registration methods: a survey. Image Vis Comput 21(11):977–1000

Part II
Organismal Dynamics: Analyzing Brain Injury and Disease

Chapter 3
High- and Low-Level Contextual Modeling for the Detection of Mild Traumatic Brain Injury

Anthony Bianchi, Bir Bhanu and Andre Obenaus

Abstract Traumatic brain injury (TBI) can lead to long-term neurological decrements. While moderate and severe TBI are readily discernable from current medical imaging modalities, such as computed tomography and magnetic resonance imaging (MRI), mild TBI (mTBI) is difficult to diagnose from current routine imaging. At the present time there no routine computational methods for the evaluation of mild traumatic brain injury (mTBI) from magnetic resonance imaging (MRI). The development of automated analyses has been hindered by the subtle nature of mTBI abnormalities, which appear as low contrast MR regions. One solution to better identify mTBI injuries from MRI is to use high-level and low-level contextual information. We describe methods and results for using high-level contextual features using a Bayesian network that simulated the evolution of the mTBI injury over time. We also utilized low-level context to obtain more spatial (within the brain) information. The low-level context utilized a classifier to identify temporal information which was then integrated into the subsequent time point being evaluated. We found that both low- and high-level context provided novel information about the mTBI injury. These were in good agreement with manual methods. Future work could combine both low- and high-level context to provide more accurate mTBI segmentation. The results reported herein could ultimately lead to better identification of regions of mTBI injury and thus when treatments become available they can be directed for improved therapy.

A. Bianchi · B. Bhanu
Department of Electrical Engineering, University of California Riverside,
Riverside, CA, USA
e-mail: abianchi@ee.ucr.edu

B. Bhanu
e-mail: bhanu@ee.ucr.edu

A. Obenaus (✉)
Department of Pediatrics, Loma Linda University, 11175 Campus St,
Coleman Pavilion Rm A1120, 92350 Loma Linda, CA, USA
e-mail: aobenaus@llu.edu

© Springer International Publishing Switzerland 2015   59
B. Bhanu and P. Talbot (eds.), *Video Bioinformatics*,
Computational Biology 22, DOI 10.1007/978-3-319-23724-4_3

3.1   Introduction

Annually, millions of people in the United States are affected by mild traumatic brain injury (mTBI) often as a consequence of an event including but not limited to sports, military activities (blast), automobile accidents, assaults, and falls [1]. Recently, the short and long-term symptoms associated with mTBI have become more apparent and include: loss of memory, loss of reasoning, neuropsychiatric alterations including decrements in social interactions [2]. Clinically, the Glasgow Coma Scale (GCS) is the current standard method for evaluating acute neurological injuries, which evaluates the patient's consciousness level through a series of mobile, verbal, and, visual stimuli. The GCS is a qualitative scale that is used extensively, in part due to its ease in delivery by a variety of health professionals. However, advances in medical imaging, such as computed tomography and magnetic resonance imaging (MRI) are used extensively to confirm the severity of injury [3]. The presence of altered brain structure(s) or small hemorrhages is a common occurrence in moderate to severe TBI but is rarely seen in mild TBI [4]. In fact, recent clinical practice guidelines define mild TBI as having no overt alterations on standard structural imaging [5]. mTBI suffers from no dramatic GCS or imaging observations, making diagnosis difficult. Thus, newer and more sophisticated methods for identifying emerging brain injury after mTBI from neuroimaging are needed.

   Clinical studies along with research in animal models clearly demonstrate no overt structural changes within the brain following mTBI at the acute time point but are now starting to identify long-term neuropsychological disruptions (i.e., anxiety, depression etc.) [6, 7]. The physiological basis in many of these mTBI cases appears to be a disruption in white matter tracts within the brain [8–10]. Computational methods and approaches can be used to identify emerging abnormalities within the brain following TBI, particularly mild injury on MRI where visual alterations are difficult to observe. Recently, we undertook such a study where we used computational analysis of the injury site to determine increased vulnerability to the brain following repeated mTBI [11]. Using these computational approaches, we reported that there was increased extravascular bleeding when animals received a second mTBI to the contralateral hemisphere within 7 days after a first mTBI. Thus, computational analyses of MRI data can identify physiological alterations that would be virtually impossible using either visual or manual methods of detection.

   Currently, visual and manual detection of abnormalities in many neurological diseases, including mTBI, is considered the "gold standard" in neuroimaging. Numerous studies have used manual detection to identify lesion location and size from MRI with correlative histology and assessment of long-term neurological effects [12–14]. Manual detection has numerous deficiencies that include, hours of analysis per scan, requires a trained operator, inter- and intra-operator error, and difficulty in multimodal analysis. These deficiencies identify the need for a rapid and consistent automated approach.

To automate mTBI detection from magnetic resonance images (MRI), we formulated the problem as a voxel-based classification problem, where we find the probabilities of a voxel being a lesion based on features of the voxel's local neighborhood. Recent studies in other neurological diseases have formulated the problem in a similar manner by utilizing the discriminatory potential of texture in MRI [15, 16]. Texture analysis allows for the identification of changes within the tissues of interest. Holli and colleagues showed significant differences in the texture of brain structures in patients with mTBI and healthy controls, suggesting that use of texture in mTBI analysis could identify alterations not found by other methods [17, 18]. Therefore, visual assessment of mTBI alone does not provide sufficient information for lesion identification or accurate segmentation of the injury site, particularly in mTBI.

To improve the segmentation of mTBI lesions, we have introduced an additional analysis feature, contextual information. We use the following definition of context, "any information that might be relevant to object detection, categorization and classification tasks, but not directly due to the physical appearance of the object, as perceived by the image acquisition system" [19]. Contextual image types that have been described include: local pixels, 2D scene gist, 3D geometric, semantic, photogrammetric, illumination, weather, geographic, temporal, and cultural [20]. Context can also be split into two types, low-level and high-level, where: (a) low-level context includes spatial relationships learned at the pixel/region level and has been used in conditional random fields [21] and autocontext [22]. (b) high-level context can be thought of as an estimation of the spatial location of an object. We have previously exploited low (voxel) level context to identify mTBI lesions, where our approach uses a contextually driven generative model to estimate the lesion location [23]. We now describe our efforts to explore a high-level context approach followed by use of our previously successful application of low-level context to improve detection of mTBI injury sites.

3.2   High-Level Contextual Modeling for MTBI

3.2.1   Description of Our High-Level Contextual Modeling for MTBI

Our proposed system uses a combination of visual and high-level contextual modeling. We utilized a database of known mTBI cases (rodent model of TBI) that includes multimodal MRI where the lesion volumes have been manually detected. These and other associated contexts were used to build the model. The visual model uses 3D texture features to build a Probabilistic Support Vector Machine (pSVM) that describes the lesion and normal appearing brain matter (NABM) space. pSVM is a discriminative model which performs well with a large amount of data and can describe complex decision spaces. A voxel-based classifier satisfies both of these conditions. The resultant output is a probability map that is combined with the contextual model, where higher probabilities signify an increased chance for a particular voxel being part of a lesion. Lesion components were further defined to consist of both blood and edema, similar to our previously published findings [11]. One strength of this approach is that if a lesion is not contiguous, that is, if the lesion is not located in only a one focal brain region but rather multiple locations, we are still be able to capture the lesion entirety as a voxel-based classifier is used. This is achieved by adding all the locations together to identify total lesion.

The contextual model uses a Bayesian network (BN) to estimate the location(s) of the mTBI lesions (Fig. 3.1). This model uses prior knowledge about the subject, where both temporal and spatial information are used to describe the development of the mTBI over time. An advantage of using a BN is that it is a generative model that is able to extrapolate based on little information when the underlying distribution and assumptions hold. Generative models are used to simulate cause and effect relationships and processes, while discriminative models such as SVM do not have this ability. Our computational system combines the advantages from both discriminative and generative models.

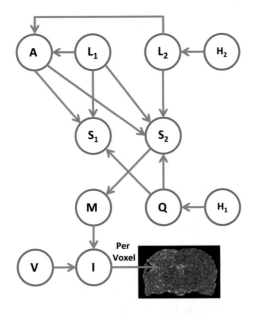

Fig. 3.1 Graphical representation of the Bayesian network showing the dependencies of random variables. Intuitive definition of distributions: A—anatomical constraints, L_1, L_2—central location of injury for injury one and two, H_I—time since the first event, H_2—time between first and second injury, Q—volume (quantity) of injury with time, S_1, S_2—spread for first and second injury, M—max operator, V—visual information, I—estimated injury. Where $I = 1$ is absolute certainty of injury and $I = 0$ is absolute certainty of NABM (normal appearing brain matter)

When an MRI data volume enters the system, the texture features are extracted and then the BN is implemented to evaluate the dependencies of the variables (Fig. 3.1). Contextual information about the mTBI is also utilized by the system including an estimated location of the focal points of impact (L_1, L_2), time since the first injury (H_1), and time interval between the two injuries (H_2) (Fig. 3.1). In this particular model, we are identifying the effects of repeated mTBI. These additional contextual inputs can be exact (e.g., 1 day since injury) or a range of values (e.g., 5–14 days since injury).

Like many other neurological injuries, mTBI evolves with time, so our contextual model (Fig. 3.1) includes multiple random variables (H, L, Q, etc.) that also evolve with time. H_1 is the distribution of the time since the first injury, which can be learned directly from the data or it can be modeled as an exponential distribution since it is most probable that a patient will arrive soon after a mild injury. The random variable Q is the volume of injury over time. This distribution can also be learned from the data directly or modeled as a lognormal distribution. This distribution follows the natural progression of the mild brain injury where there is an initial peak of an abnormality that slowly resolves over time. H_2 is the time between the first and second injury (i.e., repeated mTBI). After an initial injury, metabolic and cellular injury cascades occur which may lead to a window of increased vulnerability whereby a second mTBI may worsen ongoing pathology [24]. We have reported similar effects in mTBI previously [11]. The H_2 function can also be modeled directly from the data or using an exponential distribution from the time since first injury. The process of subsequent injuries can be thought of as a Poisson process, which has the time between injuries as an exponential distribution. H_1 and H_2 are variables that would be dependent on location and can be estimated through a regional epidemiological study, since regions will have different medical response times (i.e., time to see a physician).

The components of the BN that capture the evolution of the lesion over time are S_1 and S_2. In our model, Eq. (3.1) is used to describe the injury progression given all the contextual inputs. This function is a sigmoid like function, known as the Gompertz function [25], which has parameters (m, n) to control the shape and displacement. The Gompertz function has more flexibility than a logistic function and less number of parameters than the generalized logistic function. In Eq. (3.1) parameter m determines the displacement of the function and n determines the decay rate.

$$p(S|L, A, H, Q) = Ae^{-\frac{mH}{Q}e^{\frac{nH}{Q}d(x,L)}} \qquad (3.1)$$

The shape and displacement parameters (m, n) are not the only variables that affect the shape of the curve; Q (quantity of lesion) also determines the shape. When Q is large, the function will shift to the right and have a more gradual slope. This represents a potentially larger area for injury and increases the uncertainty of the lesion location. When Q is small, the opposite is observed, the area is small and there is more certainty in the injury location. Q is estimated by taking the average lesion size at each time point for all the volumes in the database. There are a separate set of parameters for each H_2 value because the shape of the injury can be different when the repeated injuries are at distinct times from each other (i.e., 1 day apart vs. 7 days apart) (see [11]). When the first mTBI occurs, there are many cellular cascades that take place throughout the progression of the injury potentially leaving the brain vulnerable to subsequent mTBIs [24], similar to what has been found in animal models of TBI [11, 26].

As described in Eq. (3.2), d is a distance function weighted by Σ, where Σ weights the 3D space so it accounts for rotation and scaling. Equation (3.2) also measures the distance from L (location of injury) to every other point in the 3D space. The parameters described in Eqs. (3.2–3.6) describe the rotation and scaling in the coronal space, and σ_z describes the scaling in the z space. The parameters that need to be set in these functions are σ_x and σ_y, which control the scaling in the direction of the tangent and the direction of the normal, respectively. Finding the tangent angle θ at L is done by utilizing the Fourier descriptor of the closed perimeter of the brain slice [27]. In addition, when converting to Fourier space a low-pass filter is applied that cuts off the upper twenty percent of the frequencies, which smoothens the low resolution noise at the brain boundary. Finally, the parameters, a, b, and c, control the rotation in the coronal plane using θ, to create a new "axis" for the distance function. Equation (3.1) models the progression of mTBI, since the injury is assumed to more likely spread along the perimeter of the brain than migrate into the center of the brain. This is a valid assumption as there are physical and physiological barriers within the brain to prevent significant migration into deeper brain tissues. Hence, the distance function is weighted more along the tangent of the perimeter.

$$d(x, L) = \sqrt{(x - L)^T \Sigma (x - L)} \tag{3.2}$$

$$\Sigma = \begin{bmatrix} a & b & 0 \\ b & c & 0 \\ 0 & 0 & \sigma_z \end{bmatrix} \tag{3.3}$$

$$a = \frac{\cos^2 \theta}{2\sigma_x^2} + \frac{\sin^2 \theta}{2\sigma_y^2} \tag{3.4}$$

$$b = \frac{\sin 2\theta}{4\sigma_x^2} + \frac{\sin 2\theta}{4\sigma_y^2} \tag{3.5}$$

$$c = \frac{\sin^2 \theta}{2\sigma_x^2} + \frac{\cos^2 \theta}{2\sigma_y^2} \tag{3.6}$$

The final contextual model is shown in Eq. (3.7). This function takes the maximum of the two spread functions evaluated over each value in the range of each contextual input. The contextual model output is an estimate of the lesion extent. If a contextual input is known then the probability has a value of 1. For example, if H_1 is known to be 3 days then $p(H_{1,3}) = 1$ and all other values in the distribution equal zero. Another example is when a ranged input is given. If H_1 is known to be between 5 and 14 days, the distribution becomes the priors known at those values normalized by the sum of the probabilities at those values (all other values are set to zero). When one of the contextual inputs is not known, all values in the prior distribution are considered.

$$p(M|S_1, S_2) = \sum_{\forall i | P(L_i) > 0,} \sum_{\forall j | P(Q_j) > 0,} \sum_{\forall k | P(H_{2k}) > 0}$$
$$\max(p(L_i)p(Q_j)p(H_{2k})p(S_{1ijk}),$$
$$p(L_i)p(Q_j)p(H_{2k})p(S_{2ijk})) \tag{3.7}$$

$$p(I|M, V) = p(M)p(V) \tag{3.8}$$

An estimation of the total lesion volume is finally given by Eq. (3.8). The contextual model and visual model are independent. This assumption has been made by other approaches using context [19]. The final output is a probability map giving a probability of each voxel being part of the lesion.

The visible model is the probability of an injury based on the visual appearance of a sample voxel within the MR image. While our approach lends itself to any MR imaging modality, for simplicity we have chosen to utilize a standard MR image based on physical relaxation properties within the tissues, T2-weighted imaging (T2WI). The actual physical relaxation values (in milliseconds) can be obtained from the quantitative T2 map, removing any computational issues related to use of signal intensities. T2WI allows for observation of both edema (increased signal intensities and T2 values) and the presence of extravascular blood (decreased signal intensities and T2 values). Both edema and blood are potentially at the site of injury, but are dependent upon the time post mTBI [11, 14].

Since edema and blood are known to have local tissue effects, the textures around known lesions can be used in our computational model. The texture features that were used in this study were 3D local mean, variance, skewness, kurtosis, entropy, range, gradient x, gradient y, gradient z. These local statistics were incorporated as they are able to describe small local neighborhoods at every voxel, which gives a description of the tissue type. We found that a local neighborhood feature size of $5 \times 5 \times 5$ was the most discriminative for our dataset. In our current mTBI dataset, the MRI features that were selected from the T2WI and maps were: T2 value, entropy, variance, skewness, gradient in x-direction, gradient in y-direction and the mean T2 value. A sample of the output from the combined visual and contextual models as well as their fusion is shown in Fig. 3.2.

Fig. 3.2 A typical example
of the fusion between visual
and contextual models in a rat
mTBI. In this example, the rat
experienced two mTBIs
several days apart to opposite
brain hemispheres. *Top*
Probability map after pSVM.
Middle Probability map from
the contextual model. *Bottom*
Fusion of the contextual and
visual models. Note that this
illustrates that the contextual
model for repeated injuries
progress at different rates
(compare hemispheres, where
the right hemisphere was
injured first)

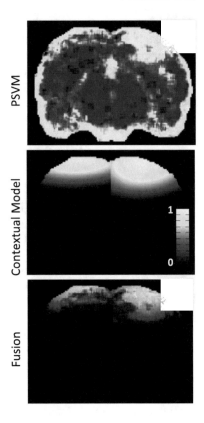

3.2.2 Acquisition of MRI Data in a Rat Model of MTBI

Mild TBI experiments were undertaken in adult Sprague Dawley rats that were
randomly assigned to three experimental groups: Single mTBI, and repeated mTBI
(rmTBI) induced 3 or 7 days apart. A mild controlled cortical impact (CCI) was
used to induce mild injury, as previously described [11, 14]. The CCI model
induces a focal injury that can be modified easily for induction of mild, moderate, or
severe traumatic injuries. Briefly, a craniectomy (5 mm) was performed over the
right hemisphere (3 mm lateral, 3 mm posterior to bregma) where a mild CCI was
induced using an electromagnetically driven piston (4 mm diameter, 0.5 mm depth,
6.0 m/s). These parameters result in a mild cortical deformation and we confirmed
that the dura is not ruptured. The overlying skin is then sutured closed after the
injury is induced followed by the appropriate postoperative care [11]. For those
animals that received a repeated mTBI at 3 or 7 days after the initial injury, a
second craniotomy on the contralateral hemisphere was performed and a CCI using
the identical parameters was induced (Fig. 3.3). Again, the overlying skin was
suture closed followed by the appropriate postoperative care. Animals in which the
dura was ruptured were excluded from the study.

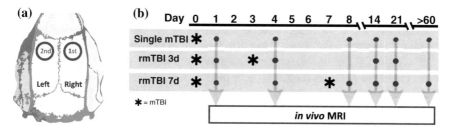

Fig. 3.3 Experimental design. **a** Illustration of the mTBI locations for the first (*right*) and second (*left*) injuries. **b** Experimental mTBI and neuroimaging timeline. A single mTBI was induced to the *right* cortex on day 0 (denoted as an *) in all animals. A second mTBI was induced to the left cortex at either 3 or 7 days later (*). MR imaging was performed 1 day post-first (1d), 1 day post-last (4 day rmTBI 3 day, 8 day rmTBI 7 day) and 14 day post-injury (17 day rmTBI 3 day, 21 day rmTBI 7 day) (*red circles*)

Neuroimaging was obtained and in vivo T2WI were collected as multiple 2D slices as previously described [14] (Fig. 3.3). T2WI data were obtained using a 4.7T Bruker Avance (Bruker Biospin, Billerica, MA). T2WI sequences were comprised of the following parameters: TR/TE = 3453 ms/20 ms, 25 slices at 1 mm thickness were collected with a field of view of 3 cm and with a 256 × 256 matrix. The resultant resolution of each voxel is 0.12 × 0.12 × 1 mm. On average 12 coronal slices covered the entire cerebrum.

MRI data from all three groups were collected at 1, 4, 8, 14, 21, 30, and 60 days after the first TBI (Fig. 3.3). After data were acquired T2 maps were computed using custom written software. Trained individuals then manually drew regions-of-interest (ROIs) using Cheshire imaging software (Haydan Image/Processing Group) to denote regions of TBI lesions on the T2 images and maps [11]. An outline of the whole brain was also drawn. Thus, ROIs were comprised of normal appearing brain matter (NABM) and cortical tissues that contained T2-observable abnormalities using criteria previously described [11]. Skull stripping was performed by drawing manual outlines to separate the brain from the skull and other tissues. Based on our dataset we obtained a total of 81 rodent brain volumes to test our computational models.

3.2.3 Results of High-Level Contextual Modeling for MTBI

Our derived experimental data on repeated mTBI were tested using contextual inputs to verify the effects they have in our model using our real MRI data. Testing was conducted using the leave-one-out approach to validate the learned parameters. The following cases were examined: all contexts known, L_1 and L_2 (location), unknown position (unkPOS), H_1 unknown time (unkTIME), L_1 L_2 H_1 unknown (unkALL), and V alone (Probabilistic SVM). ROC plots for each of the aforementioned contextual tests were obtained after thresholding the output probability maps to get a hard classification (i.e., nonprobabilistic) (Fig. 3.4). When all the contexts are known we observed an equal error rate of 0.93. Not surprisingly, with missing contextual information there is a decrease in the performance of the model.

The Dice plot results followed similar trends as found in the ROC plot (Fig. 3.4). With unknown position (unkPOS) and unknown time (unkTIME), a similar performance in the contextual model is observed since each of these cases results in a smoothing of the output. This smoothing effect is clearly seen in Fig. 3.5. Essentially, when the context is not known, the model considers a wider area of brain tissue. In the case of unknown position, smoothing is experienced along the edges of a brain as the focal point is shifted along the perimeter of the brain. The case of unknown time results in smoothing in all directions, which is due to those time points with a larger volume of TBI lesion having the largest effect on the

Fig. 3.4 a ROC plot with multiple contextual inputs. Results were evaluated after thresholding the output probability map. **b** Dice plot with multiple contextual inputs. The peak of the Dice curve (maximum dice coefficient) is used to threshold the output probability map. Legends in (**b**) are the same as in (**a**)

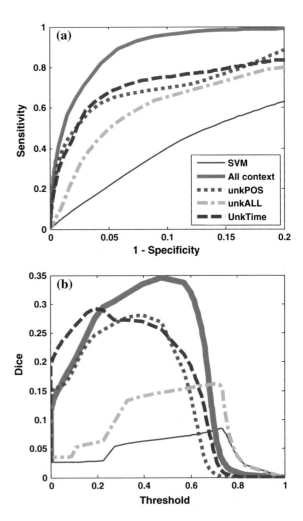

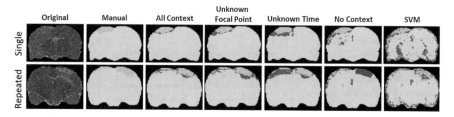

Fig. 3.5 Example outputs from the thresholded probability maps (for single and repeated mTBI) with varying levels of contextual input. The thresholds were selected at the highest point on the dice curve for each respective level of context. Each row is a coronal slice, where the *top* is from a single mTBI and the *bottom* is from a repeated mTBI. The colors represent the following: *Green* true positive, *Teal* true negative, *Red* false positive, *Brown* false negative

output from the contextual model. When all of the parameters are unknown (unkALL) smoothing due to lack of time and position occur. Thus, it is optimal to provide some context to the model, but optimal results occur when all contexts can be provided (Figs. 3.4, 3.5).

3.3 Low-Level Contextual Approach

We now extend the high-level contextual model by examining a low-level contextual model that allows the classifiers to directly estimate the posterior probability of TBI lesion and normal appearing brain matter.

3.3.1 Description

A discriminative approach is used where the classifiers estimate the probability that a voxel contains either injured tissue or NABM, only performs well when there is a large amount of training data and if it can be used for a complex decision space. A voxel-level classifier has a large amount of data to evaluate considering the 3D nature of MR images. The appearance of lesions in MRI can be very complex (blood, edema, time, location, etc.), which leads to a complex decision space. It is important to note that the ground-truth was obtained from expert segmentation of the mTBI lesion from the experimental data.

In our low-level contextual model, we use a cascade of classifiers to estimate the detected lesion at each time point. Here, we also use information from a previous time point that is propagated forward. The first classifier in the cascade estimates the lesion using only the visual features. Then context features are computed from the posterior probability map that is estimated by the classifier. These features are recalculated for each iteration in the process and results in a given number of

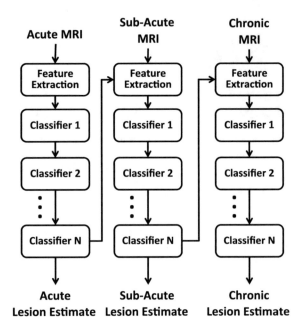

classifiers at each time point (Fig. 3.6). Spatial information is propagated by the contextual features leading to improved classification. This process has been previously used successfully for brain MR images [22].

The contextual features used by Tu and Bai [22] were point samplings of the posterior estimates. Their features demonstrated good segmentation performance on objects that were rigid in shape. However, when the shape was distorted their performance faltered. Given that MR images of the brain are often not rigid either between patients or at different imaging time points, an alternate solution is required. To overcome the shape distortion weakness, we propose two new features that generalize the static contextual information, thus allowing contextual components to work well with deformable objects.

Dynamic contextual features are calculated from the final classifier at a single time point (Fig. 3.6). Also, in our approach these features are used by the classifier at each subsequent time point where they make use of spatial and lesion growth information. Tu and Bai used spatiotemporal volumes with their basic point features in the higher dimensional space [22]. However, their approach would require extreme amounts of data and the entire sequence has to be known before segmentation. In contrast, our approach only considers pairs of brain volumes at a time, which allows for estimation at every time point.

The contextual features come from the posterior probability estimated by an already learned classifier. Previous approaches have directly sampled a dense neighborhood around an observed voxel, making each location a potential feature [22]. As noted above, this dense sampling method can lead to large feature sizes and subsequently result in over fitting due to the specific locations that are learned.

Here, we propose two new static features to overcome the problem of oversampling. One incorporates a sense of the surrounding without a predefined direction, while the other gives a general sense of direction.

The first feature gives the average posterior probability at various distances around the observed voxel. This can be thought of as a proximity feature where: what is close, medium and far away in distance is estimated. The distance function used here is the Manhattan distance allowing for a cuboidal region. These features are directionally invariant and can lead to better generalization since they describe a general area. By having a nesting of boxes the integral image can be utilized for rapid computation of the features. In 3D, only eight points are needed to find the average of a cuboidal region, using integral images [28]. Equation 3.9 provides these features, where f_{xyz} is the proximity feature, R_{1xyz}, R_{2xyz} are square neighborhoods around the voxel at xyz and where size denotes the size of the bounding box.

$$f_{xyz} = \left(\sum R_{1xyz} - \sum R_{2xyz} \right) \left(\frac{1}{size\left(R_{1xyz}\right) - size\left(R_{2xyz}\right)} \right) \qquad (3.9)$$

Directional information is important for classification since the TBI brains are rigidly registered to the brain of a naïve animal. The second contextual feature describes the posterior probability in various directions from the observed voxel (Fig. 3.7). Rays are sampled at various distance ranges and angles from the observed voxel (see Fig. 3.7b). From the distance ranges along the rays the mean is calculated leading to a refined sense of the surrounding. An example would be what is close and above the observed voxel. The integral image is also used to calculate these features. Both features can be computed at coarse or fine distance bins without a significant increase in computational time. Equation 3.9 can also be used for the direction context features, the shape of the features are changed to a rectangle with width 1.

The posterior marginal edge distance (PMED) feature is the distance a voxel is from the perimeter of objects of a class found by the maximum posterior marginal (MPM) estimate. To create this feature, first the MPM at a voxel is obtained from the output of a classifier (Eq. 3.11). This gives a binary image for each class. The

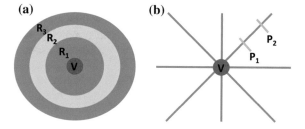

Fig. 3.7 a Illustration of the proximity feature. V is the observed voxel and the feature is the average probability of the regions (R_1, R_2, R_3). **b** Illustration of the distance features. V is the observed voxel and an example feature is the average probability between P_1 and P_2 along the 45° ray

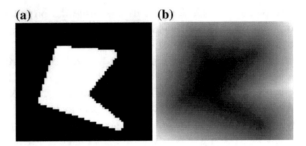

Fig. 3.8 **a** Example MPM estimate. **b** Corresponding PMED feature. Note that the values become more negative toward the center of the object and more positive farther away from the object

distance transform is applied to the image and the inverse image and the feature is given by Eq. 3.10.

$$PMED = d(MPM) - d(\sim MPM) \tag{3.10}$$

$$MPM = \underset{c}{\mathrm{argmax}}\, p(\omega = c|F) \tag{3.11}$$

Here d is the Euclidean distance transform. This gives an image that is increasing as the voxels become farther away from the edge and smaller (more negative) as the voxels get further into the object. Where ω is the estimated class, c is a specific class (lesion or normal brain in our case), and F is the feature set at a given voxel (Fig. 3.8).

3.3.2 Data

Data to test the contribution of the low-level context features used the same data set as described in Sec. 2.2 above. (see also [11, 14]). In this test, we limited the dataset to 3 MR imaging time points: acute (1st day post mTBI), subacute (8th day post-first mTBI), and chronic (14th day post-first mTBI). MRI data were acquired and processed as noted in Sec. 2.2. Right and left hemispheres and injured tissue volumes were defined as abnormal (i.e., lesion) when they included either hyper- or hypointense signal intensities within the cortex. All remaining tissues were designated as NABM. The data set was comprised of a total of six sequences, each with three time points.

3.3.3 Results

We examined the effect of our proposed features and effects of the dynamic information. For the training/testing split leave-one-out validation was used where a

whole sequence was left out (resulting in six cross-folds). Three temporally consecutive volumes remained for testing and the remaining data were used for training.
The training parameters were: 300 weak learners, learning rate 1, and 4 cascaded
classifiers. We tested three approaches: the original autocontext features [22], our
proposed approach with only the static features, and our proposed approach with the
static and dynamic features.

Our proposed dynamic approach clearly outperforms the other methods as can
be seen in the Dice coefficient curves (Fig. 3.9). These data illustrate the importance
of using dynamic information when these data types are available. The original
autocontext [22] tends to over fit due to the specific locations the features represent,
as we noted above. The same feature locations proposed by Tu and Bai [22] were
used for the original autocontext testing. During the training phase it obtained a
Dice score above 0.9, but interestingly it did not generalize well to our testing data,
perhaps due to the mild nature of the injury.

Our proposed static features give a good generalization compared to the original
autocontext [22]. Our proposed approach has a very flat Dice curve, suggesting that
it is not sensitive to a particular chosen threshold on the output probability
map. Thus, the selection of a threshold becomes less critical. The qualitative results
clearly demonstrate that our proposed low-level contextual approach works well on
small to medium size lesions (Fig. 3.10). The qualitative results also show that false
positives are only close to the lesion mass without having erroneous spatial outliers.
A caveat we found is that the proposed low-level context appears to have some
difficulties at the edges of the lesions, but we feel that this could be rectified in
future work using shape constraints on the thresholded probability maps.

Fig. 3.9 Dice coefficients
after thresholding the
posterior probability map at
the end of each cascade of
classifier (i.e., at each time
point). This is the average of
all the tests in the
leave-one-out validation

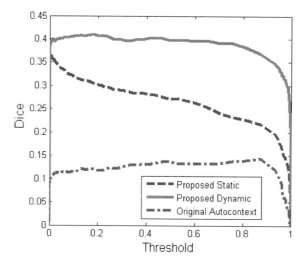

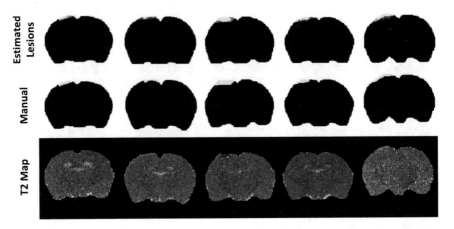

Fig. 3.10 Qualitative results of the proposed approach using dynamic and static contextual features. Each coronal slice is from a separate volume. *Color code*: *yellow* = true positive, *black* = true negative, *orange* = false negative, *brown* = false positive

3.4 Conclusions

Methods for integration of high- and low-level contextual features are demonstrated in a data set from mTBI rodents. The high-level context integration was done using a Bayesian network that simulated the evolution of lesion over time with information gathered from each patient. For the low-level context, spatial information of the lesion was gathered from a cascade of classifiers, where information from one time point is integrated into the next allowing for patient specific information to be used in segmentation of the injured area. Thus, both high- and low-level contexts could be combined leading to a more accurate segmentation. This accurate segmentation will lead to better identification of the location of the TBI which can ultimately improve the choice of treatments in brain trauma.

Acknowledgment This work was supported in part by the National Science Foundation Integrative Graduate Education and Research Traineeship (IGERT) in Video Bioinformatics (DGE-0903667). Anthony Bianchi is an IGERT Fellow.

References

1. Faul M, Xu L, Wald MM, Coronado VG (2010) Traumatic brain injury in the United States: emergency department visits, hospitalizations and deaths. Nat Cent Inj Prev Contr, Centers for Disease Control and Prevention, Atlanta, GA
2. Vaishnavi S, Rao V, Fann JR (2009) Neuropsychiatric problems after traumatic brain injury: unraveling the silent epidemic. Psychosomatics 50:198–205
3. Hunter JV, Wilde EA, Tong KA, Holshouser BA (2012) Emerging imaging tools for use with traumatic brain injury research. J Neurotrauma 29:654–671

4. Benson RR, Gattu R, Sewick B, Kou Z, Zakariah N, Cavanaugh JM, Haacke EM (2012) Detection of hemorrhagic and axonal pathology in mild traumatic brain injury using advanced MRI: implications for neurorehabilitation. Neuro Rehabil 31:261–279
5. VA/DoD (2009) Management of concussion/mild traumatic brain injury. VA/DoD Evid Based Pract: Clin Pract Guide
6. Ajao DO, Pop V, Kamper JE, Adami A, Rudobeck E, Huang L, Vlkolinsky R, Hartman RE, Ashwal S, Obenaus A, Badaut J (2012) Traumatic brain injury in young rats leads to progressive behavioral deficits coincident with altered tissue properties in adulthood. J Neurotrauma 29:2060–2074
7. Konrad C, Geburek AJ, Rist F, Blumenroth H, Fischer B, Husstedt I, Arolt V, Schiffbauer H, Lohmann H (2010) Long-term cognitive and emotional consequences of mild traumatic brain injury. Psychol Med 41:1–15
8. Donovan V, Kim C, Anugerah AK, Coats JS, Oyoyo U, Pardo AC, Obenaus A (2014) Repeated mild traumatic brain injury results in long-term white-matter disruption. J Cereb Blood Flow Metab: Off J Int Soc Cerebr Blood Flow Metab 34(4):715–723
9. Inglese M, Makani S, Johnson G, Cohen BA, Silver JA, Gonen O, Grossman RI (2005) Diffuse axonal injury in mild traumatic brain injury: a diffusion tensor imaging study. J Neurosurg 103:298–303
10. Niogi SN, Mukherjee P (2010) Diffusion tensor imaging of mild traumatic brain injury. J Head Trauma Rehabil 25:241–255
11. Donovan V, Bianchi A, Hartman R, Bhanu B, Carson MJ, Obenaus A (2012) Computational analysis reveals increased blood deposition following repeated mild traumatic brain injury. Neuro Image: Clin 1:18–28
12. Colgan NC, Cronin MM, Gobbo OL, O'Mara SM, O'Connor WT, Gilchrist MD (2010) Quantitative MRI analysis of brain volume changes due to controlled cortical impact. J Neurotrauma 27:1265–1274
13. Metting Z, Rodiger LA, De Keyser J, van der Naalt J (2007) Structural and functional neuroimaging in mild-to-moderate head injury. Lancet Neurol 6:699–710
14. Obenaus A, Robbins M, Blanco G, Galloway NR, Snissarenko E, Gillard E, Lee S, Curras-Collazo M (2007) Multi-modal magnetic resonance imaging alterations in two rat models of mild neurotrauma. J Neurotrauma 24:1147–1160
15. Ahmed S, Iftekharuddin KM, Vossough A (2011) Efficacy of texture, shape, and intensity feature fusion for posterior-fossa tumor segmentation in MRI. IEEE Trans Inf Technol Biomed: publ IEEE Eng Med Biol Soc 15:206–213
16. Kruggel F, Paul JS, Gertz HJ (2008) Texture-based segmentation of diffuse lesions of the brain's white matter. NeuroImage 39:987–996
17. Holli KK, Harrison L, Dastidar P, Waljas M, Liimatainen S, Luukkaala T, Ohman J, Soimakallio S, Eskola H (2010) Texture analysis of MR images of patients with mild traumatic brain injury. BMC Med Imaging 10(1):8–17
18. Holli KK, Waljas M, Harrison L, Liimatainen S, Luukkaala T, Ryymin P, Eskola H, Soimakallio S, Ohman J, Dastidar P (2010) Mild traumatic brain injury: tissue texture analysis correlated to neuropsychological and DTI findings. Acad Radiol 17:1096–1102
19. Marques O, Barenholtz E, Charvillat V (2011) Context modeling in computer vision: techniques, implications, and applications. Multimed Tools Appl 51:303–339
20. Divvala S, Hoiem D, Hays J, Efros A, Hebert M (2009) An empirical study of context in object detection. In: Proceedings of IEEE conference on computer vision and pattern recognition, Miami FL, pp 1271–1278
21. Lafferty J, McCallum A, Pereira F (2001) Conditional random fields: probabilistic models for segmenting and labeling sequence data. In: Proceedings of the 10th international conference on machine learning, pp 282–289
22. Tu Z, Bai X (2010) Auto-context and its application to high-level vision tasks and 3D brain image segmentation. IEEE Trans Pattern Anal Mach Intell 32:1744–1757
23. Bianchi A, Bhanu B, Donovan V, Obenaus A (2013) Visual and contextual modeling for the detection of repeated mild traumatic brain injury. IEEE Trans Med Imag 33(1):11–22

24. Vagnozzi R, Tavazzi B, Signoretti S, Amorini AM, Belli A, Cimatti M, Delfini R, Di Pietro V, Finocchiaro A, Lazzarino G (2007) Temporal window of metabolic brain vulnerability to concussions: mitochondrial-related impairment–part I. Neurosurgery 61:379–388
25. Laird AK (1964) Dynamics of tumor growth. Br J Cancer 13:490–502
26. Longhi L, Saatman KE, Fujimoto S, Raghupathi R, Meaney DF, Davis J, McMillan BSA, Conte V, Laurer HL, Stein S, Stocchetti N, McIntosh TK (2005) Temporal window of vulnerability to repetitive experimental concussive brain injury. Neurosurgery 56:364–374
27. Costa L, Cesar R (2001) Shape analysis and classification: theory and practice. CRC, Boca Raton, FL
28. Viola P, Jones M (2001) Rapid object detection using a boosted cascade of simple features. In: Proceedings of IEEE conference computer vision and pattern recognition, Kauai, Hawaii, pp 511–518

Chapter 4
Automated Identification of Injury Dynamics After Neonatal Hypoxia-Ischemia

Nirmalya Ghosh, Stephen Ashwal and Andre Obenaus

Abstract Neonatal hypoxic ischemic injury (HII) is a devastating brain disease for which hypothermia is currently the only approved treatment. As new therapeutic interventions emerge there is a significant need for noninvasive objective quantification of the spatiotemporal lesion dynamics combined with precise information about lesion constituents (ischemic core and penumbra). These metrics are important for deciding treatment parameters (type, time, site, and dose) and for monitoring injury-therapy interactions. Such information provided 'on-line' in a timely fashion to the clinician could revolutionize clinical management. Like other spatiotemporal biological processes, video bioinformatics can assist objective monitoring of injury–therapy interactions. We have been studying the efficacy of various potential treatments in translational (rodent) HII models using magnetic resonance imaging (MRI). We have developed a novel computational tool, hierarchical region splitting (HRS) to rapidly identify ischemic lesion metrics. HRS detects similar lesion volumes compared to manual detection methods and is fast, robust, and reliable compared to other computational methods. HRS also provides additional information about the location, size, and evolution of the core and penumbra, which are difficult to ascertain with manual methods. This chapter summarizes the ability of HRS to identify lesion dynamics and ischemic core-penumbra evolution following neonatal HII. In addition, we demonstrate that HRS provides information about lesion dynamics following different therapies (e.g., hypothermia, stem cell implantation). Our findings show that computational analysis of MR images using HRS provides novel quantitative approaches that can be applied to clinical and translational stroke research using data mining of standard experimental and clinical MRI data.

N. Ghosh (✉) · S. Ashwal · A. Obenaus
Department of Pediatrics School of Medicine, Loma Linda University,
Loma Linda, CA, USA
e-mail: nirmalyaghosh11@gmail.com

S. Ashwal
e-mail: sashwal@llu.edu

A. Obenaus
e-mail: aobenaus@llu.edu

© Springer International Publishing Switzerland 2015
B. Bhanu and P. Talbot (eds.), *Video Bioinformatics*,
Computational Biology 22, DOI 10.1007/978-3-319-23724-4_4

Acronyms

ADC	Apparent diffusion coefficient
CP	Core-penumbra
CVPR	Computer vision and pattern recognition
DPM	Diffusion-perfusion mismatch
DWI	Diffusion-weighted imaging
EM	Expectation-maximization
HII	Hypoxic ischemic injury
HRS	Hierarchical region splitting
HT	Hypothermia
IHC	Immunohistochemistry
MR	Magnetic resonance
MRI	Magnetic resonance imaging
MWS	Modified watershed
NIH	National Institute of Health
NSC	Neural stem cell
NT	Normothermia
PWI	Perfusion-weighted imaging
ROI	Region of interest
RPSS	Rat pup severity score
SIRG	Symmetry-integrated region growing
SWI	Susceptibility weighted imaging
T2WI	T2-weighted imaging

4.1 Introduction

Neonatal hypoxic ischemic injury (HII) is a devastating brain disease with long-term complications (e.g., cerebral palsy, cognitive impairments, epilepsy) and currently early treatment with hypothermia is the only approved therapy [23]. As new emerging treatments undergo clinical trials, it will be critically important to objectively identify and precisely quantify the spatiotemporal dynamics of the HII lesion (s) and its internal composition of salvageable (penumbra) and non-salvageable (core) tissues. Clinically, rapid computational quantification of injury from neuroimaging data, such as magnetic resonance imaging (MRI) can guide clinical therapeutic decisions about whether treatments are indicated, when such interventions should be implemented and whether or not they are effective. HII-induced neurodegeneration is a very slow biological process. For analyzing lesions, video bioinformatics with conventional 30 frames per second video data, or periodic data collected even at 1 frame per second is impractical and unnecessary. Instead, serial MRI data are acquired at salient time points (based on prior knowledge) to monitor spatiotemporal dynamics of ischemic injury (lesion, core and penumbra). Different computational tools discussed in this book for video bioinformatics applications can be then applied to such longitudinal biomedical data.

We have developed novel computational tools to identify total HII lesion, ischemic core, and penumbral volumes from standard MRI data and their quantitative parameters. Specifically, we have developed a hierarchical region splitting (HRS) approach [10], an automated MRI analysis method that has been applied to a rat pup model of neonatal HII [4, 57]. Herein, we demonstrate that HRS can outperform other methods such as watershed and asymmetry approaches that have also been applied to stroke/ischemic injury; specifically HRS is robust, generic for different MRI modalities and it extracts more information compared to others [20]. Ongoing research within the field of computational analysis of HII includes automated detection of the ischemic core and penumbra, monitoring of implanted neuronal stem cells in the ischemic brain, injury localization specific to different brain anatomical regions, and altered regional dynamics following treatment with hypothermia. We believe that computational analysis of MR images opens a vast new horizon in current preclinical and clinical translational stroke research using data mining of serial MRI data. These advancements have the ability to improve clinical management of neonatal hypoxic ischemic injury.

4.2 Ischemic Injury Dynamics

In animal studies, HII lesions observed on MR imaging modalities are generally seen as regions of abnormal signal intensity within normal brain tissues that vary in location, size, and shape depending on the duration of hypoxia [4]. Lesions also evolve spatiotemporally as the injured brain responds, and the lesion dynamics often vary depending on the individual brain's response to injury [40]. There is also an enormous clinical literature on post-HII MRI changes in term and preterm newborns [10, 22, 41]. HII leads to edema and altered water mobility inside and adjacent to the lesions that are reflected in T2—and diffusion-weighted MR images (T2WI, DWI). The quantitative computed MR physical properties—namely T2 relaxation times (in milliseconds), and apparent diffusion coefficients (ADC; in mm^2/s) can be assessed to provide inferences about underlying tissue structure/compositions [17, 18, 60]. MRI visualizes regions or boundaries of interest (ROI) and allows separation of the HII lesion from healthy brain tissues. In clinical trials, MRI indices are used as outcome measures to assess pathological changes and to monitor treatment efficacy [48, 59]. Traditionally, ROIs in MRI are manually traced and can be fraught with intra- and inter-observer variability, difficulty in replicating results, and low throughput [38]. Even if lesion detection is manually feasible, quantifying lesion evolution over time and from different brain anatomical regions is unreliable and extremely labor-intensive and objective computational methods are an absolute necessity. Such computational advances are important for (1) efficient injury diagnosis, (2) assessment of treatment effectiveness, and (3) experimental and clinical research.

4.2.1 State of the Art in Computational Methods

The first step in quantification of lesion dynamics is automated lesion detection. Computer vision and pattern recognition (CVPR) techniques have facilitated automated injury detection from MRI [9, 11, 17, 18, 25, 30, 32, 49, 56]. Most methods rely heavily on using 'a priori' models of brain tissue or specific brain diseases and 2D/3D registration between model and subjects to classify healthy tissues into different anatomical regions [8, 25]. A single normal tissue classification is compared to outlier regions that do not satisfy any normal tissue class which are then designated as abnormal (i.e. lesion) [9, 11, 56]. These a priori models require large training datasets which are often not available for age-matched neonates due to scarcity of perinatal control data and structural/maturation differences in the developing brain [17]. In addition, model-subject registration [49] generally suffers from partial-volume effects, occasional low contrast, and motion artifacts in MRI. Even after normalization [25], noisy MRI data may severely affect computational methods utilizing low-level features like image intensity [25], texture [30], and shape/connectivity [32]. Model-subject registration also suffers when injury crosses anatomical boundaries [17].

The second step is quantifying spatiotemporal changes in the HII lesion over serial MRI data. Digital image/volume subtractions to detect changes in serial MR data are occasionally utilized [34, 47]. However, accurate subtraction again relies on 2D/3D registration among serial MRI datasets [27, 63, 64]. In addition to registration-related challenges discussed above, these methods often suffer from registering MRI of the same brain from different neuroimaging time points—specifically in the rapidly maturing neonatal brains when neuroimaging time points are too far apart. Issues related to differences in structure and maturation among healthy developing brains (for comparison) as well as greater and more variable responses of the neonatal brain to HII (and many other brain injuries) result in difficulty of reliable anatomical co-registration in image-subtraction-based methods.

In summary, the majority of current lesion quantification methods depend heavily on large amounts of training data (often not available), significant preprocessing (time-consuming), complex prior models (often not reliable, specifically for neonates), model-subject registration (labor-intensive), and significant user intervention (human bias) that reduce their practical applicability in real-time medical image analysis [37]. Therefore, MRI-based automated detection of neonatal brain injury and its spatiotemporal evolution remains extremely challenging. We have addressed this challenge by undertaking a comparative study of different computational techniques that assist in identifying specific applications and/or a potential fusion of the best computational approaches to identify not only the lesion, but also lesion characteristics.

4.2.2 Lesion Dynamics: Comparative Results

We compared the performance of our HRS method with two other state-of-the-art approaches, one is based on brain-symmetry, and the other utilizes intensity topology in MRI data. None of these three methods require model-subject alignment or co-registration, thus eliminating many of the issues discussed above. It further significantly reduces computational complexity as required for practical real-time clinical applications.

The uniformity of the MR properties of the underlying brain tissue plays a key role in our hierarchical region splitting (HRS) method [16]. In brief, HRS method: (1) takes a computed map from MRI; (2) finds its global histogram; (3) fits the histogram to a bimodal distribution with two distinct and distant peaks; (4) estimates the valley-point between these two peaks as an adaptive threshold; (5) segments the entire image into two sub-images, one with the values below this threshold and the other with values above it; (6) continues this bipartite splitting recursively, to form a hierarchical tree of sub-images until the sub-images are too uniform or too small for further splitting; and finally (7) utilizes MR physical properties of the healthy and injured tissue to separate them out.

In the symmetry-integrated region growing (SIRG) method, inherent symmetry of the healthy brain is utilized to segment out the injured tissue, which is hypothesized to be asymmetric [55]. The SIRG method: (1) starts with MR intensity images; (2) finds the axis of symmetry of the brain; (3) computes a symmetry affinity matrix; (4) extracts asymmetric regions from regional skewness-kurtosis values and clustering by 3D relaxation; and finally (4) detects brain injury by an unsupervised expectation-maximization (EM) classifier that uses 3D connectivity of the regions across 2D MRI slices for outlier rejection.

A standard watershed segmentation method considers intensity images as geographical altitude maps to find local maxima–minima to segment out different regions in the image [45]. In a modified watershed (MWS) method: (1) multiple parameters are fused to compute altitude maps; (2) multiple initial locations with different altitudes are selected; (3) intensity gradients are followed as the paths taken by water droplets to reach different catchment basins (regional minima); (4) closely-located regions with similar feature are merged to counter over-segmentation; and finally (5) injured brain regions are detected by a supervised classifier trained with manually segmented examples.

4.2.2.1 Volumetric and Location Comparison

We recently reported on translational longitudinal MRI over the course of 1 year, wherein we undertook repeated MRI following neonatal HII [40]. Using these data we have compared HRS, SIRG, MWS results to manually detected ground-truth data to test for volumetric accuracy and locational overlap [19]. All three methods, HRS, SIRG, and MWS, detected lesion volume completely automatically from

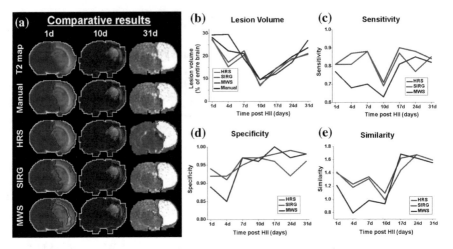

Fig. 4.1 Comparative HII lesion dynamics results between HRS, SIRG and MWS computational methods. **a** The ischemic lesion detected (*red-border*) by manual and compared methods from representative serial T2 data are similar. **b** Lesion volumes at later time points are similar, but there are clear differences at the earlier imaging time points between computational methods. **c**, **d**, and **e** Overall computational performances for lesion location are compared for sensitivity, specificity, and similarity. HRS and SIRG performances were similar while MWS was not as accurate in all respects

serial T2 data (Fig. 4.1a) and their volumetric (Fig. 4.1b) and location-overlap (Fig. 4.1c–e) performance were comparable with respect to the manually detected ground-truth data. Key results are: (1) SIRG performed marginally better than HRS in detecting lesion volumes and locations; (2) HRS had the greatest robustness (i.e., lowest standard deviation) and stability over individual 2D anterior-to-posterior MR slices across the whole brain; (3) SIRG and HRS outperformed MWS in all aspects. Overall these methods are objective and robust for real-time clinical applications as they do not require labor-intensive preprocessing or complex prior-models for model-subject alignment.

4.2.2.2 Why HRS Is Ultimately Superior

MWS [45] was outperformed (Fig. 4.1) by the other two methods due to an inherent over-segmentation problem [19]. The SIRG approach [55] utilized symmetry as a high level feature and 3D connectivity-based outlier rejection that significantly improved its power for injury discrimination. However, limitations of SIRG are: (a) extensive refining of many parameters (two region growing and five asymmetry detection criteria) which might be injury-specific, (b) inadequacy when brain structure lacks defined symmetry or the injury itself is symmetric (bilateral; [16]), and (c) dependence of T2WI intensities rather than T2 relaxation times (optimal). On the other hand, HRS [16] has (1) comparable performance with SIRG (Fig. 4.1),

(2) has fewer parameters to optimize (three region splitting and one lesion detection criteria), (3) capability of handling actual MR properties (e.g., T2 relaxation times, ADC), (4) quick and robust lesion detection over different 2D MR slices [19], (5) the added benefit of separating ischemic core from penumbra (Sect. 4.3), and (6) potential of detecting implanted iron-labeled stem cells (Sect. 4.4.2). HRS limitations including small ROI detection might be resolved in the future by utilizing asymmetry cues for small regions and 3D connectivity-based outlier rejection as implemented in SIRG [55]. Hence we have extended and utilized our HRS method for detailed lesion composition analysis and therapeutic interactions, and these results are summarized in the rest of this chapter.

4.2.3 Long-Term Lesion Dynamics: HRS Performance

Neuronal and cellular tissues impacted by an HII undergo several cellular and molecular cascades that can modify lesion volumes in a complex manner. From the manually derived lesions from serial MRI rat pup data up to 1 year post HII [40] we found that as the initial large edema event (at 1–3 days post HII) subsides, manually derived lesion volumes are reduced at 4–7 days post HII (Fig. 4.2; the sharp initial drop in injury volume). As the HII lesion "matures" normal tissues are then incorporated into the lesion as the core/penumbra expands and at later time points (>1 month), lesion volume increases and then plateaus at the final ischemic infarct size at approximately 3 months. Lesion volumes detected by HRS closely match

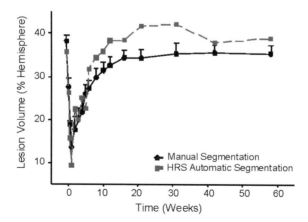

Fig. 4.2 Long term lesion dynamics from serial T2 weighted MRI data: HRS detected lesion volumes closely matched with the ground-truth of manually derived results. Lesion volume initially decreases at 4–7 days post HII (see the initial sharp drop in injury volume) and then gradually increases and reaches its final size by 15–18 weeks. Note that, HRS includes ventricular hypertrophy (5–10 %) in late imaging time-points, which human observers typically ignore

with these manually derived results (Fig. 4.2). These results also support the efficiency of HRS which detects lesions from a skull-stripped brain within 15s— making HRS potentially usable for real-time clinical applications.

4.3   Ischemic Core-Penumbra Dynamics

Generally brain tissues inside an HII lesion are not equally injured. Traditionally, the ischemic core consists of necrotic tissue that is dead or dying and is irrecoverable, while the ischemic penumbra contains more apoptotic tissue that is injured by the ischemic impact but which can potentially recover either spontaneously or in response to treatment [39, 53]. Lesion salvageability is partly related to differences in regional blood flow and mitochondria-mediated metabolism within neurons and glia [33], and interdigitated nature of (mini-) core and (mini-) penumbra changes that occur differentially over space and time [12]. Core-penumbra composition of the HII lesion is very dynamic and with time, part of the penumbral tissue undergoes necrosis and becomes core, while other parts might recover to become nearly normal healthy tissue. As the core is unsalvageable, from the therapeutic point of view, the ischemic penumbra—its presence, location, and spatiotemporal evolution is pivotal to any effective and safe intervention—specifically in a sensitive population like neonates with HII.

4.3.1   Current Methods and MRI Challenges and Potential of HRS

In translational studies, post-mortem tissue immunohistochemistry (IHC) is the best way to identify core and penumbra and their dynamics [12]. But IHC is not possible in living patients where the most feasible technique is a noninvasive method such as MRI. Presumably MRI can be used clinically if previous experimental studies have confirmed correspondence between MRI-based lesion characteristics and IHC. Ischemic stroke typically results in arterial constriction or blockade and hence blood-perfusion deficits that are first visualized in perfusion weighted MR images (PWI). Ischemic injury leads to the development of cytotoxic/vasogenic edema [7] and cellular swelling [12], which restrict water diffusivity in an HII lesion that is reflected as an early hyperintensity in DWI and hypointensity in ADC [60]. With injury progression and excessive cellular swelling, membranes finally collapse and dead cells lead to increased water content that is detectable as increased T2 relaxation times within the lesion. Thus, the ischemic core is expected to have higher T2 relaxation times and lower diffusion coefficients than the ischemic penumbra (~ 10 ms in T2 maps and $\sim 10 \times 10^{-5}$ mm^2/s in ADC maps), although these subtle signal differences are not often visually discernible. Visual segregation of core and penumbra suffers from significant inter- and intra-observer variability and potential user-fatigue-related errors [17, 18]. Diffusion-perfusion mismatch (DPM) is the standard of care in adult stroke [54]. Unfortunately DPM requires multi-modality registration (between DWI and PWI) with the associated problems as discussed above (Sect. 4.2.1) and other pitfalls [60]. More importantly, PWI is still not routinely used or available for assessments of neonates with HII due to potential side effects associated with the use of contrast infusions as well as the technical challenges in performing such studies and further underscores why serial PWI, as needed for core-penumbra evolution by DPM, are not clinically acquired in this age group.

4.3.2 HRS Based Core-Penumbra Evolution

We have searched for a single MRI-modality-based method that eliminates the need for multi-modality registration and studied HRS-based detection of core-penumbra and their evolution from a single MRI modality data (either T2 or ADC maps) of 10-day-old rat pups with HII. We have further validated the HRS results with translational ground-truth using IHC and DPM [17, 18]. HRS splits lesion regions hierarchically into subregions and prior knowledge of MR ranges are utilized to classify subregions as ischemic core and penumbra. Over time HII penumbra was found to convert into core and new tissue regions were included into penumbra due to cellular toxicity affecting neighboring healthy tissues (Fig. 4.3a). Like variations in anatomical (location) involvement, core-penumbra percentage compositions also evolve over time (Fig. 4.3b) as penumbra reduces in size and the entire lesion gradually becomes ischemic core.

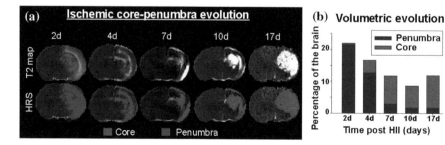

Fig. 4.3 Spatiotemporal evolution of ischemic core-penumbra. **a** Temporal HRS detections of ischemic core (*red*) and penumbra (*blue*) from serial T2-weighted MRI at representative brain depth (from anterior-to-posterior scan) demonstrate evolving anatomical involvement over time. **b** HRS extracted volumetric lesion components identifies the proportional compositions of core-penumbra within the entire lesion over time post HII. Large proportion of salvageable penumbra is quickly converted into non-salvageable necrotic core by 7 days post HII

4.3.3 Potential of Core-Penumbra Dynamics

Post-HII necrotic and apoptotic processes vary with age, severity, anatomical location and type of injury. There is further intra-population variability in anatomical structure, rate of maturation, and inherent plasticity in developing brains. Hence core-penumbra evolution is expected to vary widely following neonatal HII. This makes core-penumbra quantification even more important for candidate/treatment selection—e.g., determining whether restoration of perfusion or neuroprotective interventions might be more successful. Also, assessment of core penumbra volumes/locations is of potential significant importance in deciding treatment details—site, time, and dose. As an example, neural stem cell implantation for HII treatment could be assayed for therapeutic efficacy and critical concerns might be: (1) where to implant (near core, penumbra, or contralateral cortex/ventricle), (2) when to implant (based on significant presence of salvageable penumbra and its anatomical location for viability of stem cells), and (3) how many stem cells are to be implanted (small core and large penumbra might suggest the requirement of implanting a greater number of cells). Also, core-penumbra composition, anatomical location and their rate of evolution might alter or augment estimates of injury severity based on lesion volumes or specific locations of injury. Serial multi-modality MRI data (e.g., T2 and ADC; without need for multi-modality registration) might potentially demonstrate gradual transition of healthy tissue to ischemic penumbra and penumbra to core [17, 18] and guide selection of the site of treatment at a particular time. Thus, accurately identifying the ischemic core-penumbra and its spatiotemporal evolution are of paramount importance.

4.4 Monitoring Therapeutic Interactions in Ischemic Injury

Clinically, there are few therapies available for neonatal HII. Currently hypothermia (HT) has proven to be modestly effective in mild and moderate HII [50]. Other clinical trials are either in progress or being considered to augment the effects of hypothermia (e.g., using Xenon inhalation, topiramate, or erythropoietin as a second therapy or extending the time frame for initiating HT, etc.). In addition, there are many other pharmacological investigations in experimental models that are pursuing alternative therapies including the use of stem cells of various lineages and several studies have been published on the use of mesenchymal stem cells after neonatal HII [24, 43]. For any treatment that is to be used, translational studies can be particularly effective in examining the severity of injury and its interaction with a particular treatment with the goal of improving candidate selection for therapy and optimizing outcomes. But such research is often bottlenecked by the scarcity of effective computational tools [28].

4.4.1 Effect of Hypothermia Therapy on Core-Penumbra Evolution

Neuroinflammation is well known to play a major role in neonatal HII evolution primarily by altered inflammatory signaling of biomolecules and proteins (e.g. cytokines, chemokines). The neonatal brain often reacts differently to injury compared to adult brains, particularly in how it responds to HII [35, 58]. Post-HII hypothermia is known to reduce the inflammatory response [50] and is likely to change the course of ischemic core-penumbra evolution. It will be interesting to analyze this change between hypothermia (HT) and normothermia (NT) groups over time and assess the short-term and long-term therapeutic implications of HT in neonatal HII, whether core-penumbra compositions and their spatiotemporal evolutions correlate with biomolecular protein dynamics [3] and thus help understand injury progression—specifically as recent research reports that different injury biomarkers (e.g., cytokines) fluctuate over time post HII [2].

We recently examined in a short-term longitudinal rat pup model, the changes in cytokine/chemokine expression in two groups of animals (hypothermia vs. normothermia) for the first 24 h post HII and in addition to testing neurological behavior, we evaluated core-penumbra evolution at 0, 24, 48, and 72 h post HII using serial DWI and HRS quantification [62]. In comparison to 0 h data (immediately after HII, before HT/NT), the increase in core-penumbra volume at 24 h post HII (just after completion of HT/NT treatment) was significantly diminished in HT compared to NT treated animals (Fig. 4.4a). This reduction was maintained until 48 h post HII (24 h post HT) after which a rebound effect was observed at 72 h post HII when the core and penumbra began to gradually increase in HT animals (Fig. 4.4b, c). This result was further supported by a semi-quantitative estimation of imaging injury severity based on a rat pup severity score (RPSS) [46] from the same serial DWI data. While the RPSS cannot differentiate core-penumbra volumes it does provide an assessment of regional involvement of the entire HII lesion (core +penumbra). We have previously reported that the RPSS correlates well with HRS derived lesion volumes [16]. Concordance between RPSS and HRS, and their correlations with weight-loss, behavioral, and cytokine/chemokine data [62] demonstrate that the effect of hypothermia on the volume of ischemic injury may be transient or reversible during and after rewarming suggesting that a secondary treatment started later after treatment may be beneficial.

We also observed that pups with mild or moderate injury showed a greater potential for improvement with hypothermia. What was helpful was that when we used our HRS method on the initial MRI data (i.e., 0 h HRS lesion volume), we classified lesion severity on an 'a priori' estimate of severity. Mild injury was defined as less than 5 % and moderate injury as between 5–15 % of the entire brain. Both groups (mild, Fig. 4.4b; moderate, Fig. 4.4c) showed an improvement in the core-penumbra evolution after treatment with hypothermia. Mild HII neonates often recover with no apparent lesions on MRI while severely affected neonates often have large cystic lesions that are irreversible. Hence, it is likely that the moderate

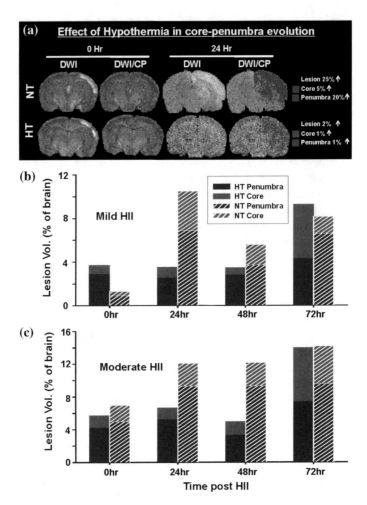

Fig. 4.4 Effect of hypothermia treatment in core-penumbra (CP) evolution. Hypothermia or normothermia (control group) treatment was applied for 24 h immediately post HII and serial DWI data was acquired to evaluate efficacy of HT therapy. **a** HRS derived core-penumbra superimposed on DWI data at 0 and 24 h clearly demonstrate that HT restricted lesion expansion. Comparative HRS/ADC result in **b** mild and **c** moderate injury show delayed injury progression in HT group with both severities until 72 h when volume of HII lesion (and ischemic core) rapidly increases and thus overall reducing potential salvageability

HII group is the group that has the greatest chance for benefiting from treatment and hence the results in Fig. 4.4c is encouraging. It was also observed that HT neuroprotection was mediated primarily by reducing penumbral injury, further supporting the notion that salvageable tissue even after HT therapy might be susceptible to inflammatory exacerbation (after 48 h post HII) without further treatment (Fig. 4.4), just like 'reperfusion injury' in clinical and translational HII [31].

In one recent clinical trial HT could be safely and reproducibly maintained during MRI acquisition suggesting that obtaining scans during treatment could better estimate the severity of injury allowing clinicians the opportunity to consider additional treatments [61]. Future clinical research is expected to increase the utility of serial neuroimaging of seriously ill newborns more readily and safely and potentially could be incorporated into the evaluation of candidate selection for treatment and for early outcome assessment of specific combinatorial therapies. Core-penumbra evolution also suggests a potential window (Fig. 4.4b, c) when a secondary treatment like stem-cell implantation might be beneficial.

4.4.2 Effect of Stem Cell Therapy on Core-Penumbra Evolution

As with many other neurological disorders, stem cell transplantation for neonatal HII has enormous potential. Neuronal stem cells (NSC) whether endogenous (already present in the brain) or exogenous (implanted) inhibit scar formation, diminish inflammation, promote angiogenesis, provide neurotrophic and neuro-protective support, and stimulate host regenerative processes within the brain [5, 52]. Behavioral and anatomical improvements (lesion volume reduction) after NSC implantation in ischemic brain have been reported and immunohistochemical data have shown that NSCs integrate into injured tissues and develop functionally active connections [42]. For clinical applications, where a noninvasive imaging method is an absolute necessity, it will be critical to analyze the core-penumbra evolution over time (post NSC implantation) using computational quantitative methods. Core-penumbra separation may be helpful in optimizing selection of where stem cells should be implanted as it may help determine the optimal environment for implantation and for testing the efficacy of different NSC implantation sites—e.g., contralateral injections at ventricle or parenchyma [40], or ipsilateral injections in lesion and peri-lesion sites [51]. The concept behind these studies is that NSCs implanted too far from HII might not react/migrate quickly enough, while those implanted too close to the lesion might die from cellular toxicity even before having any type of 'bystander effect' (i.e., paracrine type of effect on tissue repair) of the underlying brain tissues. Ischemic penumbral tissue is the primary target for rescue, where implantation either directly into the penumbra or into the contralateral ventricle (for optimal exposure to most of the brain) is a logical site. Interestingly, due to spatiotemporal evolution, existing penumbral tissues might transform into core and hence rapid automated quantification of core-penumbra dynamics (Fig. 4.3) is important.

Serial MRI data from our initial studies [40] on 10-day-old rat pups implanted with iron-labeled NSCs (into the contralateral ventricle or parenchyma) at 3 days post HII have demonstrated NSC reparative effects. When serial T2-weighted MRI data were quantitatively reevaluated with our HRS method, core-penumbra evolution was

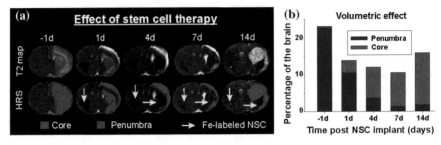

Fig. 4.5 Therapeutic effect of stem cells implanted in contralateral cortex 3 days post HII. **a** HRS derived ischemic core (*red*) and penumbra (*blue*) in serial T2-weighted MRI data demonstrate partial restriction of HII expansion as implanted NSC migrates from injection site (*downward yellow arrow*) to the tissues close to HII (*rightward yellow arrow*). **b** Volumetric summary of core-penumbra evolution suggests presence of penumbra even 7–17 days post HII (4–14 days post NSC implantation)

detected (Fig. 4.5a). Interestingly, NSC transplantation restricted lesion progression to some extent as reflected in the higher proportion of penumbral tissue even at 14 days post NSC implantation, i.e., 17 days post HII (Fig. 4.5b). In our ongoing detailed study on the therapeutic efficacy of NSC, HRS derived core-penumbra evolutions are currently being analyzed for different implantation sites (core, penumbra, and contralateral ventricle), doses (150 or 250 K cells), translational HII models (unilateral and bilateral) and injury severities (mild, moderate and severe). Noninvasive assessments of core-penumbra evolution in such studies are expected to be proven as a very important index for monitoring NSC therapeutic efficacy.

4.4.3   Potential of Combinatorial Therapies: Hypothermia + Stem Cells

Our recent HT/NT recent studies suggest that the effects of HT on core-penumbra evolution may be transient and that adding a second treatment at this time point may be helpful [62]. It appears that HT restricts penumbral volume growth as we observed a greater increase in the conversion of penumbra to core between 48 and 72 h (Fig. 4.4). Reduced penumbral growth suggests that there may be a better microenvironment for adding an additional treatment that could modify the post-HII inflammatory response. Stem cells may be of particular benefit as they could provide a multifaceted approach to treating the multiple post-HII injury cellular cascades that contribute to injury progression. Advanced imaging techniques could also be used to monitor the migration and replication of transplanted stem cells as other investigators and our own previously published studies have suggested [5, 29].

4.5   Future Research Directions

As computational methods like HRS demonstrate the potential to quantify detailed spatiotemporal evolution of neonatal HII with implanted NSC, the next logical step is to detect, quantify and track NSC and their evolution over serial MRI and then to objectively evaluate dynamic interactions between HII and NSC to better understand and monitor therapeutic effects. Finally, if one can parse the brain anatomy utilizing a brain atlas, biological information could be separated to analyze region-specific details like (1) susceptibility to HII, (2) potential tissue salvageability, (3) the migrational pathway of implanted NSCs and their degree of replication and (4) the preferred regions in which NSCs appear to have reparative effects. Without computational tools like HRS, scalable machine learning or pattern recognition techniques, handling and quantifying such complexity would be extremely difficult to accomplish [15].

4.5.1   Stem Cell Dynamics and Interactions with HII Evolution

In translational stem cell therapy, contrast-labeled NSCs [1, 40, 44] are implanted in the injured brain to boost endogenous (unlabeled) NSCs and to better understand the reparative NSC activities like migration, proliferation and differentiation by immunohistochemistry (IHC) and serial MRI [40]. These contrast agents assist in visualizing implanted stem cells for MRI which can monitor them across serial imaging data.

In our studies we have used a superparamagnetic iron-oxide (SPIO) label as the contrast agent which has been found to have no toxicity effects [6, 36]. In addition, these iron oxide labeled NSCs could be visualized for up to 1 year after HII using serial T2 and susceptibility weighted images (T2WI, SWI) [40]. The next frontier is to detect implanted stem cells computationally, track their migration from implantation site to the HII location, quantify their proliferation in response to repair signals and analyze their interactions with the HII evolution. Our earlier studies on migration and proliferation are encouraging [40], but suffer from manually derived data that might have human bias and does not consider important information like cell-density, individual NSC cluster volumes, etc. An earlier review on NSC therapy noted that the lack of computational NSC monitoring tools has hindered finding spatiotemporal characteristics of implanted cells [28].

Our initial results to quantify iron-labeled NSC clusters and its evolution using T2WI using HRS was partially successful [17, 18]. However, we also noted a weakness of HRS in detecting small ROI from T2WI and suggested that it might be improved by using symmetry cues [55]. SWI and associated quantitative maps are reported to enhance visualization by blooming effect and facilitate measuring iron content and hence number of Fe-labeled stem cells [29]. Our preliminary results on

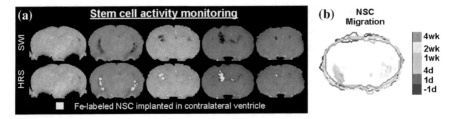

Fig. 4.6 Stem cell detection using Susceptibility weighted imaging (SWI). **a** SWI data at different brain levels (at a single imaging time-point) visualized Fe-labeled NSCs (hypointensity regions) implanted in contralateral ventricles and migrated to the lesion (*right*) site though cerebrospinal fluid. HRS could detect small NSC regions that, in future studies can be tracked in serial SWI data to monitor NSC activity. **b** Superimposed NSC locations derived by HRS from serial T2-data (in Fig. 4.5) demonstrated a clear migration of NSC from injection site in contralateral cortex towards lesion location—even as early as 4d post implantation

using HRS based NSC detection from SWI (Fig. 4.6a) are encouraging and we expect significant improvement from our previous T2-based detection methods (Fig. 4.6b)—specifically by increasing sensitivity (true positive) and reducing outliers (false positives). Beyond simple detection of NSCs and the peri-lesional locations to which they migrate (Fig. 4.6b), we are currently pursuing automated quantification of detailed NSC characteristics like leading edge, rate of migration and proliferation over time, cell-density and internal/external gradient for each NSC clusters.

Perhaps of greater importance will be quantifying dynamic interactions between HII evolution and NSC activities. This is specifically relevant when it is reported that (1) HII tissues release chemoattractant cellular and molecular cues, called 'homing signals' [42, 52] for NSC migration and activities; (2) a proportion of NSCs migrate, proliferate and differentiate to recover penumbral tissues in an attempt to restore brain function, (3) the majority of NSCs do not survive but still may have some form of reparative 'chaperone' effect [52]. Such spatiotemporal interactions are reported from qualitative and visual observations [14] but need to be objectively quantified and verified as highlighted in an earlier review [26]. We have found that NSC volumes from HRS using T2 weighted data are high when lesion volumes temporarily subsides in the sub-acute HII and thus there is more viable tissue for NSC proliferation [17, 18]. HRS quantified NSC activities from serial SWI data are expected to discover similar interactions. Unfortunately SWI data cannot effectively capture HII lesion information (Fig. 4.6a) and thus needs to be combined with serial T2-weighted data for quantifying NSC-HII interactions. Extensive computational research is needed to find relative distances between HII and NSC clusters, their relations with core-penumbra location as well as injury-gradients across the HII boundary and how these affect migratory path of NSCs.

4.5.2   Anatomy-Specific Spatiotemporal Evolutions: Implications

The final frontier is incorporating anatomical brain information and associated prior knowledge to better understand regional effects of injury and therapy. Different diseases and their translational models affect individual brain regions differently (Fig. 4.3) and the same is true for different therapies (Figs. 4.4a and 4.5a). Clinically, anatomically parcellated brain maps have been developed from age-matched healthy newborn children (control data)—specifically relevant for the developing brain [21] and then co-registered [13] with the injured brain in an attempt to quantify regional effects. Unlike those templates for the adult human brain, digital age-matched neonatal brain maps are not readily available, specifically for animals and this is one important direction of current research. Once a proper animal brain atlas is automatically co-registered using established computational tools [17, 18], HRS detected ischemic core-penumbra and NSC locations could be rapidly segregated based on brain anatomy and region-specific injury progression, NSC activity and HII–NSC interactions can be studied in translational (rodent or other species) models. Combining lesion and NSC dynamic information from serial neuroimaging with other biological information (e.g., immunohistochemistry) could lead to a large-scale information fusion that is intractable by any manually derived methods. To handle such complexity video bioinformatic methodologies from the fields of computer vision and pattern recognition need to be integrated with scalable data-mining and continuously adaptive machine learning techniques [15, 19].

4.6   Conclusions

Computational identification and quantification of lesion dynamics and core-penumbra evolution from serial neuroimaging provide a new horizon of injury assessment, deciding candidate/treatment, planning new combinatorial therapies and continuous noninvasive monitoring of therapeutic success. Post-mortem immunohistochemistry and molecular analysis at a single time point might validate computational MRI results in translational studies, but often has no direct clinical application. On the other hand, objective image/video bioinformatics from serial MRI data and computational tools like HRS have great potential to translate experimental studies to clinical use. Extending HRS based core-penumbra detection to adult stroke might also have direct clinical application for example by (a) providing an additional method to currently used selection criteria of using the time of stroke-onset for urgent treatment; (b) measuring core-penumbra composition as an alternative index of injury severity, complementing traditional National Institute of Health (NIH) stroke scale score measurements and use of methods such as diffusion-perfusion mismatch; (c) serving as a potential (online) monitoring tool for immediate recanalization post-thrombosis; and (d) becoming a reliable tool to evaluate potential of post-stroke neuroprotective agents. Specifically, for critical population like neonates with HII, computational assessment of lesion evolution before and after therapy is likely to be a significant step towards enhancing health care and reducing long-term complications of a devastating disease.

Acknowledgment This research work has been supported by grants from the Department of Pediatrics (School of Medicine, Loma Linda University) and NIH-NINDS (1RO1NS059770-01A2). The authors would also like to acknowledge kind sharing of comparative results from SIRG and MWS methods by Dr. Yu Sun and Prof. Bir Bhanu in the Center of Research in Intelligent Systems (Department of Electrical and Computer Engineering, University of California at Riverside).

References

1. Adler ED, Bystrup A, Briley-Saebo KC, Mani V, Young W, Giovanonne S, Altman P, Kattman SJ, Frank JA, Weinmann HJ, Keller GM, Fayad ZA (2009) In vivo detection of embryonic stem cell-derived cardiovascular progenitor cells using Cy3-labeled Gadofluorine M in murine myocardium. JACC Cardiovasc Imaging 2:1114–1122
2. Alonso-Alconada D, Hilario E, Alvarez FJ, Alvarez A (2012) Apoptotic cell death correlates with ROS overproduction and early cytokine expression after hypoxia-ischemia in fetal lambs. Reprod Sci 19(7):754–763
3. Aly H, Khashaba MT, El-Ayouty M, El-Sayed O, Hasanein BM (2006) IL-1beta, IL-6 and TNF-alpha and outcomes of neonatal hypoxic ischemic encephalopathy. Brain Dev 28 (3):178–182
4. Ashwal S, Tone B, Tian HR, Chong S, Obenaus A (2007) Comparison of two neonatal ischemic injury models using magnetic resonance imaging. Pediatr Res 61:9–14
5. Ashwal S, Obenaus A, Snyder EY (2009) Neuroimaging as a basis for rational stem cell therapy. Pediatr Neurol 40:227–236
6. Ashwal S, Ghosh N, Turenius CI, Dulcich M, Denham CM, Tone B, Hartman R, Snyder EY, Obenaus A (2014) The reparative effects of neural stem cells in neonatal hypoxic ischemic injury are not influenced by host gender. Pediatr Res 75(5):603–611
7. Badaut J, Ashwal S, Obenaus A (2011) Aquaporins in cerebrovascular disease: a target for treatment of brain edema? Cerebrovasc Dis 31:521–531
8. Birgani PM, Ashtiyani M, Asadi S (2008) MRI segmentation using fuzzy C-means clustering algorithm basis neural network. In: Proceedings of IEEE international conference on information & communication technology, pp 1–5
9. Corso JJ, Sharon E, Dube S, El-Saden S, Sinha U, Yuille A (2008) Efficient multilevel brain tumor segmentation with integrated bayesian model classification. IEEE Trans Med Imaging 27(5):629–640
10. Counsell SJ, Tranter SL, Rutherford MA (2010) Magnetic resonance imaging of brain injury in the high-risk term infant. Semin Perinatol 34(1):67–78
11. Cuadra MB, Pollo C, Bardera A, Cuisenaire O, Villemure JG, Thiran JP (2004) Atlas-based segmentation of pathological MR brain images using a model of lesion growth. IEEE Trans Med Imaging 23(10):1301–1314
12. del Zoppo GJ, Sharp FR, Heiss WD, Albers GW (2011) Heterogeneity in the penumbra. J Cereb Blood Flow Metab 31:1836–1851
13. Dinov ID, Van Horn JD, Lozev KM, Magsipoc R, Petrosyan P, Liu Z, Mackenzie-Graham A, Eggert P, Parker DS, Toga AW (2009) Efficient, distributed and interactive neuroimaging data analysis using the LONI pipeline. Front Neuroinform 3:22

14. Faiz M, Acarin L, Villapol S, Schulz S, Castellano B, Gonzalez B (2008) Substantial migration of SVZ cells to the cortex results in the generation of new neurons in the excitotoxically damaged immature rat brain. Mol Cell Neurosci 38:170–182
15. Ghosh N, Bhanu B (2008) How current BNs fail to represent evolvable pattern recognition problems and a proposed solution. In: Proceedings of 19th IEEE international conference on pattern recognition (ICPR), Tampa, Florida, USA, pp 3618–3621
16. Ghosh N, Recker R, Shah A, Bhanu B, Ashwal S, Obenaus A (2011) Automated ischemic lesion detection in a neonatal model of hypoxic ischemic injury. Magn Reson Imaging 33:772–781
17. Ghosh N, Sun Y, Turenius C, Bhanu B, Obenaus A, Ashwal S (2012) Computational analysis: a bridge to translational stroke treatment. In: Translational stroke research. Springer, New York, pp 881–909
18. Ghosh N, Yuan X, Turenius CI, Tone B, Ambadipudi K, Snyder EY, Obenaus A, Ashwal S (2012) Automated core-penumbra quantification in neonatal ischemic brain injury. J Cereb Blood Flow Metab 32(12):2161–2170
19. Ghosh N, Bhanu B (2014) Evolving bayesian graph for 3D vehicle model building from video. In: IEEE transaction on intelligent transportation systems (TITS), vol 15(2), pp 563–578
20. Ghosh N, Sun Y, Bhanu B, Ashwal S, Obenaus A (2014) Automated detection of brain abnormalities in neonatal hypoxia ischemic injury from MR images. Med Image Anal (MedIA). 18(7):1059–1069
21. Gousias IS, Rueckert D, Heckemann RA, Dyet LE, Boardman JP, Edwards AD, Hammers A (2008) Automatic segmentation of brain MRIs of 2-year-olds into 83 regions of interest. NeuroImage 40:672–684
22. Gutierrez LG, Rovira A, Portela LA, Leite Cda C, Lucato LT (2010) CT and MR in non-neonatal hypoxic-ischemic encephalopathy: radiological findings with pathophysiological correlations. Neuroradiology 52(11):949–976
23. Jacobs SE, Berg M, Hunt R, Tarnow-Mordi WO, Inder TE, Davis PG (2013) Cooling fornewborns with hypoxic ischaemic encephalopathy. Cochrane Database Syst Rev 1: CD003311
24. Johnston MV, Fatemi A, Wilson MA, Northington F (2011) Treatment advances in neonatal neuroprotection and neurointensive care. Lancet Neurol 10(4):372–382
25. Kabir Y, Dojat M, Scherrer B, Forbes F, Garbay C (2007) Multimodal MRI segmentation of ischemic stroke lesions. In: Conference proceedings of IEEE engineering in medicine and biology society, pp 1595–1598
26. Kim D, Hong KS, Song J (2007) The present status of cell tracking methods in animal models using magnetic resonance imaging technology. Mol Cells 23:132–137
27. Klein S, Staring M, Murphy K, Viergever MA, Pluim JP (2010) Elastix: a toolbox for intensity-based medical image registration. IEEE Trans Med Imaging 29(1):196–205
28. Kraitchman DL, Gilson WD, Lorenz CH (2008) Stem cell therapy: MRI guidance and monitoring. J Magn Reson Imaging: JMRI 27:299–310
29. Kressler B, de Rochefort L, Liu T, Spincemaille P, Jiang Q, Wang Y (2009) Nonlinear regularization for per voxel estimation of magnetic susceptibility distributions from MRI field maps. IEEE Trans Med Imaging 29:273–281
30. Kruggel F, Paul JS, Gertz HJ (2008) Texture-based segmentation of diffuse lesions of the brain's white matter. NeuroImage 39(3):987–996
31. Leger PL, Bonnin P, Lacombe P, Couture-Lepetit E, Fau S, Renolleau S, Gharib A, Baud O, Charriaut-Marlanque C (2013) Dynamic spatio-temporal imaging of early reflow in a neonatal rat stroke model. J Cereb Blood Flow Metab 33(1):137–145
32. Liu J, Udupa JK, Odhner D, Hackney D, Moonis G (2005) A system for brain tumor volume estimation via MR imaging and fuzzy connectedness. Comp Med Image Graph 29(1):21–34
33. Lo EH (2008) A new penumbra: transitioning from injury into repair after stroke. Nat Med 14:497–500

34. Manana G, Romero E, Gonzalez F (2006) A grid computing approach to subtraction radiography. In: Proceedings of IEEE international conference on image processing (ICIP), pp 3325–3328
35. McAdams RM, Juul SE (2012) The role of cytokines and inflammatory cells in perinatal brain injury. Neurol Res Int 561494
36. Muldoon LL, Sandor M, Pinkston KE, Neuwelt EA (2005) Imaging, distribution, and toxicity of superparamagnetic iron oxide magnetic resonance nanoparticles in the rat brain and intracerebral tumor. Neurosurgery 57:785–796
37. Neumann-Haefelin T, Steinmetz H (2007) Time is brain: is MRI the clock? Curr Opin Neurol 20(4):410–416
38. Niimi T, Imai K, Maeda H, Ikeda M (2007) Information loss in visual assessments of medical images. Eur J Radiol 61(2):362–366
39. Northington FJ, Zelaya ME, O'Riordan DP, Blomgren K, Flock DL, Hagberg H, Ferriero DM, Martin LJ (2007) Failure to complete apoptosis following neonatal hypoxia-ischemia manifests as "continuum" phenotype of cell death and occurs with multiple manifestations of mitochondrial dysfunction in rodent forebrain. Neuroscience 149:822–833
40. Obenaus A, Dilmac N, Tone B, Tian HR, Hartman R, Digicaylioglu M, Snyder EY, Ashwal S (2011) Long-term magnetic resonance imaging of stem cells in neonatal ischemic injury. Ann Neurol 69(2):282–291
41. Panigrahy A, Blüml S (2008) Advances in magnetic resonance imaging of the injured neonatal brain. Pediatr Ann 37(6):395–402
42. Park KI, Himes BT, Stieg PE, Tessler A, Fischer I, Snyder EY (2006) Neural stem cells may be uniquely suited for combined gene therapy and cell replacement: evidence from engraftment of Neurotrophin-3-expressing stem cells in hypoxic-ischemic brain injury. Exp Neurol 199:179–190
43. Phillips AW, Johnston MV, Fatemi A (2013) The potential for cell-based therapy in perinatal brain injuries. Transl Stroke Res 4(2):137–148
44. Qiao H, Zhang H, Zheng Y, Ponde DE, Shen D, Gao F, Bakken AB, Schmitz A, Kung HF, Ferrari VA, Zhou R (2009) Embryonic stem cell grafting in normal and infarcted myocardium: serial assessment with MR imaging and PET dual detection. Radiology 250:821–829
45. Ratan R, Sharma S, Sharma SK (2009) Brain tumor detection based on multi-parameter MRI image analysis (2009) Int J Graph Vision Image Proc (GVIP) 9(3):9–17
46. Recker R, Adami A, Tone B, Tian HR, Lalas S, Hartman RE, Obenaus A, Ashwal S (2009) Rodent neonatal bilateral carotid artery occlusion with hypoxia mimics human hypoxic-ischemic injury. J Cereb Blood Flow Metab 29(7):1305–1316
47. Sadasivan C, Cesar L, Seong J, Wakhloo AK, Lieber BB (2009) Treatment of rabbit elastase-induced aneurysm models by flow diverters: development of quantifiable indexes of device performance using digital subtraction angiography. IEEE Trans Med Imaging 28 (7):1117–1125
48. Schiemanck SK, Kwakkel G, Post MW, Prevo AJ (2006) Predictive value of ischemic lesion volume assessed with magnetic resonance imaging for neurological deficits and functional outcome poststroke: a critical review of the literature. Neurorehab Neural Repair 20:492–502
49. Schmidt M, Levner I, Greiner R, Murtha A, Bistritz A (2005) Segmenting brain tumors using alignment-based features. In: Proceedings of IEEE international conference on machine learning application, pp 215–220
50. Shankaran S, Barnes PD, Hintz SR, Laptook AR, Zaterka-Baxter KM, McDonald SA et al (2012) Brain injury following trial of hypothermia for neonatal hypoxic-ischaemic encephalopathy. Arch Dis Child Fetal Neonatal Ed 97(6):F398–F404
51. Smith EJ, Stroemer RP, Gorenkova N, Nakajima M, Crum WR, Tang E, Stevanato L, Sinden JD, Modo M (2012) Implantation site and lesion topology determine efficacy of a human neural stem cell line in a rat model of chronic stroke. Stem Cells 30(4):785–796
52. Snyder EY (2011) The intricate dance between repair and inflammation: introduction to special issue. Exp Neurol 230(1):1–2

53. Stone BS, Zhang J, Mack DW, Mori S, Martin LJ, Northington FJ (2008) Delayed neural network degeneration after neonatal hypoxia-ischemia. Ann Neurol 64:535–546
54. Straka M, Albers GW, Bammer R (2010) Real-time diffusion-perfusion mismatch analysis in acute stroke. J Magn Reson Imaging 32:1024–1037
55. Sun Y, Bhanu B, Bhanu S (2009) Automatic symmetry-integrated brain injury detection in MRI sequences. In: IEEE Wshp mathematical methods in biomedical image analysis, held in conjunction with international conference on computer vision & pattern recognition, pp 79–86
56. Van Leemput K, Maes F, Vandermeulen D, Colchester A, Suetens P (2001) Automated segmentation of multiple sclerosis lesions by model outlier detection. IEEE Trans Med Imaging 20(8):677–688
57. Vannucci RC, Vannucci SJ (2005) Perinatal hypoxic-ischemic brain damage: evolution of an animal model. Dev Neurosci 27:81–86
58. Vexler ZS, Yenari MA (2009) Does inflammation after stroke affect the developing brain differently than adult brain? Dev Neurosci 31(5):378–393
59. Vogt G, Laage R, Shuaib A, Schneider A (2012) VISTA Collaboration. Initial lesion volume is an independent predictor of clinical stroke outcome at day 90: an analysis of the Virtual International Stroke Trials Archive (VISTA) database. Stroke 43(5):1266–1272
60. Wardlaw JM (2010) Neuroimaging in acute ischaemic stroke: insights into unanswered questions of pathophysiology. J Intern Med 267:172–190
61. Wintermark P, Hansen A, Soul J, Labrecque M, Robertson RL, Warfield SK (2011) Early versus late MRI in asphyxiated newborns treated with hypothermia. Arch Dis Child Fetal Neonatal Ed 96(1):F36–F44
62. Yuan X, Ghosh N, McFadden B, Tone B, Bellinger DL, Obenaus A, Ashwal S (2014) Hypothermia modulates cytokine responses after neonatal rat hypoxic-ischemic injury and reduces brain damage. ASN Neuro. 6(6):1–15
63. Zhiguo C, Xiaoxiao L, Bo P, Yiu-Sang M (2005) DSA image registration based on multiscale gabor filters and mutual information. In: Proceedings of IEEE international conference on information acquisition, pp 105–110
64. ZitovÃ¡ B, Flusser J (2003) Image registration methods: a survey. Image vis comput 21 (11):977–1000

Chapter 5
A Real-Time Analysis of Traumatic Brain Injury from T2 Weighted Magnetic Resonance Images Using a Symmetry-Based Algorithm

Ehsan T. Esfahani, Devin W. McBride, Somayeh B. Shafiei and Andre Obenaus

Abstract In this chapter, we provide an automated computational algorithm for detection of traumatic brain injury (TBI) from T2-weighted magnetic resonance (MRI) images. The algorithm uses a combination of brain symmetry and 3D connectivity in order to detect the regions of injury. The images are preprocessed by removing all non-brain tissue components. The ability of our symmetry-based algorithm to detect the TBI lesion is compared to manual detection. While manual detection is very operator-dependent which can introduce intra- and inter-operator error, the automated detection method does not have these limitations and can perform skull stripping and lesion detection in real-time and more rapidly than manual detection. The symmetry-based algorithm was able to detect the lesion in all TBI animal groups with no false positives when it was tested versus a naive animal control group.

E.T. Esfahani · S.B. Shafiei
Department of Mechanical and Aerospace Engineering,
University at Buffalo SUNY, Buffalo, NY, USA
e-mail: ehsanesf@buffalo.edu

S.B. Shafiei
e-mail: somayehb@buffalo.edu

D.W. McBride
Department of Physiology and Pharmacology, Loma Linda University,
Loma Linda, CA, USA
e-mail: dmcbride@llu.edu

A. Obenaus (✉)
Department of Pediatrics, Loma Linda University, Loma Linda, CA, USA
e-mail: aobenaus@llu.edu

© Springer International Publishing Switzerland 2015
B. Bhanu and P. Talbot (eds.), *Video Bioinformatics*,
Computational Biology 22, DOI 10.1007/978-3-319-23724-4_5

5.1 Introduction

Traumatic brain injury (TBI) is the foremost cause of morbidity and mortality in persons between 15 and 45 years of age, with an estimated 1.7 million individuals affected each year in United State [5, 8, 34]. It is estimated that up to 75 % of all reported TBI are considered mild and/or concussive types of injuries. While often TBI symptoms may appear mild or moderate, there is convincing evidence that even subtle brain injury can lead to significant, life-long impairment in an individual's ability to function physically, cognitively, and psychologically [27, 35]. Among these incidents, approximately 7 % are fatal [8], and approximately 90 % of these fatal TBIs occur within 48 h of injury, thus successful treatment of TBI lies in early detection of brain damage after trauma [26]. At the present time there are no therapeutic interventions approved for TBI.

Clinically and in experimental models of TBI, Computed Tomography (CT) and Magnetic Resonance Imaging (MRI) are widely utilized to noninvasively diagnose and evaluate the extent of injury. Conventional MRI, such as T2 weighted imaging (T2WI) is able to identify brain injury within the brain parenchyma, based on the presence of edema and hemorrhagic contusions [17, 18]. Dependent upon the severity of injury, peak edema is typically observed 24–48 h following injury and resolves by approximately 4–7 days, while hemorrhagic lesions can occur within the first 3–4 days and resolution can take months [17, 23, 24]. Given that T2WI is sensitive to various injury components (edema, blood), it is often used for detecting abnormalities in brain tissue, as well as the severity of brain damage in human [9] and animal studies [14].

The diagnosis of TBI is traditionally based on radiological visual inspection of CT or MRI scans. In quantitative animal and clinical studies, typically, manual delineation of the boundary or region of an injury is manually selected. Although an abnormality is easily identified on imaging by a trained person, manual region drawing of an injury, to determine the exact lesion size and tissue characteristics, is arduous and time consuming. Furthermore, large deviations in lesion size can occur due to intra- and inter-operator performance. Therefore, a sensitive and reliable algorithm for assessing lesion size and classification of the TBI is an essential part in providing rapid evaluation of patient status and allowing more effective and timely treatment.

5.2 Current Algorithms for Detecting TBI Lesions

Computational assessment of TBI lesions and TBI tissue characteristics has not been well studied, in contrast to other brain injury/diseases, such as multiple sclerosis, tumors and stroke. There are only a small handful of reports that illustrate the usefulness of computational analysis of TBI imaging data in human [15] and our own work in experimental animals [4, 7]. A caveat of these reported studies is that they are not fully automated and not in real time, thus potentially limiting their immediate translation to the clinic.

Table 5.1 T2 values determined by manual detection

Tissue type	T2 value (Mean ± SD)	95 % confidence Lower T2 value (ms)	95 % confidence Upper T2 value (ms)
Grey matter	73.8 ± 8.33	$T2^L_{GM} = 70.03$	$T2^U_{GM} = 77.61$
CSF	159.7 ± 29.14	$T2^L_{CSF} = 132.8$	$T2^U_{CSF} = 186.7$
Skull	34.1 ± 0.62	$T2^L_{SK} = 38.74$	$T2^U_{SK} = 43.54$
Edema	132.2 ± 12.09	$T2^L_{ED} = 117.2$	$T2^U_{ED} = 129.2$
Blood	58.4 ± 4.78	$T2^L_{BL} = 56.23$	$T2^U_{BL} = 60.58$

Superscripts L and U represent the lower and upper T2 values
Subscripts of T2 represents the tissue type: (*GM* gray matter, *SK* skull, *ED* edema, *BL* blood, *CSF* cerebrospinal fluid)

However, there is an emerging body of literature utilizing computational analysis for identification of brain lesions, for example in stroke and multiple sclerosis. Several recent reviews provide an overview of the state of the art of computational techniques for lesion detection [4, 20, 22]. A brief state of the art in computational assessment is summarized in [4] (see Table 5.1). Many of the proposed methods are based on MRI image segmentation of anatomical brain structures [1]. For example, [28] used an EM (Expectation-Maximization) algorithm to segment anatomical brain structures where the outliers of the normal tissue models were classified as a lesion [28]. In another study, [29] developed an automatic technique to detect white matter lesions in multiple sclerosis [29]. Using 3D gradient echo T1 weighted and FLAIR images the authors identified tissue class outliers that were termed lesion beliefs. These beliefs were then expanded computationally to identify the lesion area based on a voxel-based morphometric criterion.

Recently, [10] introduced Hierarchical Region Splitting (HRS) as an automated ischemic lesion detection schema for neonatal hypoxic ischemic injury [10]. They used normalized T2 values to recursively split the MR images, wherein each segment had a uniform T2 value associated with each tissue class. This constituted either normal tissues or regions with putative ischemic lesions. Each region was then further recursively split until the region of interest (i.e. lesion) could be detected based on T2 values along with other criteria for stopping the algorithm. This method has also been compared to symmetry-integrated region growing (SIRG) [32] in human and animal magnetic resonance imaging datasets. While the SIRG method has been shown to be more accurate in detecting the lesion volume, it was more sensitive to the tuning parameters [11].

Many of the existing computational methods, irrespective of the disease state, typically rely either on large amount of training data or a prior model to identify putative lesions. Therefore, current automatic lesion detection methods are computationally intensive and not clinically useable due to the lack of real-time detection. Herein, we outline an automatic computational algorithm for detection and classification of TBI lesions from T2WI MR images. The algorithm utilizes the symmetrical structure of the brain to identify abnormalities and which is computationally inexpensive, thus allowing for real-time detection with accuracy comparable to manual detection. The performance of the proposed algorithm is evaluated in an animal model of TBI and is compared to the manual lesion detection method (gold standard).

5.3   Methods

An experimental model of TBI in adult rats was used to obtain the MRI images utilized for development and validation of our automatic lesion detection method.

5.3.1   Experimental Model of Traumatic Brain Injury

A total of 18 adult male Sprague-Dawley rats (72–93 days old at the time of TBI) were randomized into three groups: (1) mild repetitive TBI, where the injury was induced 3 days apart ($n = 10$), (2) sham animals that received a craniotomy but no TBI ($n = 6$), and (3) control animals that were without any surgical intervention ($n = 2$). All animal experiments were in compliance with federal regulations and approved by the Loma Linda University Animal Health and Safety Committees.

TBI was modeled using the well-established controlled cortical impact (CCI) model [12, 23]. Briefly, anesthetized rats (Isoflurane: 3 % induction, 1–2 % maintenance) were secured into a stereotactic frame and a midline incision exposed the skull surface. A craniotomy (5 mm diameter) was performed over the right hemisphere 3 mm posterior and 3 mm lateral from bregma, to expose the right cortical surface. The CCI was delivered using an electromagnetically driven piston with a 4 mm tip diameter at a depth of 2 mm, speed of 6.0 m/s, and contact duration of 200 ms (Leica Biosystems Inc., Richmond, Il). In the repetitive TBI group, a second CCI was delivered 3 days after the initial injury at to the same location. The animals in the sham group went through the same surgical procedure but with no CCI (surgical shams).

5.3.2   Magnetic Resonance Imaging and Manual TBI Lesion Detection

Animals were anesthetized (isoflurane: 3 % induction, 1 % maintenance) for MRI using a 30 cm bore 4.7 Tesla Bruker Advance MRI scanner (Bruker Biospin, Billerica, MA). T2-weighted imaging was obtained with the following parameters: 25 coronal slices, 1 mm thick spanning the entire brain, TR/TE = 2850 ms/20 ms, number of echoes = 6 that were 10.2 ms apart, field of view = 3×3 cm, 2 averages. T2 maps were generated using in-house software [23]. Each animal was imaged on days 1 (24 h-post first injury), 4 (72 h-post first injury and 24 h-post second injury), and 14 (14 days-post first injury) similar to our previous report [12].

Manual TBI lesions were drawn on T2WI where regions of hyper-intensities reflect increased water content (bright = edema) and hypo-intensities delineate regions of extravascular blood (dark = hemorrhage). Once these regions of interest (ROI) were delineated on the T2WI they were transferred to the quantitative T2 maps and additional visual inspection was used to confirm that the ROIs encompassed the MR observable abnormalities (Fig. 5.1a). To minimize the limitations of inter-observer manual lesion detection, a single operator was used for all analysis, including the lesion

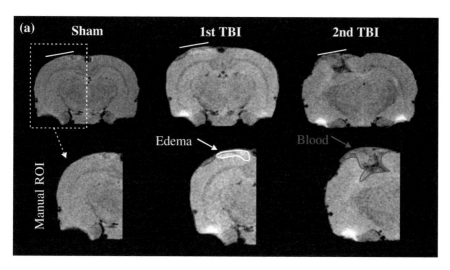

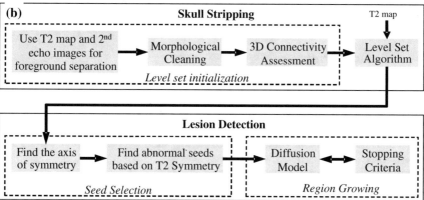

Fig. 5.1 Representative T2-weighted MR images and pseudo-code for the automatic algorithm. **a** Representative Sham (surgery, no TBI) and a repetitive TBI. The MR images were obtained 1 day after the first (1st TBI) and the second (2nd TBI) injury. The injured hemisphere (*dotted line*, sham) is expanded to illustrate the manual ROI that were used to delineate the brain injury lesion. After the first TBI there was primarily edema present but after the 2nd TBI there appeared significant amounts of blood (see also Fig. 5.5). **b** Pseudo code for our automated algorithm that is composed of two primary components, skull stripping (see Fig. 5.2) followed by lesion detection (see Fig. 5.3)

area, and was considered as the gold-standard. From the manual ROI we extracted T2 relaxation values (ms) for normal tissues and lesion components (edema, blood, see Fig. 5.1a) that were used in the belief maps for the automated computational detection algorithm we describe below. These T2 values are outlined in Table 5.1.

5.3.3 Automated Computational TBI Lesion Detection

Our proposed automatic algorithm for real-time TBI lesion detection from T2WI is created by the combination of two algorithms. The first algorithm removes the skull and all tissue not associated with the brain tissue (Skull Stripping) by comparing the T2 values. The second algorithm (Lesion Detection) detects the lesion (if present) based on the brain symmetry for the T2 values (Fig. 5.1b). Additional details of the two algorithms are outlined in the Sect. 5.4.

5.4 Analysis and Results

The proposed computational method is composed of two parts: skull stripping and lesion detection. Skull stripping is used for extracting brain-only tissues by removing the surrounding muscle and skull. In the lesion detection, first the axis of symmetry of the brain structure is found, and then using the symmetry criteria with respect to this axis, any lesion which is present is detected. The lesions are defined by a difference in the T2 values (see Table 5.1) across the axis of symmetry.

5.4.1 Skull Stripping

Numerous studies which have been developed to extract human brain tissue from an MR image using methods such as pulse coupled neural networks [21] or geometrical and statistical 3D modeling of the brain [2]. However there are still no robust and fully automatic methods for extracting rat brain tissue which can be used in real-time.

Skull-stripping methods can generally be divided into three types: intensity based, morphology based, and deformable model based [19]. In intensity-based methods the distribution of image intensity is used for multiple or iterative binary classification (thresholding) to separate brain from non-brain materials. The main drawback of intensity based methods is their sensitivity to intensity bias caused by magnetic field inhomogeneities, sequence variations, scanner drift, or random noise [36].

Morphology-based methods often consist of two steps: (1) initial classification of the brain area in each MRI slice, and (2) refining the extracted result throughout 2D or 3D connectivity-based criteria [13, 30]. intensity-based [13] implemented an intensity-based thresholding to separate high intensity regions (e.g. background, skull and cavities) from brighter regions such as brain, skin, and facial tissues. A morphological-based operation was then implemented to identify brain regions. Marr Hildreth [30] used a Marr Hildreth edge detector in combination with connectivity criteria to extract anatomic boundaries in an MR image and then select the largest central connected component as the brain region. The main drawback of these methods is their dependency on large set of parameters which are sensitive to small changes in the data.

Finally, deformable models are iterative methods where in each step an active curve evolves and deforms till it reaches the stopping criteria [33, 36]. Active contour models are very common methods for image segmentation which were introduced by Active contour [16]. The idea behind these models is to evolve a curve to detect sudden changes in gradient intensity (which describes the edges of a shape) in that image.

In this study, we approach skull stripping problem by implementing an improved level set method proposed by Li et al. [19]. This method considers an energy function defined by using image properties and a potential function that forces the zero level set function to the desired brain edge.

The level set method is an implicit interface tracking technique. Consider a contour Φ which expands with speed F in its normal direction till it reaches the edges of an object in an arbitrary image I in domain Ω. Lets denote an edge indicator g function by Eq. 5.1.

$$g_I = \frac{1}{1 + |\nabla G_\sigma * I|^2},\tag{5.1}$$

where, G_σ is Gaussian kernel with a standard deviation σ. The convolution term is used to smooth the image. Function g has its minimum value at object boundaries which ensures the stopping of curve evolution as it reaches the boundary of an object.

To calculate the evolution of the level set function $\Phi : \Omega \to R$, an energy function $E(\Phi)$ can be defined as Eq. 5.2 [19]

$$E(\Phi) = \mu R_p(\Phi) + \lambda L_g(\Phi) + \alpha A_g(\Phi),\tag{5.2}$$

where μ is the coefficient of the regularization term (R_p), λ and α are the coefficient of energy term L_g and A_g defined by Eqs. 5.3–5.5.

$$R_p(\Phi) = \int P(|\nabla\Phi|)dx,\tag{5.3}$$

$$L_g(\Phi) = \int g\delta(\Phi)|\nabla\Phi|)dx,\tag{5.4}$$

$$A_g(\Phi) = \int gH(-\Phi)dx,\tag{5.5}$$

where $P(s)$ is an energy density function represented by Eq. 5.6. H and δ are the Heaviside and Dirac delta functions, respectively.

$$P(s) = \begin{cases} \frac{1}{2\pi}(1 - \cos 2\pi s) & s \leq 1 \\ \frac{1}{2}(s - 1)^2 & s > 1 \end{cases}, \quad s = |\nabla \Phi|, \tag{5.6}$$

Based on calculus of the variance, the energy function in Eq. 5.2 can be optimized by solving the following Eq. 5.7.

$$\frac{\partial \Phi}{\partial t} = \mu \text{div}(d_p(|\nabla \Phi|)|\nabla \Phi|) + \lambda \delta(\Phi)\text{div}\left(g \frac{\nabla \Phi}{|\nabla \Phi|} + \alpha g \delta(\Phi)\right), \tag{5.7}$$

where, $d_p(s) = \frac{P'(s)}{s}$. For numerical analyses we considered the 2D level set as a function of time and position. We used a central difference method to approximate $\frac{\partial \Phi}{\partial x}, \frac{\partial \Phi}{\partial y}$ and forward difference for $\frac{\partial \Phi}{\partial t}$.

Because adding the distance regularization term, an improved level set method can be implemented using a central difference scheme which is more accurate and stable compared to a first order upwind scheme used in conventional formulation. In our numerical solution, Δx and Δy are set to be equal to 1. Therefore, numerical formulation of Eq. 5.7 can be written as:

$$\Phi_{i,j}^{k+1} = \Phi_{i,j}^k + \Delta t L(\Phi_{i,j}^k), \quad k = 0, 1, 2\ldots, \tag{5.8}$$

where, $L(\Phi_{i,j}^k)$ is the numerical approximation of the right hand side of the Eq. 5.7. In order to have a stable solution, the choice of time step should satisfy the Courant-Friedrichs-Lewy (CFL) condition: $\mu \Delta t < 0.25$.

One of the advantages of this level set formulation, besides eliminating the re-initialization necessity, is that it allows the use of a general set of initial functions as the initial level set. As shown in Fig. 5.2, the initialization of level set for skull stripping consists of three main steps: foreground separation, finding the brain bounding box and assessment of the 3D connectivity.

Step One:

Foreground separation is achieved by applying two separate thresholds, one on the quantitative T2 map and the other on the 2nd Echo image of T2WI. Let $I_{i,j}$ denote each pixel value of the gray scaled image of the 2nd Echo T2WI and $T_{i,j}$ is its associated T2 value from the T2 Map. The foreground extraction can be described by Eq. 5.9.

$$I'_{i,j} = \begin{cases} I_{i,j} & I_{i,j} > \sigma_I^2 \text{ and } 36 < T_{i,j} < 450 \\ 0 & \text{other} \end{cases}, \tag{5.9}$$

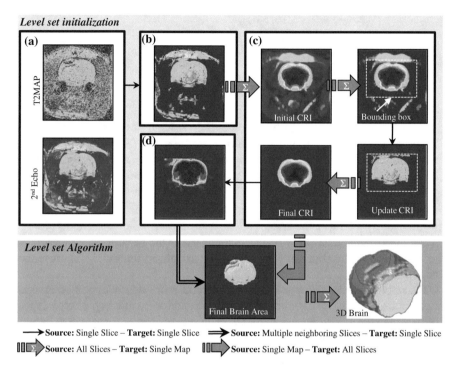

Fig. 5.2 Primary skull stripping components. Level set initialization: **a** inputs for skull stripping included both T2-weighted MR images and the quantitative T2 map (T2 values, ms). **b** Foreground separations, **c** Morphological cleaning composed of cumulative roundness index (CRI) and delineation of the brain bounding box. **d** 2D results were validated using 3D connectivity testing. Level set algorithm: here the final brain area is found for each slice and then summed to identify the brain volume in 3D (see manuscript for additional details)

where σ_I^2 is an adaptive threshold calculated based on Otsu's method [25] and $I'_{i,j}$ is the result of foreground separation. Otsos method maps an arbitary image into a binary one by selecting a threshold value which minimizes the intra-class variance. The threshold value for the T2 map is calculated based on the T2 belief map (Table 5.1). The result of foreground separation is shown in Fig. 5.2b.

Step Two:

Removing the foreground from all the MR slices, the cumulative roundness index (CRI) is calculated for the 3D image. CRI is an estimate of the likelihood of $I'_{i,j}$ to be part of the brain tissue. It combines 2D geometrical constrains (roundness) and 3D connectivity.

Lets assume that after foreground separation, the sth MR slice has M clusters in which $I'_{i,j}$ belongs to the mth cluster ($C_{m,s}$) with an area, $A_{m,s}$, and perimeter, $P_{m,s}$, associated with it. CRI can be represented by Eqs. 5.10 and 5.11.

$$R_{i,j}^s = \begin{cases} \frac{4\pi A_{m,s}}{P_{m,s}^2} & I_{i,j}' \in C_{m,s} \\ 0 & \text{other} \end{cases}, \tag{5.10}$$

$$\text{CRI}_{i,j} = \sum_s [(R_{i,j}^s)^{1.5}], \tag{5.11}$$

$R_{i,j}^s$ is the roundness index which is zero for a straight line, and one for a circle. The brain has almost a round geometry and therefore, the area with a higher roundness index has a higher probability that it is brain tissue. The power of 1.5 is used to increase the sensitivity of CRI to the roundness index. Since the clusters in $I_{i,j}'$ are calculated based on T2 values, CRI takes both the physical and the geometrical properties of the cluster into account. Hence, the CRI is a good measure for the probability of the area being in the brain tissue. After calculating CRI, a recursive adaptive threshold is implemented to select the most probable location of the brain. The termination criteria, is that the calculated threshold should remain under 0.9.

The result is a bounding box indicating the most probable location of brain tissue in each slice (Fig. 5.2c). This bounding box is used for morphological cleaning by only keeping the largest cluster of pixels in its region. The new calculated images, $I_{i,j}''$, are used to calculate the 3D connectivity and to update the CRI value. The evolution of CRI can be seen in Fig. 5.2c.

Step Three:

The level set initial template is found by applying two adaptive thresholds (σ_{CON}^2 and σ_{CRI}^2) on 3D connectivity and CRI as described by Eq. 5.12.

$$I_{i,j}'' = \begin{cases} I_{i,j}'' & \text{CON}_{i,j} > \text{sigma}_{\text{CON}}^2 \text{ and CRI}_{i,j} > \text{sigma}_{\text{CRI}}^2 \\ 0 & \text{other} \end{cases}, \tag{5.12}$$

For each brain slice, 3D connectivity ($\text{CON}_{i,j}$) considers the previous three slices and next the three slices, and is defined as the number of slices in which $I_{i,j}''$ is not zero.

The net outcome from these three steps is an initial template (zero level set contour) that is used by the level set algorithm for identification of brain-only tissues from the surrounding skull and muscle tissues (Fig. 5.2, Final Brain Area). After this step is completed and verified, the 2nd component of our algorithm, lesion detection, can be implemented.

5.4.2 Lesion Detection

Our lesion detection method consists of two main parts: seed selection and region growing. The seed selection part identifies a set of pixels which have a higher probability of being abnormal brain tissue rather than normal tissue. The growth of the lesion uses the initialized seed points to either expand or shrink the seeded region based on the distribution of seeds and the abnormality value of their neighboring pixels. The seed selection, and therefore region growing, is undertaken independently for both detection of blood and edema. The final lesion is the combination of the two (as shown in Fig. 5.3).

Correct seed point selection is a critical and basic requirement for region growing. The main criteria for selecting the seed points are the T2 values for the region (i.e. blood, edema, see Table 5.1) and T2 map asymmetry in brain structure.

In normal brain tissue there is a high degree of bilateral symmetry. This bilateral symmetry can be altered by underlying pathology, leading to deviations in symmetry. The various types of brain tissue (white matter, gray matter, cerebrospinal fluid etc.) have different T2 values, such that a uniform region likely represents a single brain tissue type. This is due to the variation in water content of the tissue types that is mediated by the underlying brain structure. Therefore, a T2 map of normal brain should have a high degree of bilateral symmetry with respect to its axis of symmetry. In our algorithm seed point selection is based on the assumption that TBI lesions are generally not symmetrical in both hemispheres, while the whole

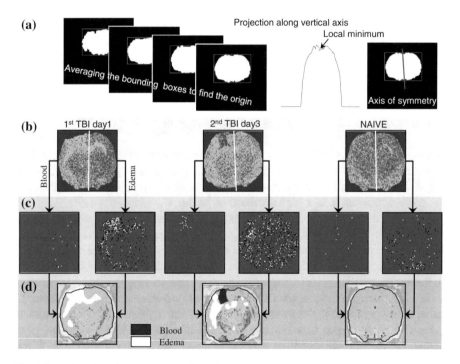

Fig. 5.3 Lesion detection components for naive, 1st TBI 1 day post injury, and 2nd TBI 3 days post injury examples. **a** Firstly, the algorithm identifies the axis of symmetry. **b** Seed point selections for edema and blood using asymmetry. **c** Region growing/shrinking based on the intensity of asymmetry for edema and blood. **d** Final lesion area(s) of brain tissues that contain edema and blood. *Black line* demarcates the brain bound found from skull stripping

brain is approximately symmetrical. Therefore, the automatic detection of these initial seed points in the brain images is the first step in detecting brain abnormalities (Fig. 5.3).

To find the axis of symmetry, the bounding box of the brain region is calculated for each slice. The averaged center point of all bounding boxes is selected as the point of origin. Finally, the projection of the brain along the vertical axis of the image is found. The local minimum within 20 pixels of the center is selected as the second point where the axis of symmetry is passed through (Fig. 5.3a). A line of symmetry is drawn using these two points.

The fact that TBI is typically associated with blood and edema, a discrete label, $L_{i,j}$, is used to separate pixels with T2 values (Table 5.1) close to blood (T2$_{BL}$) and edema (T2$_{ED}$) from that of the normal appearing brain tissue (NABT) using the criteria shown in Eq. 5.13.

$$L_{i,j} = \begin{cases} \text{Blood (BL)} & T_{i,j} < T2_{BL}^{U} \\ \text{Edema (ED)} & 0.9T2_{ED}^{L} < T_{i,j} < 1.1T2_{ED}^{U}, \\ \text{NABT} & \text{Other} \end{cases} \quad (5.13)$$

$T_{i,j}$ and $T'_{i,j}$ denote the T2 value of (i, j)th pixel and of its symmetric pixel respectively. Asymmetrical values are then equal to the absolute value of their difference, $\delta_{i,j} = |T_{i,j} - T'_{i,j}|$. The final sets of seed points for blood (S_{BL}) and edema (S_{ED}) are selected based on the following criteria (Eqs. 5.14 and 5.15):

$$S_{BL} = I_{i,j}|I_{i,j} \in BL, \delta_{i,j} < (1.1 * T2_{BL}^{U} - T2_{SK}^{L}), \quad (5.14)$$

$$S_{ED} = I_{i,j}|I_{i,j} \in ED, (1.1 * T2_{CSF}^{U} - T2_{ED}^{L}) < \delta_{i,j} < (1.1 * T2_{ED}^{U} - T2_{SK}^{L}), \quad (5.15)$$

where, in Eqs. 5.13–5.15, T2$_{ED}^{U}$ represent the upper limit of T2 value (with 95 % confidence) of edema. The selected blood and edema seed points are shown in Fig. 5.3c. In this figure, seeds are shown in gray scale, in which the brightness of the seeds corresponds to a higher difference in the T2 values.

Upon seed selection, a diffusion based region growing is then implemented to group the similar pixels in the region. The main criterion used in the region growing is region homogeneity. Pixels, or groups of pixels, are included in a region only if their characteristics are sufficiently similar to the characteristics of the pixels already in the region (seeds).

Similar to seed selection, the characteristics of interest in our approach is the T2 value of the pixel and the abnormality in brain symmetry structure which results in a significant difference between the T2 value of a pixel and the mirrored pixel across the axis of symmetry. It should be noted that in any diffusion problem, depending on the local statistics of a characteristics of interest, the diffusion technique may act as region shrinking. In this case, the selected seeds are essentially the captured noise in T2 values and are not consistent with the neighboring pixels.

The region growing algorithm allows the asymmetric value to diffuse across the neighboring pixels. Wherever, there are neighboring seeds with a higher T2 value difference, δ, (white seeds in Fig. 5.3c), the concentration of abnormal T2 values will be relatively higher. However, if the difference in the T2 value of the seeds is relatively small (dark points in Fig. 5.3c), or there are smaller seed points around them, the overall concentration of lesion values will be smaller than threshold (see below). This in turn will result in growing or shrinking of the abnormal tissue region around the seed points.

To implement two-dimensional diffusion, an implicit method is used (Eq. 5.16),

$$\delta_{i,j}^{n+1} = (1 - 4\Delta t)\delta_{i,j}^{n} + \Delta t(\delta_{i,j+1}^{n} + \delta_{i,j-1}^{n} + \delta_{i+1,j+1}^{n} + \delta_{i+1,j}^{n}); \qquad (5.16)$$

where $\delta_{i,j}^{n+1}$ is the asymmetric value evaluated at $(n + 1)$th iteration and Δt is the time step and is selected to be 0.2 s (Eq. 5.16 is numerically stable for $\Delta t < 0.25$).

The diffusion was terminated based on the number of pixels set to diffuse. This region was set as a circle with a radius of 25 pixels. Since the radial diffusion rate for Eq. 5.16 is approximately 5 pixels/s, the diffusion time is set to be 5 s.

The final lesion area is detected by comparing the final asymmetric value of blood and edema with their corresponding threshold (15 for edema and 0.6 for blood) (Fig. 5.3d). Figure 5.4 illustrates the lesion growing at different iterations

5.4.3 Validation of Our Lesion Detection Method

Our T2 symmetry method was tested on three TBI, two sham, and two naive data sets. Figure 5.4 compares the asymmetry cloud and detected lesion in a representative TBI data set. Our method is robust over the temporal course of the imaging data, in these datasets, 14 days post TBI (Fig. 5.5). Visual comparisons between the original data (left panels, Fig. 5.5) reveal excellent concordance with the automated identification of injured tissues data (right panels, Fig. 5.5). It can be readily appreciated that lesion area is correctly classified in the majority of cases prior to checking the 3D connectivity of lesion area. The lesion area determined using manual detection was compared to that of the T2 symmetry-based method in Fig. 5.5.

Validation metrics, accuracy, similarity, sensitivity, and specificity were used to determine the performance between our T2 symmetry-based algorithm and the manual method. Similarity measures the similarities between the lesion location and area for both methods. Sensitivity measures the degree to which the lesions overlap between the two methods. Specificity measures the overlap in the normal brain tissue between the two methods. The sensitivity determines the positive detection rate, while the specificity determines the negative detection rate.

Fig. 5.4 Region growing/shrinking. The seeds identified (see Fig. 5.3) from the asymmetry undergo diffusion of T2 asymmetric values for region growing/shrinking for either blood or edema. This example is from a TBI animal. The *black bar* on the color bar is the threshold

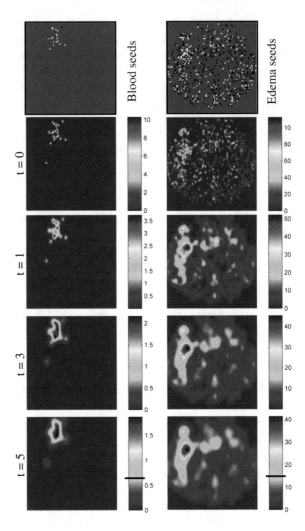

These performance indices are defined as [29, 31]:

$$\text{Similarity} = \frac{2TP}{2TP + FP + FN};\qquad(5.17)$$

$$\text{Sensitivity} = \frac{TP}{TP + FN};\qquad(5.18)$$

$$\text{Specificity} = \frac{TN}{FP + FN};\qquad(5.19)$$

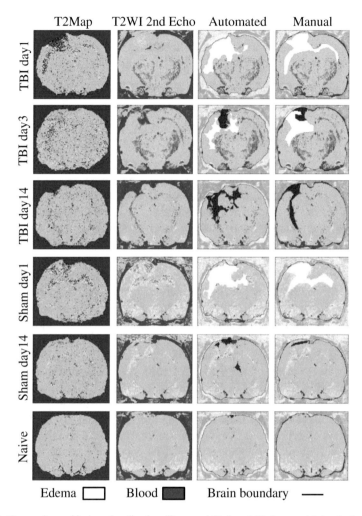

Fig. 5.5 Comparison of lesion visualization. Temporal (1, 3 and 14 days post injury) of repetitive TBI and sham animals where compared to naive animals in which no lesion was identified. Both T2 maps and T2-weighted image (T2WI) illustrate the raw data with the results of our automated algorithm shown in the two *right panels*. As can be seen, the algorithm allows accurate demarcation of evolving blood and edema in brain tissues that were validated against the gold-standard manual lesion detection

$$\text{Accuracy} = \frac{\text{TN} + \text{TP}}{\text{TN} + \text{TP} + \text{FP} + \text{FN}}; \qquad (5.20)$$

where TP and TN are true positive and true negative, respectively. FP and FN stand for false positive and false negative, respectively. The similarity index is the same as the Dice Coefficient [6]. It measures location and volume similarities between

Table 5.2 Similarity, sensitivity, specificity, and accuracy indices

	Similarity	Sensitivity	Specificity	Accuracy
1 day post-TBI	0.60 ± 0.18	0.56 ± 0.23	0.99 ± 0.00	0.99 ± 0.01
3 day post-TBI	0.40 ± 0.32	0.32 ± 0.30	0.99 ± 0.00	0.99 ± 0.00
14 day post-TBI	0.43 ± 0.22	0.50 ± 0.32	0.99 ± 0.00	0.99 ± 0.00
Total	0.48 ± 0.24	0.46 ± 0.28	0.99 ± 0.00	0.99 ± 0.00
HRS	0.40 ± 0.25	0.47 ± 0.12	0.87 ± 0.02	NA

The automatic algorithm and HRS are compared

lesions derived from both methods. Note that the similarity index actually measures the accuracy of automatic detection in comparison to the gold standard, in terms of location, size, and overlap of the detected lesion. All the measured indexes range between [0–1] and higher values indicate better performance of the algorithm. The similarity and sensitivity indexes for TBI were 0.49 ± 0.32 and 0.52 ± 0.28, respectively. The similarity, sensitivity, and specificity indices were compared to the results reported by HRS algorithm for similar brain injury cases [11] (Table 5.2).

The lesion T2 values are extremely different for the 1 day and 3 day post injury images. The 1 day post injury T2 values reflect predominately edema that develops immediately following TBI (within a day) [3, 37]. The observed edema T2 values are similar to those for CSF. After several days, the edema slowly resolves and blood within tissues becomes visible. This is observed in the 3 day post injury images. We have recently documented the occurrence of hemorrhagic progression in a similar mild TBI model [7]. Interestingly, extravascular blood within the brain parenchyma undergoes degradation into other components that may alter the T2WI and T2 Map signatures [3]. Blood from the initial TBI and any hemorrhage that occurs is thus observed in the 14 day-post injury images. The T2 values of blood are very low (due to iron composition and its dephrasing of the magnetic field) compared to the surrounding T2 values of normal appearing grey matter.

In summary, the use of our T2 symmetry method was found to accurately identify TBI lesions in all TBI animals, while no lesion was detected in the naive animals. Furthermore, the T2 symmetry method is able to detect a lesion area similar to that of the manual detection method. This computational algorithm operates in real-time which can provide immediate identification of TBI lesions to clinicians using only T2-weighted MRI.

5.5 Conclusion

Herein, a computationally inexpensive and yet effective method for automatic, real-time TBI lesion detection in T2 weighted images was developed. Our algorithm uses the T2 values of edema and blood, as well symmetrical structures of the brain, to detect all possible abnormalities. This method was evaluated in a series of TBI, sham, and naive subjects. Comparison of the results with manual detection method shows a high level of similarity, sensitivity, and specificity between the two methods. Furthermore, no false positives were observed. Although, the proposed lesion detection algorithm is evaluated on animal MRI data, it can also be used for lesion detection in human subjects.

Acknowledgements Devin McBride was supported by an NSF IGERT Video Bioinformatics Fellowship (Grant DGE 0903667). This study was supported by funding from Department of Defense (DCMRP #DR080470 to AO).

References

1. Akselrod-Ballin A, Galun M, Gomori JM, Filippi M, Valsasina P, Basri R, Brandt Achi (2009) Automatic segmentation and classification of multiple sclerosis in multichannel MRI. IEEE Trans Biomed Eng 56(10):2461–2469
2. Albert Huang A, Abugharbieh R, Tam R (2009) A hybrid geometric statistical deformable model for automated 3-D segmentation in brain MRI. IEEE Trans Biomed Eng 56(7):1838–1848
3. Belayev L, Obenaus A, Zhao W, Saul I, Busto R, Chunyan W, Vigdorchik A, Lin B, Ginsberg MD (2007) Experimental intracerebral hematoma in the rat: characterization by sequential magnetic resonance imaging, behavior, and histopathology. Effect of albumin therapy. Brain Res 1157:146–155
4. Bianchi A, Bhanu B, Donovan V, Obenaus A (2014) Visual and contextual modeling for the detection of repeated mild traumatic brain injury. IEEE Trans Med Imaging 33(1):11–22
5. Corrigan JD, Selassie AW, Langlois Orman JA (2010) The epidemiology of traumatic brain injury. J. Head Trauma Rehabil 25(2):72–80
6. Dice LR (1945) Measures of the amount of ecologic association between species. Ecology 26 (3):297–302
7. Donovan V, Bianchi A, Hartman R, Bhanu B, Carson MJ, Obenaus A (2012) Computational analysis reveals increased blood deposition following repeated mild traumatic brain injury. Neuroimage Clin 1(1):18–28
8. Finkelstein EA, Corso PS, Miller TR (2006) The incidence and economic burden of injuries in the United States. Oxford University Press
9. Gerber DJ, Weintraub AH, Cusick CP, Ricci PE, Whiteneck GG (2004) Magnetic resonance imaging of traumatic brain injury: relationship of T2 SE and T2*GE to clinical severity and outcome. Brain Inj 18(11):1083–1097
10. Ghosh N, Recker R, Shah A, Bhanu B, Ashwal S, Obenaus A (2011) Automated ischemic lesion detection in a neonatal model of hypoxic ischemic injury. J Magn Reson Imaging 33 (4):772–781
11. Ghosh N, Sun Y, Bhanu B, Ashwal S, Obenaus A (2014) Automated detection of brain abnormalities in neonatal hypoxia ischemic injury from MR images. Med Image Anal 18 (7):1059–1069
12. Huang L, Coats JS, Mohd-Yusof A, Yin Y, Assaad S, Muellner MJ, Kamper JE, Hartman RE, Dulcich M, Donovan VM, Oyoyo U, Obenaus A (2013) Tissue vulnerability is increased following repetitive mild traumatic brain injury in the rat. Brain Res 1499:109–120
13. Huh S, Ketter TA, Sohn KH, Lee C (2002) Automated cerebrum segmentation from three-dimensional sagittal brain MR images. Comput Biol Med 32(5):311–328
14. Immonen RJ, Kharatishvili I, Gröhn H, Pitkänen A, Gröhn OHJ (2009) Quantitative MRI predicts long-term structural and functional outcome after experimental traumatic brain injury. Neuroimage 45(1):1–9

15. Irimia A, Chambers MC, Alger JR, Filippou M, Prastawa MW, Wang B, Hovda DA, Gerig G, Toga AW, Kikinis R, Vespa PM, Van Horn JD (2011) Comparison of acute and chronic traumatic brain injury using semi-automatic multimodal segmentation of MR volumes. J Neurotrauma 28(11):2287–2306
16. Kass M, Witkin A, Terzopoulos D (1988) Snakes: active contour models. Int J Comput Vis 1 (4):321–331
17. Kurland D, Hong C, Aarabi B, Gerzanich V, Simard J (2012) Hemorrhagic progression of a contusion after traumatic brain injury: a review. J Neurotrauma 29(1):19–31
18. Lee B, Newberg A (2005) Neuroimaging in traumatic brain imaging. J Am Soc Exp Neurother 2(April):372–383
19. Li C, Chenyang X, Gui C, Fox MD (2010) Distance regularized level set evolution and its application to image segmentation. IEEE Trans Image Process 19(12):3243–3254
20. Lladó X, Oliver A, Cabezas M, Freixenet J, Vilanova JC, Quiles A, Valls L, Ramió-Torrentà L, Rovira À (2012) Segmentation of multiple sclerosis lesions in brain MRI: a review of automated approaches. Inf Sci (NY) 186(1):164–185
21. Murugavel M, Sullivan JM (2009) Automatic cropping of MRI rat brain volumes using pulse coupled neural networks. Neuroimage 45(3):845–854
22. Ghosh N, Sun Y, Turenius C, Bhanu B, Obenaus A, Ashwal S (2012) Computational analysis: a bridge to translational stroke treatment. In: Lapchak PA, Zhang JH (eds) Translational Stroke Research. Springer New York, pp 881–909
23. Obenaus A, Robbins M, Blanco G, Galloway NR, Snissarenko E, Gillard E, Lee S, Currás-Collazo M (2007) Multi-modal magnetic resonance imaging alterations in two rat models of mild neurotrauma. J Neurotrauma 24(7):1147–1160
24. Oehmichen M, Walter T, Meissner C, Friedrich H-J (2003) Time course of cortical hemorrhages after closed traumatic brain injury: statistical analysis of posttraumatic histomorphological alterations. J Neurotrauma 20(1):87–103
25. Nobuyuki OTSU (1979) A threshold selection method from gray-level histograms. IEEE Trans Syst Man Cybern 9(1):62–66
26. Park E, Bell JD, Baker AJ (2008) Traumatic brain injury: can the consequences be stopped? CMAJ 178(9):1163–1170
27. Reilly PR, Bullock (2005) Head injury, pathophysiology and management, vol 77, 2nd edn. Hodder Arnold Publication
28. Rouania M, Medjram, Doghmane N (2006) Brain MRI segmentation and lesions detection by EM algorithm. In: Proceedings of World Academy Science and Engineering Technology, vol 17, pp 301–304
29. Schmidt P, Gaser C, Arsic M, Buck D, Förschler A, Berthele A, Hoshi M, Ilg R, Schmid VJ, Zimmer C, Hemmer B, Mühlau M (2012) An automated tool for detection of FLAIR-hyperintense white-matter lesions in multiple sclerosis. Neuroimage 59(4):3774–3783
30. Shattuck DW, Sandor-Leahy SR, Schaper KA, Rottenberg DA, Leahy RM (2001) Magnetic resonance image tissue classification using a partial volume model. Neuroimage 13(5):856–876
31. Shen S, Szameitat AJ, Sterr A (2008) Detection of infarct lesions from single MRI modality using inconsistency between voxel intensity and spatial location: a 3-D automatic approach. IEEE Trans Inf Technol Biomed 12(4):532–540
32. Sun Y, Bhanu B, Bhanu S (2009) Automatic symmetry-integrated brain injury detection in mri sequences. In: IEEE computer society conference on computer vision and pattern recognition workshops, 2009. CVPR Workshops 2009, pp 79–86
33. Suri JS (2001) Two-dimensional fast magnetic resonance brain segmentation. IEEE Eng Med Biol Mag 20(4):84–95
34. Tagliaferri F, Compagnone C, Korsic M, Servadei F, Kraus J (2006) A systematic review of brain injury epidemiology in Europe. Acta Neurochir (Wien) 148(3):255–68 (discussion 268)
35. Vaishnavi S, Rao V, Fann JR Neuropsychiatric problems after traumatic brain injury: unraveling the silent epidemic. Psychosomatics 50(3):198–205

36. Zhuang AH, Valentino DJ, Toga AW (2006) Skull-stripping magnetic resonance brain images using a model-based level set. Neuroimage 32(1):79–92
37. Zweckberger K, Erös C, Zimmermann R, Kim S-W, Engel D, Plesnila N (2006) Effect of early and delayed decompressive craniectomy on secondary brain damage after controlled cortical impact in mice. J Neurotrauma 23(7):1083–1093

Chapter 6
Visualizing Cortical Tissue Optical Changes During Seizure Activity with Optical Coherence Tomography

M.M. Eberle, C.L. Rodriguez, J.I. Szu, Y. Wang, M.S. Hsu, D.K. Binder and B.H. Park

Abstract Optical coherence tomography (OCT) is a label-free, high resolution, minimally invasive imaging tool, which can produce millimeter depth-resolved cross-sectional images. We identified changes in the backscattered intensity of infrared light, which occurred during the development of induced seizures in vivo in mice. In a large region of interest, we observed significant decreases in the OCT intensity from cerebral cortex tissue preceding and during generalized tonic-clonic seizures induced with pentylenetetrazol (PTZ). We then leveraged the full spa-

M.M. Eberle (✉) · C.L. Rodriguez · Y. Wang · B.H. Park
Department of Bioengineering, Materials Science and Engineering 243,
University of California, 900 University Ave., Riverside, CA 92507, USA
e-mail: meber002@ucr.edu

C.L. Rodriguez
e-mail: carissa.reynolds@email.ucr.edu

Y. Wang
e-mail: ywang57@mgh.harvard.edu

B.H. Park
e-mail: hylepark@engr.ucr.edu

J.I. Szu
Translational Neuroscience Laboratory, University of California, 1247 Webber Hall,
900 University Ave., Riverside, CA 92507, USA
e-mail: jennys@ucr.edu

M.S. Hsu
Translational Neuroscience Laboratory, University of California, 1234B Genomics,
900 University Ave., Riverside, CA 92507, USA
e-mail: mike.hsu@ucr.edu

D.K. Binder
School of Medicine, University of California, 1247 Webber Hall, 900 University Ave.,
Riverside, CA 92507, USA
e-mail: devin.binder@ucr.edu

© Springer International Publishing Switzerland 2015
B. Bhanu and P. Talbot (eds.), *Video Bioinformatics*,
Computational Biology 22, DOI 10.1007/978-3-319-23724-4_6

tiotemporal resolution of OCT by studying the temporal evolution of localized changes in backscattered intensity in three dimensions and analyzed the seizure propagation in time-resolved 3D functional images. This allowed for a better understanding and visualization of this biological phenomenon.

6.1   Introduction

Imaging techniques that are high resolution and noninvasive have been vital in understanding and localizing the structure and function of the brain in vivo. However, there is a trade-off between resolution and imaging depth (Fig. 6.1). Functional MRI (fMRI) provide millimeter resolution with centimeter imaging depth and high-frequency ultrasound can obtain a higher resolution (micrometers), but with higher frequencies limiting the imaging depth to millimeters (Fig. 6.1) [1–3]. Confocal microscopy can achieve submicron resolution but its use in in vivo imaging is restricted due to the limited imaging depth [4].

Techniques are being developed to achieve functional imaging in order to capture real-time biological dynamics in vivo. Positron emission tomography (PET) and genetically encoded calcium indicators (GECIs) are two such techniques [2, 5–7]. PET can acquire millimeter resolution with functionally labeled substrates but lacks structural image information if not paired with another imaging modality such as MRI [2]. Furthermore, the labeling agents are positron-emitting radionuclides that can have toxic side effects. GECIs can achieve single-cell resolution with

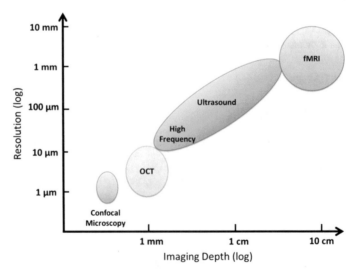

Fig. 6.1 Comparison of resolution and imaging depth for functional magnetic resonance imaging (fMRI), ultrasound, OCT, and confocal microscopy

fluorescent, confocal, or two-photon imaging and can functionally control the activation and deactivation of select groups of cells. However, degradation due to optical scattering limits the imaging depth to a few hundred micrometers and genetically engineered proteins are required altering the native biology [5–7].

Then there is a need for a label-free method for detecting neural activity. Optical coherence tomography (OCT) is a label-free, high resolution, minimally invasive imaging tool, which can produce depth-resolved, cross-sectional, three-Dimensional (3D) volumes, filling the gap between ultrasound and microscopy techniques (Fig. 6.1) [8–11]. These combined traits make OCT an ideal imaging technique for functional localization of brain activity in the cerebral cortex in vivo. This chapter describes the utilization of OCT and the development of visualization tools for studying, with high resolution, video 3D image data of the spatiotemporal changes in backscattered intensity during the progression of induced seizures.

6.2   Optical Coherence Tomography

6.2.1   Basic Principles

OCT is an optical analog of ultrasound imaging, in which the magnitude and delay of backscattered waves reflected from within a sample are used to generate images. The fact that light travels much faster than sound makes direct differentiation of light "echoes" from different depths within a biological tissue impossible. This necessitates the use of low-coherence interferometry to detect the intensity and optical path delay in OCT [11]. Interferometry creates interference patterns between light that is backscattered from a sample and light reflected from a known reference path length. In a Michelson interferometer, light is split between a reference path $E_r(t)$ and a sample path $E_s(t)$ before being reflected, interfered, and sent to a detector arm (Fig. 6.2). The intensity (I_o) at the detector as a function of ΔL, the path length difference between the sample and reference arms, [11] is given by

$$I_o \sim |E_r|^2 + |E_s|^2 + 2E_rE_s \cos(2k\Delta L)$$

Fig. 6.2 Michelson interferometer with a beam splitter

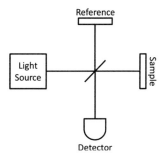

In OCT, gating in the axial direction is achieved by the use of a low-coherence, broad-bandwidth light source. In time domain-OCT (TD-OCT), the reference arm is scanned to obtain a depth profile from the sample, which is generated by demodulation of the resulting time-resolved interferogram. By collecting sequential depth profiles, cross-sectional imaging can be performed. Due to limited sensitivity and physical limitations of the mirror scanning speed, depth profile acquisition with this methodology is limited to several thousand hertz [12], making video-rate image acquisition difficult.

6.2.2   Spectral Domain Optical Coherence Tomography

Spectral domain-OCT (SD-OCT) uses a broadband, continuous wave source and the reference and sample arms are fixed at the same path length. The interference pattern is detected in a spectrally resolved manner. Fourier transformation of the interferogram yields a depth-resolved intensity profile. This has a number of significant advantages over TD-OCT detection: an order of magnitude improvement in detection sensitivity that can be used to increase acquisition speed, simultaneous acquisition of information at all depths, and improved system stability resulting from a stationary reference arm configuration [13–16]. This increase in acquisition speed is of particular importance to video bioinformatics applications as it enables video-rate visualization of cross-sectional tissue behavior.

An SD-OCT system schematic is illustrated in Fig. 6.3 with a light source composed of two superluminescent diodes (SLDs) centered at 1298 nm, the longer wavelength allowing for deeper imaging depth due to a local minimum in the water absorption spectrum. This provides an imaging depth of 2 mm with high spatial resolution [17]. The axial resolution is dependent on the coherence length of the source and the lateral resolution is dependent on the incident beam width; thus a source bandwidth of 120 nm results in a calculated axial and lateral resolution of 8 and 20 μm, respectively [18]. From the source, the light is split with an 80/20 fiber coupler between the sample and reference arm. The reflected light from both arms is then recombined and now the interfered light is sent to a spectrometer, producing spectral lines, which are incident on a 1024 pixel line scan camera (lsc, Goodrich SUI SU-LDH linear digital high speed InGaAs camera). An axial scan line (A-line) is one spectral measurement and cross-sectional images are composed of multiple A-lines, which are acquired by scanning the incident beam transversely across the tissue using galvanometers. 3D volume data sets are acquired by raster scanning the incident beam creating 3D-OCT "optical biopsy" data of tissue microstructure (Fig. 6.3a) [11]. OCT cross-sectional images of cortical tissue are displayed by plotting the intensities on a logarithmic inverse grayscale and the cerebral cortex and the corpus callosum are visible (Fig. 6.3b). After raster scanning a 3D volume rendering can be made of 200 consecutive cross-sectional images (Fig. 6.3c).

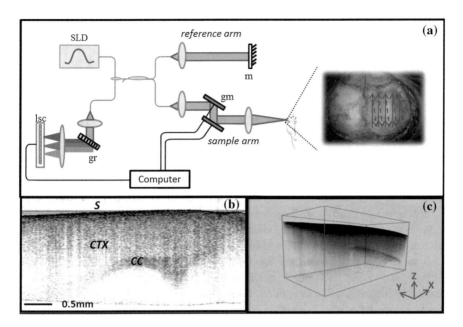

Fig. 6.3 a SD-OCT System Diagram: superluminecent diodes (SLDs), line scan camera (lsc), mirror (m), galvanometers (gm), grating (gr). Image of a mouse skull with raster scanning pattern for volume data acquisition represented by *solid* and *dashed arrows*. **b** dB grayscale OCT image: thinned skull (s), cerebral cortex (ctx), corpus callosum (cc), scale bar: 0.5 mm. **c** 3D 4 × 2 × 2 mm volume rendering of 200 grayscale images

6.2.3   Basic Image Preprocessing

In SD-OCT, each A-line, containing the structural information, is calculated from the interferogram measured by the spectrometer and the wavenumbers (k) are not evenly distributed across the pixels, which is necessary in order to perform a Fourier transform that relates physical distance (z) with wavenumber ($k = 2\pi/\lambda$) [19, 20]. An accurate A-line requires preprocessing to obtain data that is evenly spaced in k-space. Proper wavelength assignment can be achieved by introducing a perfect sinusoidal modulation as a function of k by passing light through a coverslip in the source arm. This spectral modulation is important in assigning the correct wavelength to each pixel of the lsc. The wavelength mapping is determined by minimizing the nonlinearity of the phase of the perfect sinusoid through an iterative process [21]. The initial wavelength array W is used to interpolate the spectral interference fringes to equally spaced k values. The phase of the zero-padded, k-space interpolated spectrum is determined and fitted with a third-order polynomial. The nonlinear part $\sigma(k)$ of the polynomial fit is used for correcting the wavelengths W. Thus a new k-array, k', is calculated from the previous $k = 2\pi/W$ array and $\sigma(k)$, using the equation:

$$k' = k + \sigma(k)/z_{peak}$$

where z_{peak} is given by

$$z_{peak} = 2\pi PI/(k_{max} - k_{min})$$

PI, or peak index, is the location of the coherence peak and k_{max} and k_{min} are the extremes of *k*. This correction, applied iteratively to the original spectrum, results in the wavelength array

$$W' = 2\pi/k'$$

The wavelength array W' (Fig. 6.4b) can be used to map the spectrum (Fig. 6.4a) to the correct *k* values for each A-line (Fig. 6.4c) [21]. After interpolation, the fast Fourier transform (FFT) is taken (Fig. 6.4d).

Also, in SD-OCT, the sensitivity of the system decreases as a function of depth due to the finite resolution of the spectrometer and because of this, a depth-correction function is applied, during image processing by multiplying each A-line by the calculated correction curve [22].

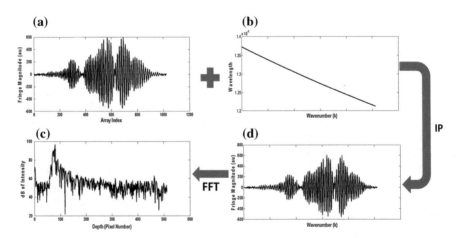

Fig. 6.4 A-line preprocessing schematic. **a** Fringe magnitude with respect to the pixel array for one A-line. **b** Wavelength array W' with respect to wavenumber used to correct wavenumber assignment in each pixel in (**a**). IP: Interpolation step of A with respect to B. **c** Resulting dB intensity A-line with depth (mirror image removed). **d** Resulting fringe magnitude plot post-IP with correct wavenumber assignment. FFT: Fast Fourier Transform

6.3   Utilizing OCT to Identify Cortical Optical Changes for In Vivo Seizure Progression Studies

Previous studies used optical techniques to study changes in backscattered light associated with changes in the electrical properties of cortical tissue [23–26]. Previous research identified that a decrease in intensity of backscattered light is observed during tissue activation, which is due to changes in the cortical tissue composition, specifically hemodynamic responses and glial cell swelling associated with the reduction in the extracellular space [23–27]. OCT has shown to be a promising method for in vivo imaging in highly scattering tissues such as the cerebral cortex and several groups have demonstrated that changes in tissue composition can be detected optically during stimulation [28–32]. OCT was used to visualize these changes in average backscattered intensity during seizure progression as well as visualize and further localize, in three dimensions (3D), regions of tissue exhibiting changes in intensity through functional OCT (fOCT) data processing of video data.

6.3.1   Detection of Average Changes in Backscattered Intensity

Utilizing the SD-OCT system described previously, it has been demonstrated that by inducing a generalized seizure, changes in the backscattered intensity from cortical tissue during seizure progression in vivo can be detected and quantified [33].

6.3.1.1   Experimental Protocol

6–8 week old, CD1 female mice (25–35 g) ($n = 4$) were initially anesthetized intraperitoneally (i.p.) with a combination of ketamine and xylazine (80 mg/kg ketamine, 10 mg/kg xylazine). Baseline Imaging was performed for 10 min and then a saline injection was administered to ensure an injection alone does not cause a change in backscattered intensity. After 25 min, a generalized tonic-clonic seizure was induced with pentylenetetrazol (PTZ) in vivo. Imaging was performed through a thinned skull cortical window with 6 mW of incident power obtaining a 2 mm imaging depth [34]. Scanning was performed 1.5 mm lateral from midline of the right hemisphere centered over the corpus callosum at a 15 kHz A-line rate with each cross-sectional image consisting of 2048 A-lines spanning a 4 mm imaging plane. Images were continuously acquired through the onset of a full seizure where the animal was sacrificed with an anesthesia overdose. All experimental procedures were approved by the University of California, Riverside Institutional Animal Care and Use Committee [33].

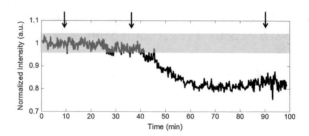

Fig. 6.5 Normalized average intensity (N) from an ROI plotted over time. The three arrows from *left* to *right* represent saline injection, PTZ injection, and initiation of full seizure, respectively. The shaded region is the 2SD of the 10 min baseline

6.3.1.2 Analysis of the Average Change in Intensity

To analyze the changes in backscattered intensity resulting from changes in tissue composition during seizure progression, the average intensity (A) from a region of interest (ROI) was calculated over a designated A-line range ($a_o:a_n$) and depth range ($z_o:z_n$) from intensity matrix (I) and included only pixels with signal above 10 dB to exclude noise.

$$A = \frac{\sum_{t=0}^{n} I(a_o : a_n, z_o : z_n)}{n}$$

$$N = A/b$$

An image was analyzed every 10 s and the average intensity, which was normalized to an average 10 min baseline matrix (b), was plotted versus time as well as its two standard deviation (2SD) interval, which was calculated from the mean of the baseline (Fig. 6.5).

There was no significant deviation in intensity out of the interval during the controls. However, once the PTZ injection was administered there was a significant 20 % decrease in intensity from baseline that is persistent through full seizure. From this study, it was observed that a decrease in intensity during seizure progression can be detected with OCT [33]. By leveraging the high spatiotemporal resolution of OCT, it is possible to visualize the localized intensity changes during seizure progression through the cerebral cortex in vivo with increased spatial specificity.

6.3.2 Spatial Localization of Backscattered Intensity Changes

To study the localized changes in backscattered intensity, SD-OCT volumetric data acquisition was performed. The previously outlined in vivo animal model was used and data acquisition was modified with raster scanning in the X and Y direction for volume collection with X representing the image length of 2048 A-lines spanning 4 mm and Y representing the 200 images in the volume spanning 2 mm (Fig. 6.3c). Volumes were acquired every 2 min and 16 baseline volumes were collected with PTZ injection occurring at 30 min. Volumes were acquired through the onset of full seizure and until the animal expired, 66 min after the start of the experiment. To spatially localize the changes in backscattered intensity during seizure progression, time-resolved, fOCT volumes were developed along with two visualization techniques.

6.3.3   Development of fOCT Volumes and Visualization Techniques

6.3.3.1   Volume Registration

Accurate volume registration was crucial in developing the fOCT volumes because, unlike the previous technique where large ROIs were analyzed, here small localized changes in intensity are being considered and it is extremely important that the same localized areas are being compared from one volume to the next. The first step used to register the volumes was volumetric cross-correlation. The 3D volume matrices were all registered to the first acquired volume through the following algorithm: Initially, the correlation matrices were calculated for each preceding volume matrix G with the initial volume matrix F (r_C) and with its self ($r_F r_G$) using the Fourier transform ($\mathcal{F}$) method for matrix F and the complex conjugate matrix $\bar{G}$.

$$r_c = \mathcal{F}^{-1}[\mathcal{F}(F) * \mathcal{F}(\bar{G})]$$

Next, the matrix location of the maximum of the correlation matrix r_C was found, the volumes were shifted in all three dimensions with respect to the location of the maximum, and the correlation coefficient was calculated for each cross-correlation.

$$R = r_C / \sqrt{r_F * r_G}$$

Once the volumes were correlated in all three dimensions, a surface correction was applied to correct for surface tilt during the experiment. The skull surface was mapped using a threshold algorithm where the maximum intensity point was subtracted from the minimum intensity point in each A-line and then was multiplied by a defined threshold value. The calculated value was then subtracted from each A-line and the location of the value that was greater than zero became the surface location. Once the volume surface was identified, it was subtracted from an average baseline surface and each A-line was shifted with respect to the difference from the average baseline ensuring the surface remained at its initial location and that the pixels in each A-line represented the same area of tissue throughout the experiment.

Once the series of volumes were registered, a median filter of 16 × 64 pixels was applied maintaining the correct aspect ratio as well as the structural information. They were then analyzed for functional intensity changes.

6.3.3.2   Calculation of Spatially Resolved *f*OCT Volumes

In order to determine the changes in backscattered intensity during seizure progression from a pre-seizure state for each pixel, an average baseline volume matrix (B) was calculated (average of 16 volumes). Pixel intensity ratio matrices (R) were then calculated by dividing each volume matrix (F) by the average baseline volume.

$$R = (\log(F/B))^2$$

The square of the logarithm of the fractional changes were then computed in order to better visualize all changes in intensity from baseline, with the sign of the original logarithm reapplied after squaring. To visualize the changes from baseline intensity, color scales were assigned to the ratio values in each pixel. Blue represented ratios below baseline and red represented ratios above baseline. The color was scaled from black to full color saturation at a predetermined threshold ratio value representing a greater than ±0.5 or 50 % change from baseline (Fig. 6.6). Visualization techniques were then developed to assist in analyzing the changes in intensity in each pixel during seizure progression in time as well as throughout the 3D sampled volume of the cerebral cortex.

6.3.3.3   Maximum Intensity Projection (MIP) of *f*OCT Volumes

Maximum intensity projection (MIP) images of the functional map volumes is a technique developed to visualize the overall intensity changes and their spatial specificity with regard to the X and Y orientation of the cerebral cortex. The MIPs

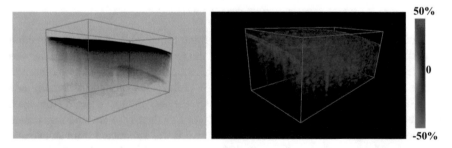

Fig. 6.6 SD-OCT grayscale intensity volume (4 × 2 × 2 mm) of mouse cortical tissue (*left*) and its *f*OCT volume (48 min) (*right*). *Color bar* represents ratio values ± 50 % away from a 0 % change at baseline

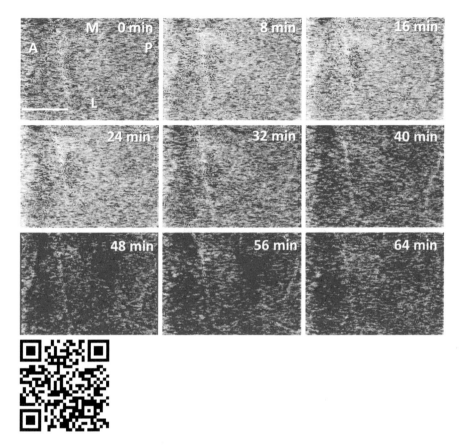

Fig. 6.7 MIP frames of cortical volumes. Numbers are time (min) of volume. 0–32: Control, 4–56: Seizure activity, 64: post-overdose. Color is scaled to the *color bar* in Fig. 6.6 and represents ratio values ±50 % away from a 0 % change at baseline. *Scale bar* 1 mm. See Media 1: https://www.youtube.com/watch?v=3ylu3oHZEC4

were created by finding the largest change from baseline in each A-line across the entire volume and assigning a pixel the corresponding color scale value located at that A-line position in a 2048 × 200 pixel image. This collapses the z-axis of a 512 × 2048 × 200 pixel volume into a 2D *en face* image representing the largest change from baseline at each pixel (Fig. 6.7). However, this technique does limit the full spatial resolution provided by SD-OCT volume acquisition.

6.3.3.4 3D Visualization of *f*OCT Volumes

To get a full 3D visualization of the changes in intensity through the cerebral cortex during seizure progression, Amira, a 3D visualization tool from FEI Visualization Sciences Group, was used for high resolution volume rendering of the calculated

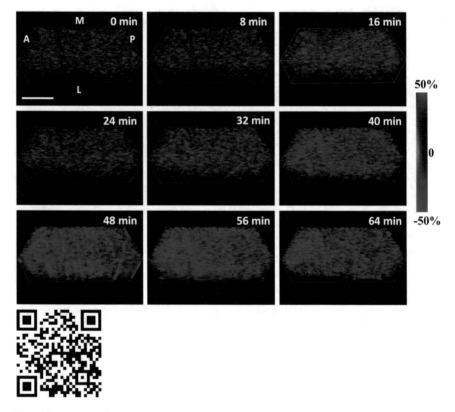

Fig. 6.8 Frames of cortical tissue volumes with functional maps applied. Volumes are 4 × 2 × 2 mm Numbers are time (min) of volume. 0–32: Control, 40–56: Seizure activity, 64: post-overdose. *Color bar* percent change from baseline (0 %) saturating at ±50 %. *Scale bar* 1 mm. See Media 2: https://www.youtube.com/watch?v=sfNhuK9JSxA

functional volumes. After the first five baseline volumes, there is a large decrease in backscattered intensity, which is apparent due to the accumulation of blue pixels (Figs. 6.7 and 6.8). The global nature of the decrease verifies the generalized nature of PTZ seizure progression.

Both visualization techniques were developed to display the calculated ratio ƒOCT volumes in a way that preserved spatial specificity and accurately conveyed the intensity trends that occur during seizure progression. In both techniques, there was a consensus that a large decreasing trend in the backscattered intensity occurred during seizure progression throughout the cortical tissue. This coincided well with the large decrease found in the average ROI method [33]. Although the PTZ animal model displays a global trend, these results are a preliminary demonstration that more localized detection and visualization of intensity change is possible in other less severe pathologies of neural activity.

6.4 Conclusion

This chapter demonstrated the capability of OCT to identify intensity changes during seizure progression in vivo with spatial specificity. Functional processing was applied to the time-resolved, volumetric data enabling the development of visualization techniques in order to accurately convey the intensity trends occurring during seizure progression in three dimensions in time. Large decreases in intensity were found in all of the analysis methods presented and the visualization techniques displayed the global nature of PTZ onset through large decreases in intensity throughout the imaged volume of cortical tissue. Video analysis of these 3D data sets allows for a far better understanding and visualization of these biological phenomena would be possible with isolated volumes.

SD-OCT is a powerful imaging technology that can achieve micrometer resolution with millimeter imaging depth, producing depth-resolved, volumetric image data. It is a versatile tool with many additional processing techniques such as Doppler OCT [35], which allow for simultaneous imaging of tissue structure and blood flow, which can be incorporated with no system modifications. Due to its many advantages and continued advancements, OCT is dynamic tool with many promising applications in furthering fundamental research and clinical medicine.

Acknowledgments This research was sponsored at UC Riverside by the National Institutes of Health R00-EB007241, K08-NS059674, and R01-NS081243; the National Science Foundation IGERT Video Bioinformatics DGE 0903667; and the UC Discovery Grant #213073.

References

1. Matthews PM et al. (2006) Applications of fMRI in translational medicine and clinical practice. Rev Neuroimage 7:732
2. Catana C et al. (2012) PET/MRI for Neurologic Applications. J Nucl Med 53(12):1916
3. Hedrick WR, Hykes DL, Starchman DE (2005) Ultrasound physics and instrumentation, 4th edn. Elsevier, Mosby
4. Conchello JA, Lichtman JW (2005) Optical sectioning microscopy. Nat Methods 2:920
5. Townsend DW (2006) In: Valk PE et al (eds) Positron emission tomography. Springer, New York, pp 1–16
6. Looger LL, Griesbeck O (2012) Genetically encoded neural activity indicators. Curr Op Neurobiol 22:18
7. Deisseroth K (2011) Optogenetics. Nat Methods 8:26
8. Schmitt JM (1999) Optical coherence tomography (OCT): a review. IEEE J Sel Top Quant Electron 5(4):1205
9. Huang D et al. (1991) Optical coherence tomography. Science 254:1178
10. Izatt JA et al. (1996) Optical coherence tomography and microscopy in gastrointestinal tissues. IEEE J Sel Top Quant Electron 2(4):1017
11. Fujimoto J (2008) In: Drexler W, Fujimoto J (eds) Optical coherence tomography technology and applications. Springer, New York, pp 1–45
12. Rollins AM et al. (1998) In vivo video rate optical coherence tomography. Opt Exp 3(6):219

13. de Boer JF et al. (2003) Improved signal-to-noise ratio in spectral-domain compared with time-domain optical coherence tomography. Opt Lett 28, 2067
14. Yun SH et al. (2003) High-speed spectral domain optical coherence tomography at 1.3 μm wavelength. Opt Exp 11:3598
15. Choma MA et al. (2003) Sensitivity advantage of swept source and Fourier domain optical coherence tomography. Opt Exp 11:2183
16. Leitgeb RA et al. (2004) Ultra high resolution Fourier domain optical coherence tomography. Opt Exp 12(11):2156
17. Bizheva K et al. (2004) Imaging in vitro brain morphology in animal models using ultrahigh resolution optical coherence tomography. J B O 9, 719
18. Wang Y et al. (2012) GPU accelerated real-time multi-functional spectral-domain optical coherence tomography system at 1300nm. Opt Exp 20:14797
19. Wojtkowski M et al. (2002) In vivo human retinal imaging by Fourier domain optical coherence tomography. J Biomed Opt 7(3):457
20. de Boer JF (2008) In: Drexler W, Fujimoto J (eds) Optical coherence tomography technology and applications. Springer, New York, pp 147–175
21. Mujat M et al. (2007) Autocalibration of spectral-domain optical coherence tomography spectrometers for in vivo quantitative retinal nerve fiber layer birefringence determination. J B O 12(4):041205
22. Yun SH et al. (2003) High-speed spectral domain optical coherence tomography at 1.3 μm wavelength. Opt Exp 11:3598
23. Weber J R et al (2010) Conf Biomed Opt (CD), BSuD110p, (OSA)
24. Binder DK et al. (2004) In vivo measurement of brain extracellular space diffusion by cortical surface photobleaching J Neurosci :8049
25. Rajneesh KF et al. (2010) Optical detection of the pre-seizure state in-vivo. J Neuosurg Abs 113:A422
26. Holthoff K et al. (1998) Intrinsic optical signals in vitro: a tool to measure alterations in extracellular space with two-dimensional resolution. Brain Res Bull 47(6):649
27. Jacqueline A et al. (2013) Glial cell changes in epilepsy: Overview of the clinical problem and therapeutic opportunities. Neurochem Intern 63(7):638
28. Satomura Y et al. (2004) In vivo imaging of the rat cerebral microvessels with optical coherence tomography. Clin Hem Micro 31:31
29. Aguirre AD et al. (2006) Depth-resolved imaging of functional activation in the rat cerebral cortex. Opt Lett 31:3459
30. Chen Y et al. (2009) Optical coherence tomography (OCT) reveals depth-resolved dynamics during functional brain activation. J Neurosci Methods 178:162
31. Rajagopalan UM, Tanifuji M (2007) Functional optical coherence tomography reveals localized layer-specific activations in cat primary visual cortex in vivo. Opt Lett 32:2614–2616
32. Tsytsarev V et al. (2013) Photoacoustic and optical coherence tomography of epilepsy with high temporal and spatial resolution and dual optical contrasts. J Neuro Meth 216:142
33. . Eberle MM et al. (2012) In vivo detection of cortical optical changes associated with seizure activity with optical coherence tomography. Bio Opt Exp 3(11):2700
34. Szu JI et al. (2012) Thinned-skull Cortical Window Technique for In Vivo Optical Coherence Tomography Imaging. J V Exp 69, e50053. doi:10.3791/50053
35. White BR et al. (2003) In vivo dynamic human retinal blood flow imaging using ultra-high-speed spectral domain optical Doppler tomography. Opt Exp 11(25):3490

Part III
Dynamics of Stem Cells

Chapter 7
Bio-Inspired Segmentation and Detection Methods for Human Embryonic Stem Cells

Benjamin X. Guan, Bir Bhanu, Prue Talbot and Nikki Jo-Hao Weng

Abstract This paper is a review on the bio-inspired human embryonic stem cell (hESC) segmentation and detection methods. Five different morphological types of hESC have been identified: (1) unattached; (2) substrate-attached; (3) dynamically blebbing; (4) apoptotically blebbing; and (5) apoptotic. Each type has distinguishing image properties. Within each type, cells are also different in size and shape. Three automatic approaches for hESC region segmentation and one method for unattached stem cell detection are introduced to assist biologists in analysis of hESC cell health and for application in drug testing and toxicological studies.

7.1 Introduction

In recent years, human embryonic stem cells (hESCs) have been used to assay toxicity of environmental chemicals [1–5]. The hESCs are one of the best models currently available for evaluating the effects of chemicals and drugs on human prenatal development [6]. The hESCs also have the potential to be a valuable model

B.X. Guan (✉) · B. Bhanu
Center for Research in Intelligent Systems, University of California,
Winston Chung Hall Suite 216, 900 University Ave., Riverside, CA 92507, USA
e-mail: xguan001@ucr.edu

B. Bhanu
e-mail: bhanu@ee.ucr.edu

P. Talbot
Department of Cell Biology and Neuroscience, University of California,
2320 Spieth Hall, 900 University Avenue, Riverside, CA 92507, USA
e-mail: talbot@ucr.edu

N.J.-H. Weng
Department of Cell Biology and Neuroscience, University of California,
2313 Spieth Hall, 900 University Avenue, Riverside, CA 92507, USA
e-mail: jweng002@ucr.edu

© Springer International Publishing Switzerland 2015
B. Bhanu and P. Talbot (eds.), *Video Bioinformatics*,
Computational Biology 22, DOI 10.1007/978-3-319-23724-4_7

(a)   (b)   (c)

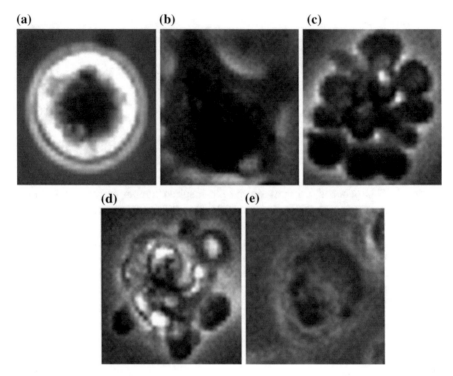

(d)   (e)

Fig. 7.1 a Unattached single stem cell; **b** substrate-attached single stem cell; **c** dynamically blebbing stem cell; **d** apoptotically blebbing stem cell; **e** apoptotic stem cell

for testing drugs before clinical trial. Equally important, the hESCs can potentially be used to treat cancer and degenerative diseases such as Parkinson's disease, Huntington's disease, and type 1 diabetes mellitus [6, 7].

Figure 7.1 shows five different morphological types of hESCs: (1) unattached single stem cell; (2) substrate-attached single stem cell; (3) dynamically blebbing stem cell; (4) apoptotically blebbing stem cell; and (5) apoptotic stem cell. As shown in Fig. 7.1, the inconsistency in intensities and shapes make the segmentation difficult when more than two types of cells are present in a field. Therefore, we have used bio-inspired segmentation and detection method to automate the analysis of hESCs.

Despite the enormous potential benefits of the hESC model, large-scale analysis of hESC experiments presents a challenge. Figure 7.2 shows an example of stem cell images at different magnifications. The analysis of hESCs is either semi-automated or manual [8]. CL-Quant software (DRVision Technologies) is an example of current state-of-the-art software for cell analysis, and it offers a semi-automatic approach. It requires users to develop protocols which outline or mask the positive and negative samples of stem cells using a soft matching procedure [9]. It usually takes an expert biologist 5–6 min of protocol making. The

Fig. 7.2 **a** Image of hESCs
taken with 10× objective;
b Image of hESCs taken with
20× objective; **c** Image of
hESCs taken with 40×
objective

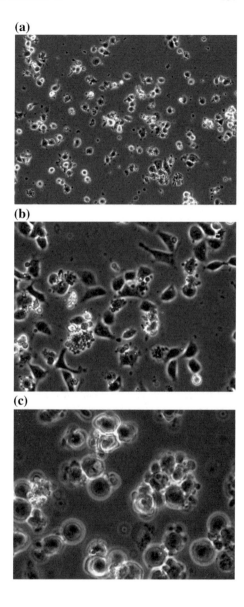

(a)

(b)

(c)

protocols are generally more accurate in the sample videos where the positive and negative samples are derived [8]. The segmentation accuracy of the software is also depended on the number of positive and negative samples. Although the software requires users to make protocols, it is a better alternative than analyzing the stem cells manually. Since there is no existing software for bleb detection and bleb counts, biologists often analyze frame by frame to determine the number of blebs in a video. Therefore, automatic segmentation and detection methods are essential for the future development of fast quantifiable analysis of hESCs.

In this paper, we will review bio-inspired automatic hESC region segmentation and unattached stem cell detection methods. Segmentation is a gateway to automated analysis of hESCs in biological assays. For example, the segmented results from those methods make automatic unattached single stem cell detection possible [10]. The detection of unattached stem cells is important in determining whether the cell is reactive to test chemicals. If the percentage of unattached stem cells is higher than attached stem cells, then the test chemical inhibited attachment and probably killed the cells that are unattached.

7.2   Related Work

There are two previous works for cell region detection in phase contrast images [11, 12]. Ambriz-Colin et al. [11] discuss two segmentation methods in this paper: (1) segmentation by pixels intensity variance (PIV); and (2) segmentation by gray level morphological gradient (GLMG). The PIV method performs pixel classification on the normalized image. It recognizes the probable cell regions and labels the rest as the background in the normalized image. The GLMG method detects the cell regions using morphological gradient that is calculated from the dilation and erosion operations, and by a threshold that separates the pixels belonging to a cell and to the background. Li et al. [12] use a histogram-based Bayesian classifier (HBBC) for segmentation. Li et al. [12] discuss a combined use of morphological rolling-ball filtering and a Bayesian classifier that is based on the estimated cell and background gray scale histograms to classify the image pixels into either the cell regions or the background.

In this paper, we also touch upon segmentation methods such as k-means and mixture of Gaussians by expectation–maximization (EM) algorithm. These approaches are widely used techniques in image segmentation. The k-means segmentation by Tatiraju et al. [13] considers each pixel intensity value as an individual observation. Each observation is assigned to a cluster with the mean intensity value nearest to the observation [14, 15]. The intensity distribution of its clusters is not considered in the partition process. However, the mixture of Gaussians segmentation method by the EM (MGEM) algorithm proposed by Farnoosh et al. [16] uses intensity distribution models for segmentation. The MGEM method uses multiple Gaussians to represent intensity distribution of an image [7, 13, 17, 18]. However, it does not take into account the neighborhood information. As the result, segmented regions obtained by these algorithms lack connectivity with pixels within their neighborhoods. Their lack of connectivity with pixels within their neighborhoods is due to the following challenges:

(1) incomplete halo around the cell body.
(2) intensity similarity in cell body and substrate.
(3) intensity sparsity in cell regions.

Our proposed method is intended to solve these problems using spatial information of the image data. We evolve the cell regions based on the spatial information until the optimal optimization metric value is obtained. The proposed methods are developed with three criteria: (1) robustness; (2) speed; and (3) accuracy.

The existing papers often detect cells in phase contrast images in a simple environment [19, 20]. Eom et al. [19] discuss two methods for cell detection: detection by circular Hough transform (CHT) and detection by correlation. However, those methods do not work on the complex environment where the occurrence of dynamic blebbing and existence of overlapping cells are prevalent. The CHT is sensitive to the shape variations, and the correlation method in [20] does not work on cell clusters. Therefore, a detection method based on the features derived from inner cell region is developed to solve the aforementioned problem in unattached single stem cells detection.

7.3 Technical Approach

The segmentation approaches reviewed in this paper range from simple to complex and efficient to inefficient. Even though spatial information is used in all three segmentation methods, the derivation of spatial information in each method is different. In addition, we also provide a review on the unattached single stem cell detection method. The segmentation and detection methods are discussed in the following order.

1. Gradient magnitude distribution-based approach
2. Entropy-based k-means approach
3. Median filter-induced texture-based approach
4. Unattached single stem cell detection

7.3.1 Gradient Magnitude Distribution-Based Approach

The approach in [21] iteratively optimizes a metric that is based on foreground (F) and background (B) intensity statistics. The foreground/hESC region, F, is a high intensity variation region while the background/substrate region, B, is a low intensity variation region. Therefore, we can use the magnitude of gradients of the image to segment out the cell region from the substrate region. Equations (7.1) and (7.2) show the calculation of spatial information.

$$G = \left(\frac{dI}{dx}\right)^2 + \left(\frac{dI}{dy}\right)^2 \tag{7.1}$$

$$I_{mg} = \log_e\left(\frac{(-1+e^1) \times G}{\max(G)} + 1\right) \times 255 \tag{7.2}$$

where G is the squared gradient magnitude of image, I. $\frac{dI}{dx}$ and $\frac{dI}{dy}$ are gradients of image, I, in the x and y directions. I_{mg} is the spatial information and the log transform further emphasizes the difference between cell and substrate region. Equation (7.2) normalizes G as well as transforms the image into a bimodal image. Therefore, a single threshold can segment the transformed image into cell region and substrate region.

The proposed approach uses a mean filter on I_{mg} iteratively to evolve the cell regions. It is able to group the interior cell region pixels together based on the local information. The method updates I_{mg} and evolves the cell region until optimization metric in [21] is maximized. The window size of the mean filter contributes to how fast the cell region is evolved. The transformed image is iteratively thresholded by Otsu's algorithm. The parameters for the optimization metric are also updated iteratively.

7.3.2 Entropy-Based K-Means Approach

The approach mentioned in [22] utilizes the k-mean algorithm with weighted entropy. The cell regions generally have higher entropy values than the substrate regions due to their biological properties. However, not all cell regions have high entropy values. Therefore, the approach exploits intensity feature to solve the following problems: (1) intensity uniformity in some stem cell bodies; and (2) intensity homogeneity in stem cell halos. Since high entropy values happen in areas with high varying intensities, the aforementioned properties can greatly affect the cell region detection. Therefore, the method in [22] proposed a weighted entropy formulation. The approach uses the fact that stem cell image has an intensity histogram similar to a Gaussian distribution. Moreover, the stem cell consists of two essential parts: the cell body and the halo [6]. The cell bodies and halos' intensity values are located at the left and right end of the histogram distribution, respectively. As the result, regions with low or high intensity values have a higher weight. The background distribution is represented by the following equation:

$$D_{bg} \sim N_{256}(\mu, \sigma^2) \tag{7.3}$$

where D_{bg} is a Gaussian distribution of background with mean, μ, and variance, σ^2, and $D_{bg} \in \mathcal{R}^{256}$. The foreground distribution is shown in Eq. (7.4).

$$W = \text{Max}(D) - D \tag{7.4}$$

The weighted entropy is calculated by Eq. (7.5):

$$I_{\textbf{we}}(r,c) = \log\left(1 + E(r,c) \times \sum_{(y,x)\in W} W(I(y,x))^2\right) \tag{7.5}$$

where $I_{\text{we}}(r,c)$ is the weighted entropy at the location (r,c) with $r,c \in \mathcal{R}$. $E(r,c)$ is the un-weighted entropy value of the image's gradient magnitude at location (r,c), and $I(y,x)$ is the intensity value at location (y,x). $x,y \in \mathcal{R}$, and W is the set of neighboring coordinates of (r,c). $\sum_{(y,x)\in W} W(I(y,x))^2$ is the spatial energy term. Equation (7.5) enhances the separation between cell/foreground and substrate/background regions. The normalization of I_{we} back into a 8-bit image is shown in Eq. (7.6):

$$I_{\text{wen}} = \frac{(I_{\text{we}} - \min(I_{\text{we}}))}{(\max(I_{\text{we}}) - \min(I_{\text{we}}))} \times 255 \tag{7.6}$$

7.3.3 Median Filter-Induced Texture-Based Approach

The segmentation method by Guan et al. [23] combines the hESC spatial information and mixture of Gaussians for better segmentation. The mixture of Gaussians alone does not able to segment the cell regions from the substrate due to its lack of spatial consistency. The cell region intensities lie on both left and right end of the image histogram. However, the mixture of Gaussians can accurately detect fragments of the cell regions. Therefore, it can serve as a good template to find the optimal threshold, T_{opt}, and filter window size, m, in the spatial transformed image for cell region detection. As a result, we exploited the spatial information due to the biological properties of the cells in the image and used the result from the mixture of Gaussians as a comparison template [23].

With the spatial information, we can easily distinguish the cell region from the substrate region. The spatial information is generated with the combination of median filtering. A spatial information image I_S at scale m in [23] is calculated by the following equation:

$$I_{\text{mf}}(m) = med(|I - med(I,m)|, m+2) \tag{7.7}$$

where $med(\cdot, m)$ denotes the median filtering operation with window size m. The operation $|I - med(I,m)|$ yields low values in the substrate region and high values in the cell regions. The larger median filter window connects the interior cell regions while preserving the edges. The transformation of the image, I, by Eq. (7.7) generates a bimodal image from an original image that contains three intensity modes. Therefore, we can use the result from the mixture of Gaussians to find T_{opt} and m_{opt}. The cell region detection is done by finding the maximum correlation coefficient value between the results from the mixture of Gaussian and the spatial information at various T and m. The filter window size, m, varies from 3 to 25 with a step size of 2. The threshold T is from the minimum to the maximum of the spatial information image in steps of 0.5.

7.3.4 Unattached Single Stem Cell Detection

The unattached single stem cell detection by Guan et al. [10] is a feature-based classification that utilizes detected inner cell region features. The inner cell region is derived from thresholding the normalized probability map. The feature vector contains area size, eccentricity, and convexity of the inner cell region. We use the Euclidean distance as the classification measure. The Euclidean distances of the target's feature vector and the feature vectors in the training data are calculated by

$$K_f(i) = \begin{cases} \frac{1}{J}\sqrt{\sum_j K_{Coef}(i,j)^2} & \sqrt{\sum_j K_{Coef}(i,j)^2} \leq J \\ 1 & \text{else} \end{cases} \quad (7.8)$$

The symbol $K_f(i)$ is the Euclidean distance of the target feature vector and feature vectors from the training data where $i \in \mathbb{R}$. $K_{Coef}(i,j)$ is a matrix that contains the differences of the target feature vector and feature vectors in the training data. Variable J is equal to $\sqrt{3}$ and $j \in \{1,2,3\}$ since we have three features in our classification method.

7.4 Experimental Result

In this subsection, we briefly discuss the experimental results for each aforementioned approach.

7.4.1 Data

All time lapse videos were obtained with BioStation IM [1]. The frames in the videos are phase contrast images with 600×800 resolutions. The videos are acquired using three different objectives: $10\times$, $20\times$, and $40\times$.

7.4.2   Metrics

The true positive, TP, is the overlapped region between the binary image of detected cell regions and its ground-truth. True negative, TN, is the overlapped region between the binary image of detected background region and its ground-truth. The false positive, FP, is the detected background in the background binary image that is falsely identified as part of the cell region in cell region binary image. The false negative, FN, is the detected cell region in cell region binary image that is falsely identified as part of the background in the background binary image [24].

The true positive rate or sensitivity, TPR, measures the proportion of actual positives which are correctly identified.

$$TPR = \frac{TP}{(TP + FN)} \tag{7.9}$$

The false positive rate, FPR, measures the proportion of false positives which are incorrectly identified.

$$FPR = \frac{FP}{(FP + TN)} \tag{7.10}$$

7.4.3   Gradient Magnitude Distribution-Based Approach

The gradient magnitude distribution approach has more than 90 % average sensitivity and less than 15 % average false positive rate for all datasets that are not corrupted by noise [21]. The high performance remained after filtering is done on the noisy dataset. The dataset taken with 40× objective is corrupted by noise, and the proposed method has lower performance without filtering as shown in Tables 7.1 and 7.2. Tables 7.1 and 7.2 also show comparison of the proposal

Table 7.1 Average sensitivity

Method	10× (%)	20× (%)	40× (%)	40×[a] (%)
Guan [21]	95.85	95.65	75.09	90.42
KM	51.92	51.83	79.66	40.26
M2G	79.27	80.54	63.56	80.99

[a]filtered data

Table 7.2 Average false positive rate

Method	10× (%)	20× (%)	40× (%)	40×[a] (%)
Guan [21]	14.64	13.33	3.18	7.22
KM	4.41	5.95	30.18	0.46
M2G	11.57	14.10	3.63	9.37

[a]filtered data

method with k-means (KM) and mixture of two Gaussians (M2G). KM method performs the worst in all three methods. The proposed method's average sensitivity is higher than M2G by at least 10 % in all datasets.

7.4.4 Entropy-Based K-Means Approach

In this approach, we also compare the proposed method with k-means. However, the initial means for each cluster are assigned based on our knowledge for each cluster. Therefore, KM in this approach has significant lower false positive rate in all datasets. The entropy and k-means based on segmentation has more than 80 % average sensitivity and less than 16 % average false positive rate [22]. Table 7.3 shows that the proposed method has above 96 % average sensitivity except for the unfiltered 40× dataset. Table 7.4 shows that KM has lower average false positive rate, but it is due to the fact that it detect all pixels as background.

7.4.5 Median Filter-Induced Texture-Based Approach

The median filter-induced texture-based approach has tested only on the six datasets taken with 20× objective. The approach has more than 92 % average sensitivity and less than 7 % average false positive rate [23]. The means of all six datasets in average sensitivity and average false positive rate are 96.24 and 5.28 %. As shown in Table 7.5, mixture of two Gaussians and mixture of three Gaussians methods have the lowest average sensitivities. Even though other methods have the mean average sensitivities above 94 %, they have high false positive rate as shown in Table 7.6. Their mean average false positive rates are above 17 %.

7.4.6 Unattached Single Stem Cell Detection

The unattached single stem cell detection approach was tested on four datasets taken under 20× objective [10]. Figure 7.3 shows an ROC plots for four different

Table 7.3 Average sensitivity

Method	10× (%)	20× (%)	40× (%)	40×[a] (%)
Guan [22]	98.09	96.73	80.82	96.18
KM	37.76	38.37	40.28	40.28

[a]filtered data

Table 7.4 Average false positive rate

Method	10× (%)	20× (%)	40× (%)	40×[a] (%)
Guan [22]	19.21	17.81	5.83	15.68
KM	1.77	2.51	0.61	0.61

[a]filtered data

Table 7.5 Average sensitivity

Video	HBBC (%)	GLMG (%)	PIV (%)	M2G (%)	M3G (%)	Guan [23] (%)
1	92.08	97.08	98.90	43.52	53.16	94.30
2	84.95	93.80	97.37	29.88	42.18	92.31
3	98.46	98.89	99.80	70.46	78.14	97.67
4	99.27	98.78	99.54	79.85	87.33	98.38
5	97.07	97.66	98.82	77.38	84.40	97.25
6	97.81	98.17	99.23	74.88	81.15	97.54

Table 7.6 Average false positive rate

Video	HBBC (%)	GLMG (%)	PIV (%)	M2G (%)	M3G (%)	Guan [23] (%)
1	19.71	22.47	19.86	2.83	5.16	4.41
2	26.67	21.82	18.83	2.65	4.39	6.57
3	16.62	22.47	18.50	4.80	7.22	4.78
4	14.19	15.91	16.69	4.77	7.52	5.74
5	12.34	16.42	15.10	4.39	6.86	5.51
6	16.33	22.07	20.33	5.06	8.09	4.69

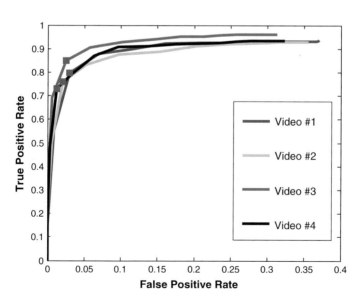

Fig. 7.3 True positive rate versus false positive rate (ROC curves) (*Note* the *red squares* are the optimal results of the proposed method)

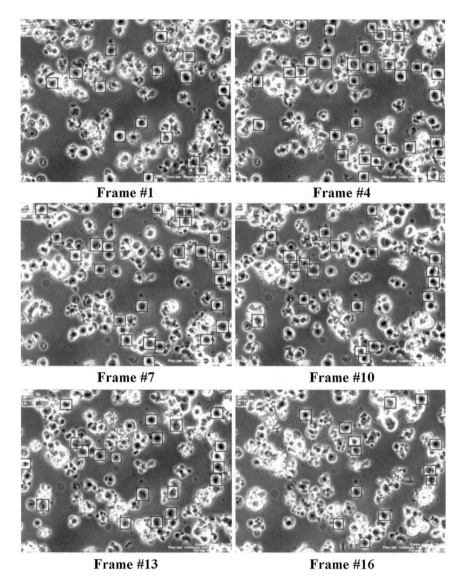

Fig. 7.4 A sample of results of unattached single stem cell detection approach

experiments. The proposed method achieves above 70 % in true positive while keeping the false positive rate below 2.5 %. Figure 7.4 shows the result of the proposed method on a sequence of images. The proposed method captured majority of the unattached single stem cells.

7.5 Performance Analysis

In this subsection, we evaluate the three aforementioned segmentation approaches mentioned above in terms of robustness, speed, and accuracy.

7.5.1 Robustness

The median filter-induced texture-based method in [23] was tested only on a 20× dataset with a set of parameters. Since the parameters are tuned for the 20× datasets, it is not reliable for datasets collected using different objectives as shown in Fig. 7.5. Most importantly, its performance is heavily depended on initial segmentation result of the mixture of Gaussians. The entropy-based k-means approach in [22] and the gradient magnitude distribution-based approach in [21] are more

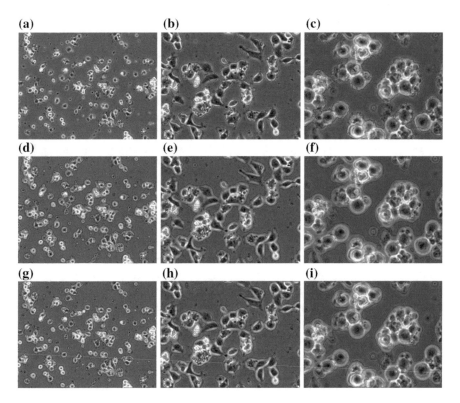

Fig. 7.5 **a–c** Median filter-induced texture-based method results for 10×, 20×, and 40× images respectively; (**d–f**) entropy-based k-means method results for 10×, 20×, and 40× images, respectively; (**g–i**) gradient magnitude distribution-based method results for 10×, 20×, and 40× images, respectively

robust with different objectives. Since both methods do not need a template to obtain the optimal segmentation, they are robust under different objectives with a same set of parameters.

7.5.2 Speed

The entropy-based k-means in [22] and gradient magnitude distribution-based in [21] approaches have only one optimization step. The median filter-induced texture-based method in [23] requires two optimization steps. However, the entropy-based k-means approach has higher time complexity than the other methods. The reason it is slow is because the method performs k-means clustering and entropy filtering for segmentation. The operations of k-means clustering and entropy filtering are extremely time consuming. Consequently, the approach's processing time is about 28 s/frame.

Since the median filter-induced texture-based method has two optimization steps, it also has high time complexity. The first optimization step is done to determine the initial segmentation result by the mixture of Gaussians. The second optimization step is done with the mixture of Gaussians result to obtain the final segmentation. Therefore, the approach's processing time is about 26 s/frame.

The gradient magnitude distribution-based approach only uses mean filtering and the Otsu's algorithm. Since mean filtering and Otsu's algorithm are faster operations, the method is the fastest among the aforementioned approaches. The gradient magnitude distribution-based segmentation only requires 1.2 s/frame of processing time. All experiments for each method are done on a laptop with an Intel (R) Core™ 2 Duo CPU processor that run at 2.53 GHz.

7.5.3 Accuracy

The entropy-based k-means approach has the highest average true positive rate (TPR) as well as the highest average false positive rate (FPR) [24]. It has a mean of about 97 % average TPR and near 18 % average FPR. Both median filter-induced texture-based and gradient magnitude distribution-based approaches have less than 12 % average false positive rate. However, the median filter-induced texture-based approach was only tested on the dataset taken with a 20× objective, and it had a mean of 96.24 % in average TPR. The gradient magnitude distribution-based method has a mean near 94 % in average TPR. Both entropy-based k-means and gradient magnitude distribution-based approaches were tested on datasets under different objectives.

7.6 Conclusion

This paper provides a brief review of three existing bio-inspired segmentation methods as well as a concise discussion on the unattached single stem cell detection method. All three segmentation approaches mentioned in this paper have above 90 % average true positive rate and less than 18 % average false positive rate in segmentation. However, the gradient magnitude distribution-based approach outperforms the other methods in the overall measures of robustness, speed, and accuracy. The gradient magnitude distribution-based approach has good detection results for images collected with all three objectives. It has smoother detected result than the other methods as shown in Fig. 7.5. In terms of speed, the method requires processing time of 1.2 s/frame. Compared to its counterparts, it is about 20 times faster when run in the same machine. In terms of accuracy, the gradient magnitude distribution-based method still has a mean near 94 % in average TPR and less than 12 % in average FPR. The unattached single stem cell detection makes possible with the accurate cell region segmentation. Therefore, the gradient magnitude distribution-based method is significant for the future development of fast quantifiable analysis of stem cells. The methods described in this chapter could be used to evaluate the health and viability of hESC cultures and the response of hESC to environmental changes or toxicants.

Acknowledgment This research was supported by National Science Foundation-Integrated Graduate Education Research and Training (NSF-IGERT): Video Bioinformatics Grant DGE 0903667 and by Tobacco-Related Disease Research Program (TRDRP): Grant 20XT-0118.

References

1. Stojkovic M, Lako M, Strachan T, Murdoch A (2004) Derivation, growth and applications of human embryonic stem cells. Reproduction 128:259–267
2. Lin S et al (2010) Comparison of the toxicity of smoke from conventional and harm reduction cigarettes using human embryonic stem cells. Toxicol Sci 118:202–212
3. Behar RZ, Bahl V, Wang Y, Lin S, Davis B, Talbot P (2012) A method for rapid dose-response screening of environmental chemicals using human embryonic stem cells. J Pharmacol Toxicol Methods. doi:10.1016/j.vascn.2012.07.003
4. Bahl BV, Lin S, Xu N, Davis B, Wang Y, Talbot P (2012) Comparison of electronic cigarette refill fluid cytotoxicity using embryonic and adult models. Reprod Toxicol 34(4):529–537
5. Behar RZ, Bahl V, Wang Y, Weng J, Lin S, Talbot P Adaptation of stem cells to 96-well plate assays: use of human embryonic and mouse neural stem cells in the MTT assay, Current Protocols, Chapter 1:Unit1C.13. doi:10.1002/9780470151808.sc01c13s23
6. Talbot P, Lin S (2011) Mouse and human embryonic stem cells: can they improve human health by preventing disease? Curr Top Med Chem 11(13):1638–1652
7. Lin S, Talbot P (2011) Methods for culturing mouse and human embryonic stem cells. Methods Mol Biol 690:31–56
8. Lin S et al (2010) Video bioinformatics analysis of human embryonic stem cell colony growth. JOVE 39, May 2010

9. Nikon (2013) CL-Quant. http://www.nikoninstruments.com/News/US-News/Nikon-Instruments-Introduces-CL-Quant-Automated-Image-Analysis-Software. Accessed July 2013
10. Guan BX, Bhanu B, Talbot P, Lin S (2012) Detection of non-dynamic blebbing single unattached human embryonic stem cells. In: IEEE international conference on image processing, Orlando, FL
11. Ambriz-Colin F, Torres-Cisneros M, Avina-Cervantes J, Saavedra-Martinez J, Debeir O, Sanchez-Mondragon J (2006) Detection of biological cells in phase-contrast microscopy images. In: Fifth mexican international conference on artificial intelligence, 2006. MICAI '06, pp 68 –77
12. Li K, Chen M, Kanade T (2007) Cell population tracking and lineage construction with spatiotemporal context. In: Proceedings of the 10th international conference on medical image computing and computer-assisted intervention (MICCAI), pp 295–302
13. Tatiraju S, Mehta A (2008) Image segmentation using k-means clustering, EM and normalized cuts, UC Irvine
14. Alsabti K, Ranka S, Singh V (1998) A efficient k-means clustering algorithm. In: Proceedings of first workshop on high performance data mining
15. Kanungo T, Mount DM, Netanyahu NS, Piatko CD, Silverman R, Wu AY (2002) An efficient k-means clustering algorithm: analysis and implementation. IEEE Trans PAMI 881–892
16. Farnoosh R, Zarpak B (2008) Image segmentation using Gaussian mixture model. Int J Eng Sci 19:29–32
17. Xu L, Jordan MI (1996) On convergence properties of the EM algorithm for Gaussian mixture. Neural Comp 129–151
18. Gopinath S, Wen Q, Thakoor N, Luby-Phelps K, Gao JX (2008) A statistical approach for intensity loss compensation of confocal microscopy images. J Microsc 230(1):143–159
19. Eom S, Bise R, Kanade T (2010) Detection of hematopoietic stem cells in microscopy images using a bank of ring filters. In: Proceedings of 7th IEEE international symposium on biomedical imaging, Rotterdam, Netherlands, pp 137–140
20. Miroslaw L, Chorazyczewski A, Buchholz F, Kittler R (2005) Correlation-based method for automatic mitotic cell detection in phase contrast microscopy. Adv Intell Soft Comput 30:627–634
21. Guan BX, Bhanu B, Talbot P, Lin S (2012) Automated human embryonic stem cell detection. In: IEEE 2nd international conference on healthcare informatics, imaging and systems biology, pp 75–82
22. Guan BX, Bhanu B, Thakoor N, Talbot P, Lin S (2013) Automatic cell region detection by K-means with weighted entropy. In: International symposium on biomedical imaging: From Nano to Macro, pp 418–421
23. Guan BX, Bhanu B, Thakoor N, Talbot P, Lin S (2011) Human embryonic stem cell detection by spatial information and mixture of Gaussians. In: Proceedings of 1st IEEE international conference on health informatics, imaging and system biology, San Jose, CA, pp 307–314
24. Pepe M, Longton GM, Janes H (2008) Comparison of receiver operating characteristics curves, UW Biostatistics Working Paper Series- Working Paper 323 [eLetter] January 2008. http://biostats.bepress.com/uwbiostat/paper323

Chapter 8
A Video Bioinformatics Method to Quantify Cell Spreading and Its Application to Cells Treated with Rho-Associated Protein Kinase and Blebbistatin

Nikki Jo-Hao Weng, Rattapol Phandthong and Prue Talbot

Abstract Commercial software is available for performing video bioinformatics analysis on cultured cells. Such software is convenient and can often be used to create suitable protocols for quantitative analysis of video data with relatively little background in image processing. This chapter demonstrates that CL-Quant software, a commercial program produced by DRVision, can be used to automatically analyze cell spreading in time-lapse videos of human embryonic stem cells (hESC). Two cell spreading protocols were developed and tested. One was professionally created by engineers at DRVision and adapted to this project. The other was created by an undergraduate student with 1 month of experience using CL-Quant. Both protocols successfully segmented small spreading colonies of hESC, and, in general, were in good agreement with the ground-truth which was measured using ImageJ. Overall the professional protocol performed better segmentation, while the user-generated protocol demonstrated that someone who had relatively little background with CL-Quant can successfully create protocols. The protocols were applied to hESC that had been treated with ROCK inhibitors or blebbistatin, which

N.J.-H. Weng (✉)
Department of Cell Biology & Neuroscience, University of California,
2313 Spieth Hall, 900 University Ave., Riverside, CA 92507, USA
e-mail: jweng002@ucr.edu

R. Phandthong
Department Cell, Molecular, and Developmental Biology, University of California,
2320 Spieth Hall, 900 University Ave., Riverside, CA 92507, USA
e-mail: rphan005@ucr.edu

P. Talbot
Department of Cell Biology & Neuroscience, University of California,
2320 Spieth Hall, 900 University Ave., Riverside, CA 92507, USA
e-mail: talbot@ucr.edu

© Springer International Publishing Switzerland 2015
B. Bhanu and P. Talbot (eds.), *Video Bioinformatics*,
Computational Biology 22, DOI 10.1007/978-3-319-23724-4_8

tend to cause rapid attachment and spreading of hESC colonies. All treatments enabled hESC to attach rapidly. Cells treated with the ROCK inhibitors or blebbistatin spread more than controls and often looked stressed. The use of the spreading analysis protocol can provide a very rapid method to evaluate the cytotoxicity of chemical treatment and reveal effects on the cytoskeleton of the cell. While hESC are presented in this chapter, other cell types could also be used in conjunction with the spreading protocol.

8.1   Introduction

Live cell imaging has been widely used in our laboratory for many years to study dynamic cell processes [9, 18, 20, 40–42] and has more recently been applied to toxicological problems [10, 12, 22, 23, 27, 32–36, 39]. Analysis of dynamic events, such as cell attachment, migration, division, and apoptosis, can provide mechanistic insight into normal cellular processes [28, 37] as well as how toxicants affect cells [2, 5, 23, 42]. Collection of video data has recently improved due to the introduction of commercial incubators with built-in microscopes and cameras for collecting time-lapse data during short- and long-term experiments [8, 30, 37].

After videos are collected, it is important to extract quantitative data from them. A challenging but important issue until recently has been how to analyze large complex data sets that are produced during live cell imaging. When video data analyses are done manually by humans, many hours of personnel time are usually required to complete a project, and manual analysis by humans is subject to variation in interpretation and error. Video bioinformatics software can be used to speed the analysis of large data sets collected during video imaging of cells and can also improve the accuracy and repeatability of analyses [37]. Video bioinformatics, which involves the use of computer software to mine specific data from video images, is concerned with the automated processing, analysis, understanding, visualization, and knowledge extracted from microscopic videos. Several free video bioinformatics software packages are available online such as ImageJ and Gradientech Tracking Tool (http://gradientech.se/tracking-tool-pro/). Also, some advanced video bioinformatics software packages, such as CL-Quant, Amira, and Cell IQ, are now commercially available and can be used to generate customized protocols or libraries to analyze video data and determine quantitatively how cells behave during experimental conditions.

This chapter presents a new application of CL-Quant software to automatically analyze cell spreading in time-lapse videos of human embryonic stem cells (hESC) (WiCell, Madison, WI). While hESC are presented in this chapter, other cell types could also be used in conjunction with these protocols.

8.2 Instrumentation Used to Collect Live Cell Video Data

8.2.1 BioStation IM

Data were collected with a BioStation IM. The BioStation IM or its newer version the IM-Q, manufactured by Nikon, is a bench top instrument that houses a motorized inverted microscope, an incubator with a built-in high sensitivity cooled CCD camera, and software for controlling exposures, objectives, and the type of imaging (e.g., phase contrast or fluorescence). The components of this instrument are fully integrated and easy to set up. In a BioStation IM, cells are easily maintained at a constant temperature (37 °C) and relative humidity (85 %) in a 5 % CO_2 atmosphere. The BioStation IM enables time-lapse data to be collected reliably over hours or days without focus or image drift. In time-lapse experiments, images can be collected as quickly as 12 frames/s. In a BioStation IM, imaging can be also performed in the X-, Y-, and Z-direction. The unit comes with a well-designed software package and a GUI for controlling the instrument and all experimental parameters.

The BioStation IM is available in two models, the BioStation II and BioStation II-P, optimized for either glass bottom or plastic bottom culture dishes, respectively, and the magnification range is different in the two models. Both models accommodate 35 and 60 mm culture dishes. A four-chambered culture dish, the Hi-Q4 sold by Nikon, can be used for examining four different conditions of culture in the same experiment.

8.3 Software Used for Video Bioinformatics Analysis of Stem Cell Morphology and Dynamics

8.3.1 CL-Quant

CL-Quant (DRVision, Seattle, WA) provides tools for developing protocols for recognition and quantitative analysis of images and video data [1]. The software is easy to learn and does not require an extensive knowledge of image processing. CL-Quant can be used to detect, segment, measure, analyze, and discover cellular behaviors in video data. It can be used with both phase contrast and fluorescent images. Several basic protocols for cell counting, cell proliferation, wound healing, and cell migration have been created by DRVision engineers and can be obtained when purchasing CL-Quant software. Protocols can be created by DRVision at a user's request, or users can create their own protocols. Later in this chapter, we will describe how to create a protocol in CL-Quant, show the difference between a professional protocol and user-generated protocol, and also show an example in which CL-Quant was used to measure hESC spreading in experimental time-lapse videos.

8.3.2   ImageJ

ImageJ is a public domain Java-based image-processing program, which was developed at the National Institutes of Health. ImageJ was designed with an open architecture that provides extensibility via plug-ins and recordable macros. A number of tutorials are available on YouTube and are helpful for beginners learning to use this software. ImageJ is compatible with major operating systems (Linux, Mac OS X, and Windows), works with 8-bit color and grayscale, 16-bit integer, and 32-bit floating point images. It is able to read many image formats, and it supports time-lapse or z-stacks. There are numerous plug-ins that can be added to ImageJ to help solve many imaging processing and analysis problems. ImageJ is able to perform numerous standard image-processing operations that may be useful in labs dealing with image analysis and video bioinformatics. For example, researchers have used ImageJ to quantify bands in western blots and also to quantify the fluorescent intensity on the images. One of the advantages of using ImageJ is that it enables rapid conversion of images to different formats. For example, ImageJ can convert tif images to avi, it can create 3D images from z-stacks with 360° rotation, and it can be used to obtain ground-truth information when setting up a new video bioinformatics protocol.

8.4   Protocols for Cell Attachment and Spreading

In toxicological assays using stem cells, endpoints of interest often occur days or sometimes a month after the experiment begins [27, 37]. There has been interest in shortening such assays, often by using molecular biomarkers to obtain endpoints more rapidly [6]. In the mouse embryonic stem cells test, endpoints can now be obtained in 7 days using biomarkers for heart development. While this significantly reduces the time to reach an endpoint, it is still a relatively long time to obtain data on cytotoxicity.

Pluripotent hESC model the epiblast stage of development [31] and are accordingly a valuable resource for examining the potential effects of chemicals at an early stage of human prenatal development [38]. We are developing video assays to evaluate cellular processes in hESC in short-term cultures, and we then use these assays to identify chemicals that are cytotoxic to young embryos. One hESC-based assay involves evaluation of cell spreading, a dynamic process dependent on the cytoskeleton. When a treatment alters the cytoskeleton or its associated proteins, cells are not able to attach and spread normally. We have used two video bioinformatics protocols to analyze these parameters during 4 h of in vitro culture. At the beginning of cell plating, hESC are round and unattached. Usually, hESC attach to their substrate and begin spreading within 1 h of plating. As cells attach and start spreading, their area increases, and this can be measured in time-lapse images using video bioformatics tools, thereby providing a rapid method to evaluate a process dependent on the cytoskeleton. Two parameters can be derived from the time-lapse data: rate of cell spreading (slope) and fold increase in cell area. These two parameters were compared in control and treated groups using the linear regression and 2-way ANOVA analysis (GraphPad Prism, San Diego).

We will compare two protocols for measuring cell spreading in this chapter. Both protocols are performed using CL-Quant software. One, which we term a professional protocol, was created by DRVision engineers for quantifying cell proliferation. Even though the professional protocol was not created specifically for hESC spreading, we were able to use the segmentation portion of the protocol for measuring spreading (area) of hESC during attachment to Matrigel. In addition, we created our own protocol using CL-Quant for analyzing the area of hESC colonies during attachment and spreading. Our method, which we refer to as the user-generated protocol, was created by an undergraduate student with a basic background in biology and 1 month of experience with CL-Quant software.

hESC were cultured using methods described in detail previously [29]. To create time-lapse videos for this project, small colonies of hESC were incubated in mTeSR medium at 37 °C and 5 % CO_2 in a BioStation IM for 4 h. Frames were captured every minute from 4–5 different fields. When applying either the professional or user-generated protocol, the first step was to segment the image so as to select mainly hESC. During segmentation, we used the DRVision's soft matching procedure, which allowed us to identify the objects of interest. The second step was to remove noise and small particles/debris that were masked during segmentation. In the third step, the area of all cells in a field was measured in pixels. The protocol was applied to all images in time-lapse videos to obtain the area occupied by hESC when plated on Matrigel and incubated for 4 h. Because the survival efficiency is low for single cells [43], hESC were plated as small colonies, which normally attach, spread, and survive well. Figure 8.1a shows phase contrast images of several small hESC colonies plated on Matrigel at different times over 4 h. In the first frame, the cells were unattached, as indicated by the bright halo around the periphery of some cells. During 4 h of incubation, the cells attached to the Matrigel and spread out. By the last frame, all cells in the field have started to spread. Figure 8.1a also shows the same images after segmentation, enhancement, and masking using the professional protocol supplied by DRVision. Comparison of the phase contrast and segmented sequences shows that the masks fit each cell well.

To determine if the measurement data obtained from the professional CL-Quant protocol were accurate, ground-truth was obtained by tracing each hESC in all of the video images using the freehand selection tool in ImageJ, and then measuring the pixel area for each frame. The ground-truth (dotted line) and CL-Quant derived area were very similar for hESC grown in control conditions (mTeSR medium) (Fig. 8.1b). For most videos, CL-Quant slightly overestimated the area of the cells due to difficulty in fitting a perfect mask to each cell.

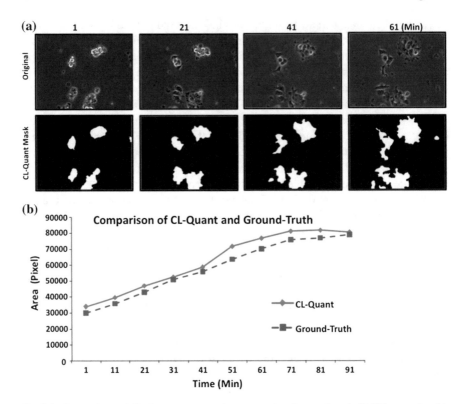

Fig. 8.1 Comparison of CL-Quant segmentation protocol and ground-truth. hESC were plated in mTeSR medium in a 35 mm dish and incubated in a BioStation IM for 4 h. **a** Phase contrast images modified with ImageJ to remove text labels and the same images with masks applied using the professional CL-Quant protocol. **b** Graph showing cell area (spreading) in pixels for CL-Quant derived data and the ground-truth. The areas obtained from the two methods were in good agreement

8.5 Application of Video Bioinformatics Protocols to hESC Cell Spreading in the Presence of Rock Inhibitors and Blebbistatin

In 2007, ROCK inhibitor (Y27632) was shown to increase the efficiency of hESC survival in culture [43]. However, ROCK is involved in numerous signaling pathways, and therefore may affect many cell properties [25]. ROCK inhibitors decrease non-muscle myosin II activity, and this decrease helps attachment of hESC. However, when single hESC are plated with ROCK inhibitor, they do not adhere to each other due to downregulation of e-cadherin [19]. In addition, hESC treated with Y27632, a potent ROCK inhibitor, appeared stressed and less healthy than untreated controls [3]. Finally, the use of ROCK inhibitor (Y27632) in a toxicological study with methyl mercury decreased the IC_{50} [11]. Blebbistatin, an inhibitor of myosin II which is downstream of ROCK, can also be used to increase attachment of hESC and thereby improve plating efficiency [19] and may also alter cell morphology [21].

In this study, time-lapse data were collected on hESC treated with different ROCK inhibitors or blebbistatin, and two CL-Quant protocols were used to quantitatively compare spreading of treated cells to controls. H9 hESC were seeded on Hi-Q4 dishes coated with Matrigel and incubated using different treatment conditions. Cells were then placed in a BioStation IM and cultured as described above for 4 h during which time images were taken at three or four different fields in each treatment/control group at 1 min intervals.

The protocols described in Sect. 1.4 for quantifying hESC area were used to compare spreading of cells subjected to different ROCK inhibitors (Y27632, H1152) or to blebbistatin. First, the written data that was stamped on each image by the BioStation software was removed using the remove outlier's feature in ImageJ. The professional CL-Quant protocol was applied to the resulting time-lapse videos. Examples of phase contrast and masked images are shown in Fig. 8.2 for treatment with the two ROCK inhibitors and blebbistatin. Cells in each group were masked accurately by the segmentation protocol, and even thin surface cell projections were masked with reasonable accuracy. The measurement protocol was then applied to each frame to determine the area (pixels) of the masked cells. To establish the accuracy of this protocol, the ground-truth for control and treated groups was determined using ImageJ in two separate experiments (Fig. 8.3a, b). Each point in Fig. 8.3 is the mean of 3 or 4 videos. As shown in Fig. 8.3, the ground-truth and the CL-Quant derived data were in good agreement for all groups in both experiments.

In the first experiment, the fold increase in spread area was elevated by Y27632 and blebbistatin relative to the control ($p < 0.0001$ for Y27632 and $p < 0.05$ for blebbistatin 2-way ANOVA, Graphpad Prism), while H1152 was not significantly different than the control ($p > 0.05$). The rate of spreading, as determined by the slope for each group, was greater in the three treated groups than in the control; however, only the slope for Y27632 was significantly different than the control ($p < 0.0001$) (slopes = 0.124 control; 0.135 H1152; 0.142 blebbistatin; 0.245 Y27632). In this experiment, the Y27632 group was distinct from all other groups in both its rate of spreading and fold increase in spread area. The morphology of the control cells was normal; cells had smooth surfaces with relatively few projections. In contrast, all treated cells had irregular shapes and more projections than the controls (Fig. 8.3a).

In the second experiment, the three treated groups spread faster and more extensively than the control group. All three treated groups were significantly different than the control with respect to fold change in spread area (by 2-way ANOVA $p < 0.05$ for Y27632 and H1152; $p < 0.0001$ for blebbistatin). In contrast to the first experiment, the Y27632 group was similar to the other two treatments (Fig. 8.3d). The rate of spreading was greater in the three treated groups than in the control (slope = 0.084 control; 0.117 H1152; 0.150 blebbistatin; 0.136 Y27632), and both Y27632 ($p < 0.01$) and blebbistatin ($p < 0.001$) were significantly different than the control. As seen in the first experiment, all treated cells were

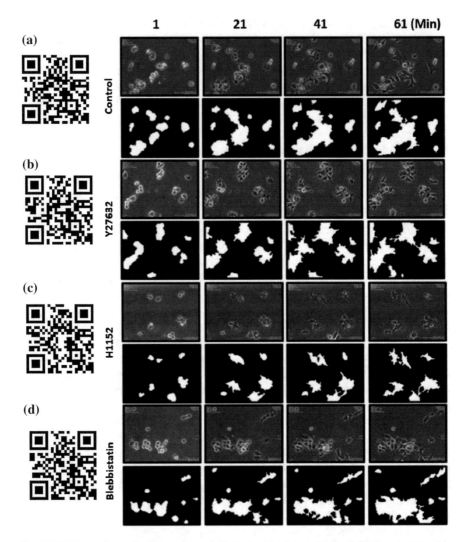

Fig. 8.2 Cell area (spreading) was successfully masked by the professional CL-Quant protocol in different experimental conditions. hESC were treated with ROCK inhibitors (Y27632 and H1152) or blebbistatin, incubated in a BioStation IM for 4 h, and imaged at 1 min intervals. Phase contrast images and the corresponding masked images are shown for hESC treated with: **a** control medium, **b** Y27632, **c** H1152, and **d** blebbistatin

morphologically distinct from the controls. Treated cells appeared attenuated and had long thin projections extending from their surfaces, while control cells were compact and had smooth surfaces (Fig. 8.3c). It is possible that CL-Quant underestimated the area of the Y27632 inhibitor treated cells in the second experiment due to the attenuation of the surface projections, which were more extensive than in the first experiment and were difficult to mask accurately.

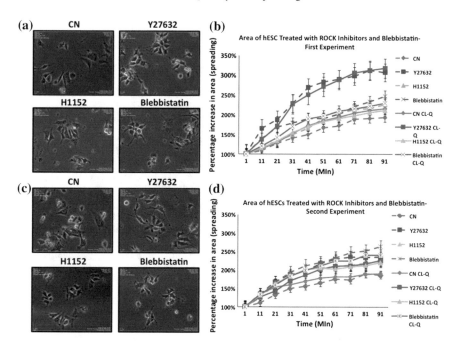

Fig. 8.3 The morphology and spreading of hESC was affected by treatment with ROCK inhibitors (Y27632 and H1152) and blebbistatin in two experiments. Spreading was measured using the professional CL-Quant protocol. **a** Phase contrast images from the first experiment showed that treated cells were morphologically different than the control. **b** The rate of spreading and the fold increase in spread area was greater in Y27632 and blebbistatin treated cells than in controls. **c** Phase contrast images of control and treated cells in the second experiment showed morphological changes in the treated groups. **d** The fold increase in spread area was greater in the treated cells than in the controls in the second experiment; however, the effect of Y27632 was not as great as previously seen. Data in (**b**) and (**d**) are plotted as a percentage of the area in the first frame. Each point is the mean ± the SEM

Videos showing the effect of ROCK inhibitors and blebbistatinon hESC spreading can be viewed by scanning the bar code.

8.6   Comparison of Professional and User-Generated Protocols

We also compared the professional and user-generated protocols to each other. The cells in both the control and treated groups were well masked by the professional protocol, and the mask included the thin surface projections characteristic of the treated group (Fig. 8.4a). To validate the quantitative data obtained with the professional protocol, ground-truth was determined using ImageJ (Fig. 8.4b). Both the

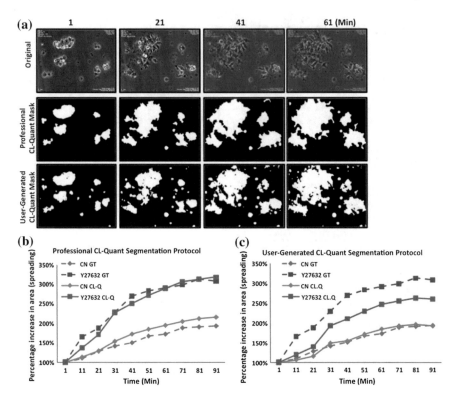

Fig. 8.4 Comparison of the professional and user-generated cell spreading protocols. **a** Phase contrast micrographs of hESC treated with Y27632 and the corresponding masks created with the professional and user-generated protocols. **b** Comparison of ground-truth to area (spreading) data obtained with the professional protocol in control and treated groups. **c** Comparison of ground-truth to area (spreading) data obtained with the user-generated protocol in control and treated groups

control and treated groups were in good agreement with the ground-truth (Fig. 8.4b).

An advantage CL-Quant is that users can generate their own protocols without a programming background. Our user-generated protocol, which was created by an undergraduate student with 1 month of experience using CL-Quant, was applied to the control and treated groups. The resulting masks did not cover the surface projections of the treated group as well as the professional protocol; however, the user-generated protocol did include single cells, some of which were filtered out by the professional protocol (Fig. 8.4a). The user-generated protocol did not filter out the small debris as well as the professional protocol (Fig. 8.4a). The data obtained with the user-generated protocol were close to ground-truth for the control group, but not for the treated group (Fig. 8.4c). Phase contrast images showed that the cells treated with Y27632 had many more attenuated surface projections than the control cells (Fig. 8.5a, d). Our user-generated protocol and the professional protocol were

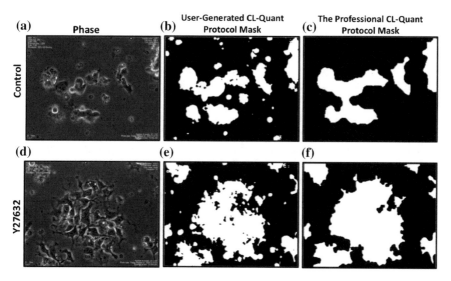

Fig. 8.5 Differences in cell morphology showing why treated hESC are more difficult to segment than control cells. **a** Phase contrast image of hESC colonies taken at 60 min of incubation. Segmentation of the image in "(**a**)" created with the user-generated protocol (**b**) and the professional protocol (**c**). **d** Phase contrast image of hESC colonies treated with Y27632 for 60 min. The cells have many thin surface projections not present on controls. Segmentation of the image in "(**c**)" with the user-generated protocol (**e**) and the professional protocol (**f**)

able to mask the control cells well (Fig. 8.5b, c). However, the user-generated protocol was not able to mask the thin projections on treated cells as well as the professional protocol (Fig. 8.5e, f). Neither protocol recognized gaps between cells in the treated group. Overall, the professional protocol was more similar to the ground-truth than the user-generated protocol in this experiment; however, with more experience, the user could improve the protocol to include surface projections more accurately.

8.7 Discussion

In this chapter, we introduced a video bioinformatics protocol to quantify cell spreading in time-lapse videos using CL-Quant image analysis software, and we validated it against ground-truth. A professionally developed version of the protocol was then compared to a protocol developed by a novice user of CL-Quant. We also applied this protocol to hESC treated with blebbistatin and ROCK inhibitors, which are commonly used during in vitro passaging of hESC [43].

Most evaluations of cells in culture have involved processes such as cell division, confluency and motility, but not spreading. Cell attachment and spreading depend on the interaction between cells and the extracellular matrix to which they attach. When cells are plated on a substrate, they first attach, then flatten and spread. At the molecular level, spreading depends on the interaction of membrane-based integrins with their extracellular matrix and the engagement of the cytoskeleton, which initiates a complex cascade of signaling events [7]. This in turn leads to the morphological changes observed during spreading and enables cultured cells to flatten out and migrate to form colonies [24, 44].

Changes in cell behavior that depend on the cytoskeleton often indicate that cell health is being compromised by environmental conditions [2, 27]. Because spreading depends on the cytoskeleton, it can be used to evaluate the effect of chemical treatments on cytoskeletal health. Cell spreading is a particularly attractive endpoint in toxicological studies as it occurs soon after plating and does not require days or weeks of treatment to observe. Many types of cells, such as fibroblasts can attach and spread in 10–15 min. hESC require about 1–2 h to spread, and therefore can be plated and data collected in 4 h or less time depending on the experimental design.

The application of the spreading protocol to hESC enables chemical treatments to be studied using cells that model a very early stage of postimplantation development. hESC are derived by isolating and culturing the inner cell mass from human blastocysts. As these cells adapt to culture, they take on the characteristics of epiblast cells which are found in young post-implantation embryos [31]. Embryonic cells are often more sensitive to environmental chemicals than differentiated adult cells, and it has been argued that risk assessment of environmental chemicals should be based on their effects on embryonic cells, as these represent the most vulnerable stage of the life cycle [13]. The sensitivity of hESC to environmental toxicants may be due to mitochondrial priming which occurs in hESC and makes them more prone to apoptosis than their differentiated counterparts [26].

A major advantage of the cell spreading assay is that it requires relatively little time to perform. A complete spreading assay can be done in as little as 4 h, while other endpoints such as cell division and differentiation require days or weeks to evaluate. The rapidity of the spreading assay makes it valuable in basic research on the cytoskeleton or in toxicological studies involving drugs or environmental chemicals. Moreover, this tool could be used in the future as a quality control check when evaluating stem cell health for clinical applications.

The cell spreading assay introduced in this chapter provides a rapid method for evaluating the cytoskeleton and assessing the quality of cells. Using bioinformatics tools to analyze the video data significantly reduces the time for data analysis [14–17]. If an experiment is done for 4 h with 1 min intervals between frames, 240 frames would be collected for each video by the end of the experiment. Each group could have 4–10 different videos. Before we used bioinformatics tools to analyze our data, cell spreading was analyzed by measuring cell area manually. ImageJ was used to calculate cell area for each frame. The total time for cell spreading analysis for each video was about 24 h. The CL-Quant bioinformatics protocol, which we introduced in this chapter, greatly reduces this time. In general, it takes about 30–60 min to create a protocol for a specific purpose. Once the protocol is created, it takes 5 min to run the protocol using CL-Quant. The protocol can be batch run on multiple videos without requiring users to tie up valuable time performing analysis of spreading.

Our data show that the professionally developed protocol performed better than the one developed by the novice user. However, the novice was able to rapidly learn to use CL-Quant software, and he obtained accurate control data as shown by comparison to the ground-truth. It was clear that the novice had difficulty masking the fine projections on the hESC surfaces, but with additional training, he would likely be able to create a better segmentation protocol and achieve more accurate masking of the treated group data.

We also examined the effect of ROCK inhibitors and blebbistatin on hESC spreading. The ROCK inhibitors and blebbistatin improved cell attachment and spreading, as reported by others [19, 43]. However, those hESC treated with ROCK inhibitors or blebbistatin appeared stressed, had thin attenuated projections off their surfaces, and did not seem as healthy as control cells. ROCK inhibitor or blebbistatin are often used to improve cell survival, especially when plating single cells. ROCK inhibitor and blebbistatin allow cells to be accurately counted and plated. Cell survival is also improved by the efficient attachment observed when these ROCK inhibitors and blebbistatin are used. Cells are stressed during nucleofection, so ROCK inhibitors are often used to improve cell survival when nucleofected cells are ready to plate. ROCK inhibitors and blebbistatin inhibit ROCK protein and downregulate myosin II activity which accelerates cell attachment and cell spreading [43]. Our analysis shows that both ROCK inhibitors and blebbistatin alter the morphology of spreading cells. Moreover, Y27632 and blebbistatin significantly increased the rate of spreading and the fold change in spread area when compared to untreated controls, which may be a factor in why they are more commonly used than H1152. While the full significance of the above changes in cell morphology and behavior are not yet known, these inhibitors do either indirectly or directly decrease myosin II activity, which may lead to the stressed appearance of the treated cells. It has not yet been established why decreasing myosin II activity leads to more rapid and extensive spreading of hESC.

Use of ROCK inhibitors is not recommended in toxicological applications of hESC as they can alter IC_{50}s [11]. A spectrophotometer method has been established to determine cell number when hESC are in small colonies [4]. This method enables an accurate number of cells to be plated without the use of ROCK inhibitors and may generally be applicable to hESC culture and would avoid the morphological stress observed in ROCK inhibitor treated cells.

The protocols reported in this chapter can be used to quantify two parameters of cell spreading, rate and fold change. The use of spreading as an assay for cytoskeletal responses and cell health is attractive as it takes relatively little time to collect data and analyze data with the application of video bioformatics tools. In the future, the segmentation and filtering aspects of these protocols may be improved to gather more accurate data on the challenging cell surface projections, but these protocols in their current form can be reliably applied to spreading of hESC colonies. In the future, improvements in software could include more advanced preprocessing tools, tools for analyzing 3-dimensional data, and tools for analyzing different cell morphologies.

Acknowledgments We thank DRVision, Sam Alworth, Ned Jastromb, Randy Myers, and Mike Allegro of Nikon Inc. for their invaluable help with the BioStation IM and CL-Quant software training. Work in this chapter was supported by an NSF IGERT grant on Video Bioinformatics (#DGE 093667), the California Institute for Regenerative Medicine (CIRM NE-A0005A-1E), and a grant from the Tobacco-Related Disease Research Program of California (TRDRP 22RT-0127).

References

1. Alworth SV, Watanabe H, Lee JS (2010) Teachable, high-content analytics for live-cell, phase contrast movies. J Biomol Screen 15(8):968–977. doi:10.1177/1087057110373546
2. Bahl V, Lin S, Xu N, Davis B, Wang Y, Talbot P (2012) Comparison of electronic cigarette refill fluid cytotoxicity using embryonic and adult models. Reprod Toxicol 34(4):529–537. doi:10.1016/j.reprotox.2012.08.001
3. Behar RZ, Bahl V, Wang Y, Weng J, Lin SC, Talbot P (2012) Adaptation of stem cells to 96-well plate assays: use of human embryonic and mouse neural stem cells in the MTT assay. Curr Protocols Stem Cell Biol Chapter 1(Unit1C):13
4. Behar RZ, Bahl V, Wang Y, Lin S, Xu N, Davis B, Talbot P (2012) A method for rapid dose-response screening of environmental chemicals using human embryonic stem cells. J Pharmacol Toxicol Methods 66:238–245. doi:10.1016/j.vascn.2012.07.003
5. Behar RZ, Davis B, Wang Y, Bahl V, Lin S, Talbot P (2014) Identification of toxicants in cinnamon-flavored electronic cigarette refill fluids. Toxicol In Vitro. doi:10.1016/j.tiv.2013.10.006
6. Buesen Roland, Genschow Elke, Slawik Birgitta, Visan Anke, Spielmann Horst, Luch Andreas, Seiler Andrea (2009) Embryonic stem cell test remastered: comparison between the validated EST and the new molecular FACS-EST for assessing developmental toxicity in vitro. Toxicol Sci 108(2):389–400
7. Cuvelier D, Thery M, Chu YS, Thiery JP, Bornens M, Nassory P, Mahadevan L (2007) The universal dynamics of cell spreading. Curr Biol 17(8): 694–699
8. Cervinka M, Cervinkova Z, Rudolf E (2008) The role of time-lapse fluorescent microscopy in the characterization of toxic effects in cell populations cultivated in vitro. Toxicol In Vitro 22 (5):1382–1386. doi:10.1016/j.tiv.2008.03.011. S0887-2333(08)00082-9 [pii]
9. DiCarlantonio G, Shaoulian R, Knoll M, Magerts T, Talbot P (1995) Analysis of ciliary beat frequencies in hamster oviducal explants. J Exp Zool 272(2):142–152
10. DiCarlantonio G, Talbot P (1999) Inhalation of mainstream and sidestream cigarette smoke retards embryo transport and slows muscle contraction in oviducts of hamsters (Mesocricetus auratus). Biol Reprod 61(3):651–656
11. Fujimura M, Usuki F, Kawamura M, Izumo S (2011) Inhibition of the Rho/ROCK pathway prevents neuronal degeneration in vitro and in vivo following methylmercury exposure. Toxicol Appl Pharmacol 250(1):1–9
12. Gieseke C, Talbot P (2005) Cigarette smoke inhibits hamster oocyte pickup by increasing adhesion between the oocyte cumulus complex and oviductal cilia. Biol Reprod 73(3):443–451
13. Grandjean P, Bellinger D, Bergman A, Cordier S et al (2007) The Faroes statement: human health effects of developmental exposure to chemicals in our environment. Basic Clin Pharmacol 102:73–75

14. Guan BX, Bhanu B, Thakoor N, Talbot P, Lin S (2011) Human embryonic stem cell detection by spatial information and mixture of Gaussians. In: IEEE first international conference on healthcare informatics, imaging and systems biology, pp 307–314

15. Guan BX, Bhanu B, Talbot P, Lin S (2012) Detection of non-dynamic blebbing single unattached human embryonic stem cells. IEEE Int Conf Image Process 2293–2296

16. Guan BX, Bhanu B, Talbot P, Lin S (2012) Automated human embryonic stem cell detection. In: IEEE second international conference on healthcare informatics, imaging and systems biology (HISB). doi:10.1109/HISB.2012.25

17. Guan BX, Bhanu B, Thakoor N, Talbot P, Lin S (2013) Automatic cell region detection by K-means with weighted entropy. In: International symposium on biomedical imaging: from nano to macro, San Francisco, CA

18. Howard DR, Talbot P (1992) In vitro contraction of lobster (Homarus) ovarian muscle: methods for assaying contraction and effects of biogenic amines. J Exp Zool 263(4):356–366

19. Harb N, Archer TK, Sato N (2008) The Rho-ROCK-Myosin signaling axis determines cell-cell integrity of self-renewing pluripotent stem cells. PLos One 3(8):e3001

20. Huang S, Driessen N, Knoll M, Talbot P (1997) In vitro analysis of oocyte cumulus complex pick-up rate in the hamster Mesocricetus auratus. Molec Reprod Develop 47:312–322

21. Holm F, Nikdin H, Kjartansdottir K et al (2013) Passaging techniques and ROCK inhibitor exert reversible effects on morphology and pluripotency marker gene expression of human embryonic stem cell lines. Stems Cells Dev 22:1883–1892

22. Knoll M, Talbot P (1998) Cigarette smoke inhibits oocyte cumulus complex pick-up by the oviduct in vitro independent of ciliary beat frequency. Reprod Toxicol 12(1):57–68

23. Knoll M, Shaoulian R, Magers T, Talbot P (1995) Ciliary beat frequency of hamster oviducts is decreased in vitro by exposure to solutions of mainstream and sidestream cigarette smoke. Biol Reprod 53(1):29–37

24. Lauffenburger DA, Horwitz AF (1996) Cell migration: a physically integrated molecular process. Cell 84:359–369. doi:10.1016/S0092-8674(00)81280-5

25. Liao JK, Seto M, Noma K (2007) Rho Kinase (ROCK) inhibitors. J Cardiovasc Pharmacol 50 (1):17–24

26. Liu JC, Guan X, Ryan JA, Rivera AG, Mock C, Agarwal V, Letai A, Lerou PH, Lahav G (2013) High mitochondrial priming sensitizes hESCs to DNA-damage-induced apoptosis. Cell Stem Cell 13(4):483–491

27. Lin S, Fonteno S, Weng J-H, Talbot P (2010) Comparison of the toxicity of smoke from conventional and harm reduction cigarettes using human embryonic stem cells. Toxicol Sci 118:202–212

28. Lin S, Fonteno S, Satish S, Bhanu B, Talbot P (2010) Video bioinformatics analysis of human embryonic stem cell colony growth. J Vis Exp. http://www.jove.com/index/details.stp?id=1933

29. Lin S, Talbot P (2010) Methods for culturing mouse and human embryonic stem cells. Embryonic stem cell therapy for osteodegenerative disease. Humana Press, pp 31–56

30. Lin SC, Yip H, Phandthong G, Davis B, Talbot P (2014) Evaluation of cell behavior and health using video bioinformatics tools. In: Bhanu B, Talbot P (eds) Video bioinformatics, Chapter 9. Springer, New York

31. Nichols J, Smith A (2009) Naïve and primed pluripotent states. Cell Stem Cell 4(6):487–492. doi:10.1016/j.stem.2009.05.015

32. Riveles K, Roza R, Arey J, Talbot P (2004) Pyrazine derivatives in cigarette smoke inhibit hamster oviductal functioning. Reprod Biol Endocrinol 2(1):23

33. Riveles K, Roza R, Talbot P (2005) Phenols, quinolines, indoles, benzene, and 2-cyclopenten-1-ones are oviductal toxicants in cigarette smoke. Toxicol Sci 86(1):141–151

34. Riveles K, Iv M, Arey J, Talbot P (2003) Pyridines in cigarette smoke inhibit hamster oviductal functioning in picomolar doses. Reprod Toxicol 17(2):191–202

35. Riveles K, Tran V, Roza R, Kwan D, Talbot P (2007) Smoke from traditional commercial, harm reduction and research brand cigarettes impairs oviductal functioning in hamsters (Mesocricetus auratus) in vitro. Hum Reprod 22(2):346–355

36. Talbot P, Lin S (2010) Cigarette smoke's effect on fertilization and pre-implantation development: assessment using animal models, clinical data, and stem cells. J Biol Res 44:189–194

37. Talbot P, zur Nieden N, Lin S, Martinez I, Guan B, Bhanu B (2014) Use of video bioinformatics tools in stem cell toxicology. Handbook of nanomedicine, nanotoxicology and stem cell use in toxicology

38. Talbot P, Lin S (2011) Mouse and human embryonic stem cells: can they improve human health by preventing disease? Curr Top Med Chem 11:1638–1652

39. Talbot P, DiCarlantonio G, Knoll M, Gomez C (1998) Identification of cigarette smoke components that alter functioning of hamster (Mesocricetus auratus) oviducts in vitro. Biol Reprod 58(4):1047–1053

40. Tsai KL, Talbot P (1993) Video microscopic analysis of ionophore induced acrosome reactions of lobster (Homarus americanus) sperm. Mol Reprod Dev 36(4):454–461

41. Talbot P, Gieske C, Knoll M (1999) Oocyte pickup by the mammalian oviduct. Mol Biol Cell 10(1):5–8

42. Talbot P (1983) Videotape analysis of hamster ovulation in vitro. J Exp Zool 225(1):141–148

43. Watanabe K, Ueno M, Kamiyq D, Nishiyama A, Matsumura M, Wataya T, Takahashi JB, Nishikawa S, Muguruma K, Sasai Y (2007) A ROCK inhibitor permits survival of dissociated human embryonic stem cells. Nat Biotechnol 25(6):681–686

44. Woodhouse EX, Chuaqui RF, Liotta LA (1997) General mechanisms of metastasis. Cancer 80:1529–1537

Chapter 9
Evaluation of Dynamic Cell Processes and Behavior Using Video Bioinformatics Tools

Sabrina C. Lin, Henry Yip, Rattapol Phandthong, Barbara Davis and Prue Talbot

Abstract Just as body language can reveal a person's state of well-being, dynamic changes in cell behavior and morphology can be used to monitor processes in cultured cells. This chapter discusses how CL-Quant software, a commercially available video bioinformatics tool, can be used to extract quantitative data on: (1) growth/proliferation, (2) cell and colony migration, (3) reactive oxygen species (ROS) production, and (4) neural differentiation. Protocols created using CL-Quant were used to analyze both single cells and colonies. Time-lapse experiments in which different cell types were subjected to various chemical exposures were done using Nikon BioStations. Proliferation rate was measured in human embryonic stem cell colonies by quantifying colony area (pixels) and in single cells by measuring confluency (pixels). Colony and single cell migration were studied by measuring total displacement (distance between the starting and ending points) and total distance traveled by the colonies/cells. To quantify ROS production, cells were pre-loaded with MitoSOX Red™, a mitochondrial ROS (superoxide) indicator, treated with various chemicals, then total intensity of the red fluorescence was measured in each frame. Lastly, neural stem cells were incubated in differentiation medium for 12 days, and time lapse images were collected daily. Differentiation of neural stem cells was quantified using a protocol that detects young neurons. CL-Quant software can be used to evaluate biological processes in living cells, and the protocols developed in this project can be applied to basic research and toxicological studies, or to monitor quality control in culture facilities.

S.C. Lin · H. Yip · R. Phandthong · B. Davis · P. Talbot (✉)
UCR Stem Cell Center, Department of Cell Biology and Neuroscience,
University of California, Riverside 92521, USA
e-mail: talbot@ucr.edu

© Springer International Publishing Switzerland 2015
B. Bhanu and P. Talbot (eds.), *Video Bioinformatics*,
Computational Biology 22, DOI 10.1007/978-3-319-23724-4_9

9.1 Introduction

Evaluation of dynamic cell processes and behavior is important in basic research [11, 40, 43, 46], in the application of stem cell biology to regenerative medicine [29, 41], and in studies involving the toxicity of drug candidates and environmental chemicals [13, 24, 25, 31, 32, 35–38, 42]. Prior work in basic and toxicological research has often involved microscopic observation of cells or assays that evaluate single endpoints after chemical exposure (e.g., [4, 6–8, 25, 33]). However, much additional insight can be learned about a cells response to its environment by comparing dynamic processes, such as cell growth and motility, in treated and control cells [30, 31, 44]. Just as human body language can reveal information about human mood and well-being, cellular dynamics can often reveal information about the mode of action and the cellular targets of chemical exposure. For example, impairment of cell motility would likely be correlated with an adverse effect on the cytoskeleton. Such an effect can be quantified in video data without using any labels or genetic transformation of the cells [25, 31, 36, 49]. In addition, fluorescent labels can be used to report the condition of cells in time-lapse data thereby revealing more information about a treatment than a single endpoint assay [27]. Finally, multiple endpoints can be multiplexed and mined from video data to gain additional insight from a single experiment [2, 34].

The interest and importance of video data in cellular studies has led to the commercialization of a number of instruments (e.g., BioStation CT/IM, Cell IQ, Tokai Hit) optimized for collecting live cell images over time [10, 44]. Videos can now be made for hours, days, or even months using conditions that support in vitro cell culture and experimentation. However, while dynamic video data are rich with information about cell health and cell processes, they are often difficult to analyze quantitatively. This is due to the complexity of the data and the generally large size of the data sets. Moreover, video analysis can be very time-consuming and is error-prone due to subjectivity of the human(s) performing the analysis. The recent interest in live cell imaging has been accompanied by a need for software tools for extracting information from video data. This field of study has been termed "video bioinformatics" (http://www.cris.ucr.edu/IGERT/index.php). Video bioinformatics includes the development and application of software tools for extraction and mining of information and knowledge from video data. The advantages of using video bioinformatics tools are enormous. Tremendous amounts of time can be saved, and when properly applied, video bioinformatics tools will extract more accurate reproducible data than would generally be the case for a human performing the same task. Video bioinformatics tools are available commercially [3] and are also being developed in research laboratories to solve specific problems such as quantification of cells in colonies, cell identification, and prediction of successful development of human embryos to the blastocyst stage [14–17, 21, 51].

In this chapter, four applications of video bioinformatics tools to toxicological problems are presented. First, cell colony and individual cell growth were monitored using time-lapse data. Second, single cell and colony migration were analyzed to provide information on rate of migration, distance traveled, and total displacement. Third, a method is presented for direct observation and quantification of ROS production in cultured cells. Finally, quantification of differentiating neurons was accomplished by evaluating time-lapse videos collected over a period of 10 days. Video data were collected in either a BioStation CT or BioStation IM, both available from Nikon. Analyses were done using protocols created using a commercial software package (CL-Quant). Each application can be used with either single cells or colonies.

9.2 Collection of Time-Lapse Data

The BioStation IM is a fully motorized, automated, environmentally controlled microscope and imaging system that captures images using a cooled monochrome CCD camera. It was designed to enable live cell imaging using optimal in vitro conditions. It can accommodate 35 mm culture dishes including HiQ4 dishes (Nikon Instruments Inc., Melville, NY) that allow four different treatments to be monitored in a single experiment. Cells are incubated at 37 °C in a CO_2 controllable atmosphere with a high relative humidity. Multiple magnifications are possible for capturing phase contrast and/or fluorescence images using software that controls point selection and collection of data. Perfusion is an option to allow for real-time addition or subtraction of cell culture media and to enable longer-term observation. The BioStation IM robotics are capable of precise cell registration so the resultant movies can be analyzed quantitatively.

The BioStation CT is a much larger incubation unit that can perform high content work ideal for live cell screening. The culture conditions inside the BioStation CT can be regulated. While our unit is usually operated at 5 % CO_2, 85 % relative humidity and 37 °C, hypoxic conditions are also possible if needed. The BioStation CT is especially suitable for data collection in long-term experiments, in which cells are studied over weeks or months. It has a robotic arm for transfer of plates to and from a microscope stage which enables complete automation of the time-lapse experiment. The BioStation CT holds up to 30 experimental samples in various plate formats (6, 12, 24, 48, 96 well plate formats, 35, 60 and 100 mm dish formats, and 25 and 75 cm^2 flask formats). A cooled monochrome CCD camera collects phase and/or fluorescence images at defined intervals and points of interests. Large montages of the entire well area can be taken over the magnification range of 2×–40× which allows for complete cell characterization over the life of the cell culturing period.

9.3 CL-Quant Software

All video analyses were performed using CL-Quant software, a live-cell image analysis program produced for Nikon by DRVision Technologies (Bellevue, Washington). It can either be purchased from Nikon as CL-Quant or from DRVision under the name SVCell. The current version of the software is user friendly, features an intuitive GUI to manage high content imaging experiments, and comes with webinar instruction. All ground-truth evaluations of CL-Quant were done using either ImageJ or Photoshop.

CL-Quant comes with several modules professionally developed by DRVision for basic processing of videos. For example, bioinformatics tools for measuring cell confluency, cell migration, and cell counting can be obtained from Nikon and applied to users' videos. CL-Quant also provides tools that end users can work with to develop protocols for recognition and quantitative analysis of microscopic video data [3]. CL-Quant protocols can be applied with user directed learning and do not require image processing knowledge. Although the software has great depth, basic analyses can be done with relatively little training. CL-Quant can be used to detect, segment, measure, classify, analyze, and discover cellular phenotypes in video data. Preconfigured modules are available for some applications such as cell counting, confluency, cell division, wound healing, cell motility, cell tracking, and measuring neurite outgrowths. Moreover, the software has significant depth and can be configured for other more complex applications by the user.

In this chapter, examples will be shown for adapting CL-Quant to measure cell/colony growth rate, cell/colony migration, ROS production, and neural differentiation. Protocols, developed by DRVision and Nikon software engineers and those created by novices learning to use the CL-Quant software, will be compared and used to study cell behavior. The above parameters can be useful in toxicological studies, in work that requires knowledge of cell health, in clinical applications of stem cells to regenerative medicine, or in basic studies of cell biology.

9.4 Cell and Colony Growth

9.4.1 Growth of Human Induced Pluripotent Stem Cells (hiPSC)

Human-induced pluripotent stem cells (hiPSC; RivCAG-GFP), created in the UCR Stem Cell Core Facility and grown in 12-well plates as described previously [28] were either incubated in control medium (mTeSR, Stem Cell Technologies) or in mTeSR containing 0.1 puff equivalents (PE) of sidestream cigarette smoke (PE = the amount of smoke in one puff that dissolves in 1 ml). This concentration was shown previously to inhibit human embryonic stem cells (hESC) colony growth [31]. Cells were imaged at 10× magnification for 48 at 6 h intervals in a

BioStation CT maintained at 37 °C, 5 % CO_2 and a relative humidity of 85–90 %. Phase contrast images show the growth of a single control and treated colony over 48 h (Fig. 9.1a–c and g–i). All colonies analyzed were selected to be relatively close in size before treatment. During incubation, the sizes of the treated colonies appeared to be smaller than the control colonies. Some treated colonies did not grow, and some eventually died due to treatment (as shown in Fig. 9.1i). To obtain quantitative information on colony growth rates, a protocol, which was developed in our lab with the CL-Quant software (version 3.0) and used previously with hESC [31], was applied to these iPSC video data. The protocol first segmented images of control and treated colonies and overlaid each colony with a mask (Fig. 9.1d–f and j–l). The fidelity of the mask was excellent for both control and treated groups. After images were segmented, small objects, dead cells, and debris were removed with an enhancement module, and finally the size of each colony was measured in pixels in each frame of each video. Because each colony is slightly different at the start of an experiment, resulting data for each set of videos were normalized to the size of the colony in frame 1, then data were averaged and growth curves were graphed (Fig. 9.1m). Results showed a clear difference in growth rates between the control and treated colonies. In fact, the treated colonies decreased in size and appeared not to grow over the 48 h incubation period. Treatment was significantly different than the control (2-way ANOVA, $p \leq 0.05$), and the iPSC were more sensitive to sidestream smoke treatment than the hESC studied previously [32]. The protocol used for this analysis had previously been compared to ground-truth derived using Adobe Photoshop, and excellent agreement was found between the data obtained with CL-Quant analysis and the ground-truth [31].

The data shown in Fig. 9.1 involved analysis of 60 images. To perform this analysis by hand would require approximately 3–4 h. CL-Quant was able to perform this analysis in about 1 h, and it can be run in a large batch so that the users' time is not occupied during processing. With a larger experiment having more frames, the difference between CL-Quant and manual analysis would be much greater.

Video examples of iPSC colony growth and CL-Quant masking can be viewed by scanning the bar codes.

9.4.2 Growth of Mouse Neural Stem Cells (mNSC)

Monitoring the growth of single cells can be more challenging than monitoring hiPSC or hESC colony growth. Some single adherent cells grow very flat and do not differ much in contrast from the background making segmentation difficult. However, images can be enhanced by adjusting the brightness, the contrast, and/or the gamma parameters using CL-Quant software or other image processing software (e.g., Photoshop and ImageJ). CL-Quant comes with some professionally developed modules for use with some types of single cells. Investigators can try these protocols to see if one works well with their cell type or, alternatively, they

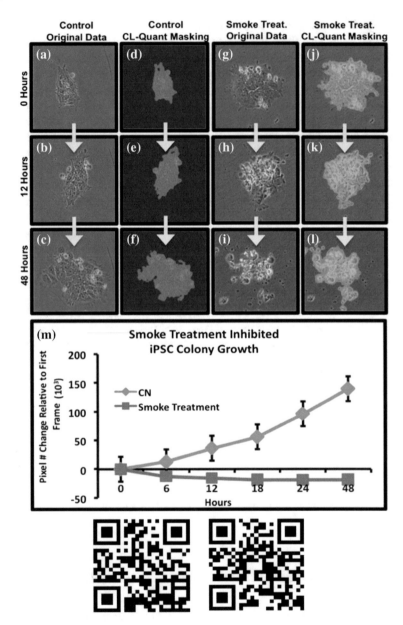

Fig. 9.1 Growth of hiPSC colonies over 48 h. **a–c** Phase contrast images of control iPSC colonies at various times during growth. **d–f** The same control images segmented using a CL-Quant protocol developed in our lab. **g–i** Phase contrast images of smoke treated iPSC colonies at various times during growth. **j–l** The same treatment group images masked using the same CL-Quant protocol as applied to control colonies. **m** Graph of control and treated cells showing growth rate. Data are means and standard errors of three experiments. *CN* control

can use the CL-Quant software to develop their own protocol and tailor it to the specific requirements of the cells they are using. However, one should not assume that the protocols are accurate and should check them against ground-truth.

In our experiments, the effect of cigarette smoke treatment on mNSC proliferation was examined. mNSC were plated in 12-well plates at 2,500 cells/well, and cells were allowed to attach for 24 h before treatment, incubation, and imaging. Various fields of interests were imaged over 48 h in the BioStation CT in 5 % CO_2 and 37 °C. The collected video data (Fig. 9.2a–c) were then processed and analyzed

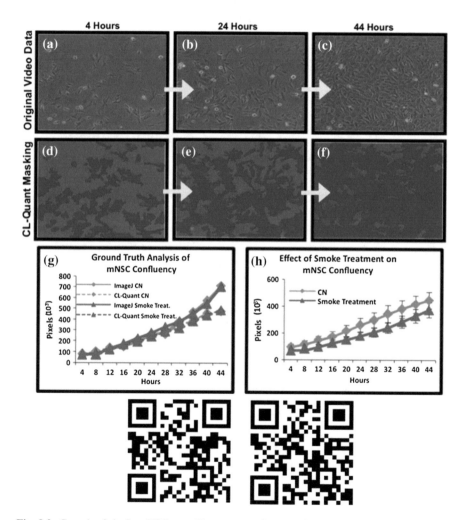

Fig. 9.2 Growth of single mNSC. **a–c** Phase contrast images of control mNSC at various times during growth over 48 h. **d–f** The same images after segmentation using a protocol developed by DR Vision. **g** Graph showing analysis for growth rate of control and treated cells (*solid lines*) and ImageJ ground-truth for each group (*dotted lines*). **h** Graph of control and treated mNSC showing confluency rate. Data are means and standard errors of three experiments. *CN* control

using the confluency module provided by DRVision Technologies for use with CL-Quant software. The confluency module masked the cells in each field, and mNSC growth was determined by measuring the number of pixels in each frame (Fig. 9.2d–f). Ground-truth was obtained to verify the validity of the CL-Quant confluency analysis tool using ImageJ. To obtain ground-truth, each cell was carefully outlined and colored to measure area (pixels), and a comparison of CL-Quant and ImageJ data showed CL-Quant was reliable over the first 34 h with some divergence at the latest times (Fig. 9.2g). A complete mNSC growth experiment was analyzed in which the growth of cigarette smoke treated cells was compared to nontreated control cells (Fig. 9.2h). Smoke treatment of mNSC significantly inhibited their proliferation from 20 to 44 h (2-way ANOVA, $p \leq 0.001$).

Video examples of mNSC proliferation (control and smoke treatment) can be viewed by scanning the bar code.

9.5   Cell Migration

9.5.1   Migration of hESCColonies

Evaluation of cell motility can be important in determining if the cytoskeleton of treated cells has been affected. The data in Fig. 9.3 were collected using hESC colonies that were incubated in a BioStation CT for 48 h, and images were collected of each colony at 10 min intervals. Colonies, which were grown on Matrigel, were incubated in either control medium (mTeSR) or mTeSR containing cigarette smoke. Normally, hESC show motility when grown on Matrigel.

CL-Quant provides a number of readouts for motility. The two that were most useful are total distance traveled and total displacement. Total distance traveled is the measurement of how far the colony has migrated over time, and total displacement is the difference in distance between the beginning point and the endpoint. Figure 9.3a–c shows examples of hESC colonies that have been masked and tracked by a motility protocol. Within a population of hESC colonies, three behaviors were observed: (1) growing, (2) shrinking, and (3) dying. The tracking module traces the path of the colonies, and those that showed growth had longer paths than colonies that were shrinking or dying.

In Fig. 9.3d, e, displacement and total distance traveled were measured for individual colonies in control and cigarette smoke treatment groups. All colonies in the control group were healthy and growing, but 2 of 12 treated colonies (red circles) were dead by the end of 48 h. For total distance traveled, measurements for control colonies appeared to be clustered, while the treated colonies were more variable, in part due to the presence of two dead colonies. For both the control and treated groups, the total displacement was quite variable, suggesting there was no directional movement in either group. A t test was performed on both parameters, after removing measurements of dying colonies, and the results showed that the

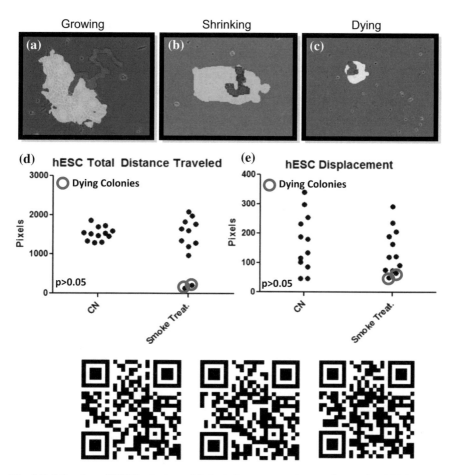

Fig. 9.3 Migration of hESC colonies. **a** Masked phase contrast image of a growing hESC colony during migration. **b** Masked phase contrast image of a shrinking hESC colony during migration. **c** Masked phase contrast image of a dying hESC colony during migration. **d, e** Graphs showing total displacement/distance traveled for each control and treated colonies. All CL-Quant masking and tracking of colonies were done by applying a tracking recipe developed by our lab. *CN* control

distance traveled and displacement of control and treated colonies were not significantly different ($p > 0.05$).

hESC migration is an important process during development, as derivatives of these cells must migrate during gastrulation to form the three germ layers properly [22]. Therefore, although our cigarette smoke treatment did not affect migration of hESC colonies, these two parameters can be useful in determining the effects of other toxicants on pluripotent cell migration. Observed effects on total distance traveled and displacement of colonies can be the first signs of potent chemical effects on the cytoskeletal integrity of the cells.

Video examples of hESC colony migration and CL-Quant masking can be viewed by scanning the bar codes.

9.5.2 Migration and Gap Closure of mNSC and NTERA2 Cells

Single cell migration can be analyzed using a gap closure assay. This assay is performed by growing a monolayer of cells, creating a gap in the middle of the monolayer, and monitoring the time required for cells to migrate into the gap and close it. The gap can be made using a pipette to remove a band of cells, but the sizes of the gaps are not always uniform. As a result, the rate of closure may not be as accurate and comparable among control and treated groups. We have used Ibidi wound healing culture inserts (Fig. 9.4, Ibidi, cat#80241, Verona, WI) to make uniform gaps. First, inserts were adhered to culture plates. Second, cells were plated in each well of the insert, and when cells in the wells were confluent, the insert was removed leaving a uniform gap (500 μm) between cells in each well. This method works well with cells grown on plastic or glass, but not with cells grown on wet coating substrates (e.g., Matrigel, poly-D-lysine, poly-L-lysine, and laminin) because the adhesive at the bottom of the inserts will not stick.

Experiments using two types of cells, mNSC and NTERA2, are shown in Fig. 9.5. Both cell types can be grown on plastic, and time-lapse videos of both cell types were collected in the BioStation CT for 44 h. Images were collected every 4 h and CL-Quant analysis was done for each frame by measuring the number of pixels in the gap. The gap closure module, developed by our lab using the CL-Quant software, includes a segmentation recipe to identify the gap between the two populations of cells and a measurement recipe that counts the number of pixels in the gap. An example of mNSC gap closure images is shown in Fig. 9.5a–c, and CL-Quant masking of the same gap is shown in Fig. 9.5d–f. In the filmstrip, the gap became smaller as the cells migrated toward each other. Gap closure analysis using the CL-Quant software was validated using ground-truth obtained from the ImageJ software for control and treated NTERA2 (Fig. 9.5g, h). While CL-Quant tended to

Fig. 9.4 Diagram of Ibidi gap closure culture inserts

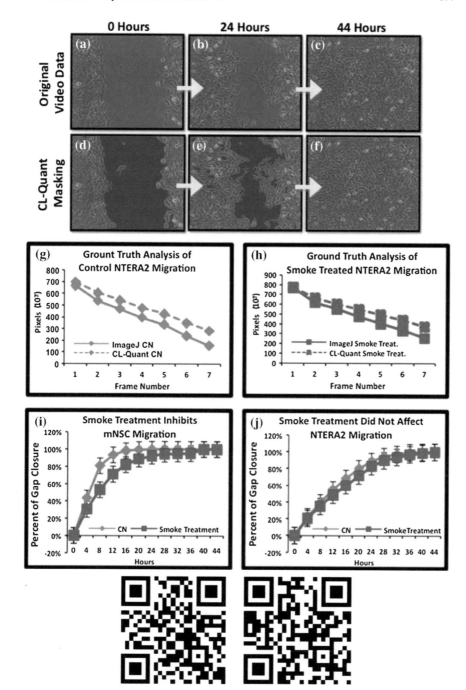

◄ **Fig. 9.5** Gap closure for mNSC and NTERA2 cells. **a–c** Phase contrast images of mNSC at various 3 times during gap closure. **d–f** The same images after segmentation using a protocol developed in our lab with CL-Quant software. **g, h** Graph showing rate of gap closure for control (*blue*) and treated NTERA2 cells (*red*) and the corresponding ground-truth (*dotted lines*) obtained using ImageJ. **i** Graph of mNSC migration by monitoring percent of gap closure over 44 h. **j** Graph of NTERA2 cell migration by monitoring percent of gap closure over 44 h. Data are means and standard errors of three experiments

overestimate area slightly due to some extension of the mask beyond the gap, ImageJ and CL-Quant analyses produced similar results for both control and treated groups. mNSC and NTERA2 cell migration experiments were analyzed with the CL-Quant gap closure module (Fig. 9.5i, j). Gap closure was completed in about 16 h for the mNSC, while the NTERA2 cells required about 40 h to completely close the gap. Migration of mNSC, but not the NTERA2, was significantly inhibited by cigarette smoke ($p \leq 0.001$ for 2-way ANOVA of mNSC data).

Although our gap closure analysis was done by measuring the pixels within the gap, it can also be monitored by masking the cells. For certain cells that produce a clear phase contrast image, this option may be easier and more accurate than monitoring the gap.

Video examples of control and smoke treated single cell migration can be viewed by scanning the bar codes.

9.6 Detection of Reactive Oxygen Species (ROS) in Human Pulmonary Fibroblasts (HPF)

Exposure to environmental chemicals can lead to stress [9, 12, 26, 47], and ROS are often produced in stressed cells [1, 19, 45]. ROS can damage macromolecules in cells including proteins and DNA, and any factor that increases ROS would be potentially damaging to a cell. It is possible to observe the production of ROS in cells using fluorescent probes such as MitoSOX Red™ (Life Technologies, Grand Island, NY). MitoSOX Red™ readily enters cells and is rapidly targeted to the mitochondria. When oxidized by superoxide, it emits red fluorescence (absorption/emission maxima = 510/580). MitoSOX Red™ can be preloaded in cells and will fluoresce as levels of superoxide increase. Its fluorescent intensity is related to the amount of superoxide in the cells.

In this example, hPF were disassociated from a culture vessel using 0.05 % trypsin and then plated in the HiQ4 dishes coated with poly-D-lysine. After hPF were allowed to attach for 24 h, cells were preloaded with 5 μM MitoSOX Red™ for 10 min at 37 °C in a cell culture incubator. Preloaded cells were washed with culture medium and then either treated with cigarette smoke which induces ROS production [47] or were left untreated (control). Dishes were placed in a BioStation IM, which was programmed to capture images every 4 min for 10 h using both the phase and red fluorescence channels.

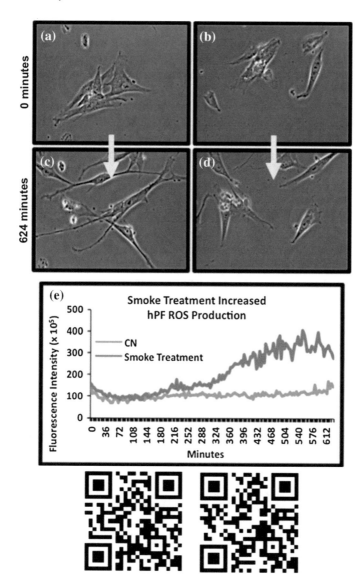

Fig. 9.6 Production of reactive oxygen species in hPF. **a–d** Merged phase contrast and fluorescent images at various times during incubation of control and treated hPF with MitoSox Red. **e** Graph showing fluorescence intensity in control and treated cells over time. *CN* control

A protocol was developed using CL-Quant to analyze the level of MitoSOX Red™ fluorescence in living cells. This was done by first developing a segmentation procedure to identify fluorescence. An enhancement program was then used to remove all debris and dead cells. The dead cells are highly fluorescent but round and easily excluded from the analysis with a size-based enhancement filter.

The background was flattened, and the mask was applied and observed to determine if it accurately covered each cell in the entire video. If the mask was not accurate, the segmentation was refined until masking accurately covered all living cells in each frame. CL-Quant was then used to measure the level of fluorescence in each field of each video.

The above protocol was applied to time-lapse videos of control and cigarette smoke treated hPF that were preloaded with MitoSOX Red™. Merged phase contrast and fluorescent images of control and treated cells are shown at various times in Fig. 9.6a–d. There are usually some highly fluorescent dead cells present in each field at the start of an experiment. It is important to filter out the dead cells before performing the analysis as they would contribute significantly to the intensity measurements. The graph shows the intensity of the MitoSOX Red™ fluorescence in control and treated cells over 10 h of incubation (Fig. 9.6e). Control levels remained low and relatively constant throughout incubation in agreement with direct observation of the videos. In contrast, fluorescence increased significantly in the treated cells. This increase begins at about 300 min of incubation and continues until the end of the experiment. This method is useful for direct monitoring of ROS production in time-lapse images. It reports which cells produce ROS, the relative amounts of ROS in control and treated groups, and the time at which ROS is elevated. Statistical analysis showed that cigarette smoke treatment significantly increased hPF ROS production over time (2-way ANOVA, $p \leq 0.001$).

Video examples of hPF ROS production in control and treated cells can be viewed by scanning the bar codes.

9.7 Detection and Quantification of Spontaneous Neural Differentiation

Differentiation of stem cell populations is an essential and important process of normal development. Many in vitro differentiation protocols have been established to derive various cell types that can then be used for degenerative disease therapy, organ regeneration, and models for drug testing and toxicology research [42]. In all cases, the health, morphology, and differentiation efficiency of the cells are important parameters that should be closely observed and evaluated. Here, we provide an example of how differentiating mNSC are monitored over time and derived neurons are quantified using the CL-Quant software. mNSC were plated in 12-well plates in NeuroCult™ Differentiation Medium (Stem Cell Technologies, Vancouver, Canada) for 12 days. The plate was incubated in the Nikon BioStation CT, and several fields of interest were randomly chosen for imaging every 24 h. The NeuroCult™ Differentiation Medium supports the differentiation of three brain cell types: (1) neurons, (2) astrocytes, and (3) oligodendrocytes. The morphologies of these three cell types are very different in that neurons have long axons and small cell bodies, and astrocytes and oligodendrocytes are flatter in appearance. As seen

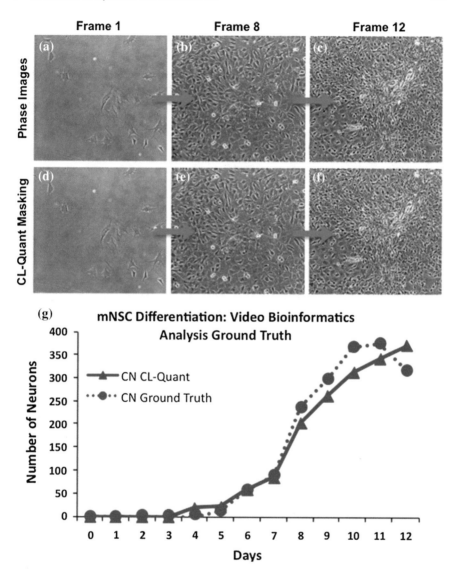

Fig. 9.7 Quantification of neurons in neural differentiation assay. **a–c** Phase contrast images of mNSC at various times during incubation. **d–f** CL-Quant software masking of mNSC phase contrast images to identify the neurons within each frame. **g** Graph showing quantification results obtained using the CL-Quant software was similar to the ground-truth obtained using the ImageJ software

in Fig. 9.7a–c, phase contrast microscopy of the differentiating neural stem cell population showed small, dark neurons sitting on top of a layer of flat cells. The stark morphological differences can be used to quantify the number of neurons in each frame. A segmentation recipe was developed using the CL-Quant software to identify the darker and smaller neurons (Fig. 9.7d–f), and a measurement recipe was used to count the number of neurons in each frame. We further validated our recipe with ground-truth generated using the ImageJ software (Fig. 9.7g), and the number of neurons identified in each frame using the video bioinformatics tool agreed closely to the ground-truth data. Automation of the identification process is a critical component for making future stem cell research more efficient and effective.

9.8 Discussion

This chapter gives four examples of how video bioinformatics tools can be applied to experimental time-lapse data thereby enabling quantification of dynamic cellular processes with attached cells that grow as colonies or single cells. Live cell imaging is easily obtainable with modern instrumentation designed for culturing cells in incubators equipped with microscopes [44]. Analysis of such data, as shown above, can be done using professionally developed software tools [3] or tools developed by end users with software such as CL-Quant [31, 44, 49]. However, any commercial software may be limited in its ability to segment difficult subjects, in which case custom software would need to be created [14–16, 21, 51]. In all of the examples mentioned, use of video bioinformatics tools significantly reduced the time for analysis and provided greater reproducibility than would normally be obtained with manual human analysis. Although not demonstrated in this chapter, the power of live cell imaging can be increased by multiplexing several endpoints together in one experiment. For example, hESC or iPSC colony growth and migration can be evaluated from the same set of video data, thereby conserving time and resources.

We demonstrated how video bioinformatics tools were used to evaluate the effects of cigarette smoke on dynamic processes (growth, migration, and ROS production) in cultured cells. Many in vitro toxicological assays (e.g., MTT, neutral red, and lactic dehydrogenase assays) are useful and effective in evaluating chemical potency. Several recent studies from our lab effectively used the MTT assay to screen the toxicity of various electronic cigarette fluids with embryonic stem cells, mNSC, and hPF [4, 6, 7, 50]. Due to the sensitive nature of hESC cultures, a new 96-well plate MTT protocol was also established for pluripotent cells, which allows exact numbers of cells (in small clumps) to be plated from well to well [6, 7]. While these in vitro assays are relatively quick and efficient, they provide a single endpoint at one time/experiment, and the cells are often killed to obtain the endpoint. As a result, dynamic changes in cell behavior and morphology are not observed, and potential data are lost. In both basic research and toxicological applications, examination of video data can reveal changes in cell dynamics as well as rate data that are not gathered by single time point analysis. By determining specific processes that are altered during treatment, the mode of action and cellular targets may be identified. As an example, motility of mNSC, but not of NTERT-2 cells, was affected by cigarette smoke, suggesting that the cytoskeleton is more sensitive to smoke exposure in the former cells. Observations made from time-lapse video data also provide insight on when during exposure chemicals affect dynamic cellular processes.

Although this chapter presented toxicological applications of video bioinformatics tools, other biological disciplines can benefit from this approach. For example, Auxogyn, Inc. has established a method to determine the health of early human embryos using time-lapse microscopy and an automated embryo stage classification procedure [48, 51]. The protocol employs a set of learned embryo features that allow 88 % classification accuracy of embryos that will develop to the blastocyst stage. This advancement is being used in in vitro fertilization (IVF) clinics to help physicians transfer only healthy embryos with the capacity to develop into blastocysts. This not only increases IVF success rates, but decreases the chance for multiple births that often result in unhealthy children. In June 2013, Auxogyn announced the birth of the first baby to be born in an IVF clinic that used the "Early Embryo Viability Assessment" (Eeva) test to select the best embryos for transfer (http://www.auxogyn.com/news.2013-06-14.first-auxogyn-baby-born-in-scotland.php).

The use of video bioinformatics tools will also be important when monitoring the health of cells that will eventually be used in stem cell therapy. In the future, stem cells grown for transfer to patients will be cultured over long periods during passaging and differentiation making them costly in time and resources. Therefore, it is important to monitor the culturing process using time-lapse data to verify that cells are healthy and robust throughout in vitro culture and differentiation. It will be important to have noninvasive monitoring systems for stem cell applications in regenerative medicine. If a problem develops during expansion and culturing of cells used in therapy, experiments can be terminated and restarted to assure that only cells of excellent quality are transferred to patients.

Time-lapse data are also used in basic studies of cell biology. Qualitative and quantitative analysis of video data have revealed information on dynamic cellular processes [18, 20, 27], such as spindle formation during mitosis, actin protein dynamics in cells, and gamete fusion [5, 23, 39]. Video data can also be used to study cell processes that occur rapidly and are not easily understood by direct observation, such as the acrosome reaction of lobster sperm [46]. Frame-by-frame analysis of the acrosome reaction enabled each step during acrosomal eversion to be analyzed and further enabled quantitative measurement of the forward movement of sperm during the reaction. Time-lapse video microscopy has also been used to study the process and rate of oocyte cumulus complexes pick-up by explants of hamster oviducts [43].

New instrumentation, such as the Nikon BioStation CT/IM, provide long-term stable incubation conditions for live cell imaging and enable acquisition of better quality data than possible in the past. Improved methods for live cell imaging coupled with video bioinformatics tools provide a new technology applicable to numerous fields in the life sciences.

Acknowledgments Work presented in this chapter was supported by the following grants: TRDRP 22RT-0127, CIRM NE-A0005A-1E, NSF IGERT DGE 093667, a Cornelius Hopper Award from TRDRP, and a TRDRP postdoctoral fellowship 20FT-0084. We thank Dr. Evan Snyder for providing the mNSC and Randy Myers and Ned Jastromb for their help with the BioStations and CL-Quant.

References

1. Adler V, Yin Z, Tew KD, Ronai Z (1999) Role of redox potential and reactive oxygen species in stress signaling. Oncogene 18:6104–6111
2. Albrecht DR, Underhill GH, Resnikoff J, Mendelson A, Bhatiaacde SN, Shah JV (2010) Microfluidics-integrated time-lapse imaging for analysis of cellular dynamics. Integr Biol 2:278–287
3. Alworth SV, Watanabe H, Lee JS (2010) Teachable, high-content analytics for live-cell, phase contrast movies. J Biomol Screen 15(8):968–977. 1087057110373546 [pii]
4. Bahl V, Lin S, Xu N, Davis B, Wang Y, Talbot P (2012) Comparison of electronic cigarette refill fluid cytotoxicity using embryonic and adult models. Reprod Toxicol 34(4):529–537. doi:10.1016/j.reprotox.2012.08.001
5. Ballestrem C, Wehrie-Haller B, Imhof BA (1998) Actin dynamics in living mammalian cells. J Cell Sci 111:1649–1658
6. Behar RZ, Bahl V, Wang Y, Weng J, Lin SC, Talbot P (2012) Adaptation of stem cells to 96-Well plate assays: use of human embryonic and mouse neural stem cells in the MTT assay. Curr Protoc Stem Cell Biol Chapter 1: Unit1C 13
7. Behar RZ, Bahl V, Wang Y, Lin S, Xu N, Davis B, Talbot P (2012) A method for rapid dose-response screening of environmental chemicals using human embryonic stem cells. J Pharmacol Toxicol Methods 66:238–245. doi:10.1016/j.vascn.2012.07.003
8. Behar RZ, Davis B, Wang Y, Bahl V, Lin S, Talbot P (2013) Identification of toxicants in cinnamon-flavored electronic cigarette refill fluids. Toxicol In Vitro. doi:10.1016/j.tiv.2013.10.006
9. Carlson C, Hussain SM, Schrand AM, Braydich-Stolle LK, Hess KL, Jones RL, Schlager JJ (2008) Unique cellular interaction of silver nanoparticles: size-dependent generation of reactive oxygen species. J Phys Chem 112:13608–13619
10. Cervinka M, Cervinkova Z, Rudolf E (2008) The role of time-lapse fluorescent microscopy in the characterization of toxic effects in cell populations cultivated in vitro. Toxicol In Vitro 22 (5):1382–1386. doi:10.1016/j.tiv.2008.03.011. S0887-2333(08)00082-9 [pii]
11. DiCarlantonio G, Shaoulian R, Knoll M, Magers T, Talbot P (1995) Anaysis of ciliary beat frequencies in hamster oviductal explants. J Exp Zool 272(2):142–152
12. Drechsel DA, Patel M (2008) Role of reactive oxygen species in the neurotoxicity of environmental agents implicated in Parkinson's disease. Free Raic Biol Med 44:1873–1886
13. Gieseke C, Talbot P (2005) Cigarette smoke inhibits hamster oocyte pickup by increasing adhesion between the oocyte cumulus complex and oviductal cilia. Biol Reprod 73(3):443–451
14. Guan BX, Bhanu B, Thakoor N, Talbot P, Lin S (2011) Human embryonic stem cell detection by spatial information and mixture of Gaussians. In: IEEE first international conference on healthcare informatics, imaging and systems biology, pp 307–314
15. Guan BX, Bhanu B, Talbot P, Lin S (2012) Detection of non-dynamic blebbing single unattached human embryonic stem cells. In: International conference on image processing. IEEE, pp 2293–2296
16. Guan BX, Bhanu B, Thakoor NS, Talbot P, Lin S (2013) Automatic cell region detection by k-means with weighted entropy. In: 10th international symposium biomedical imaging (ISBI). IEEE, pp 418–421

17. Guan BX, Bhanu B, Talbot P, Lin S (2014) Bio-driven cell region detection in human embryonic stem cell assay. IEEE/ACM Trans Comput Biol Bioinform 11(3):604–611. doi:10. 1109/TCBB.2014.2306836
18. Haraguchi T (2002) Live cell imaging: approaches for studying protein dynamics in living cells. Cell Struct Funct 27:333–334
19. Held P (2010) An introduction to reactive oxygen species. BioTek Application Guide. http:// www.biotek.com/resources/articles/reactive-oxygen-species.html
20. Hinchcliffe E (2005) Using long-term time-lapse imaging of mammalian cell cycle progression for laboratory instruction and analysis. Cell Biol Educ 4:284–290
21. Huth J, Buchholz M, Kraus JM, Schmucker M, Wichert GV, Krndija D, Seufferlein T, Gress TM, Kestler HA (2010) Significantly improved precision of cell migration analysis in time-lapse video microscopy through use of a fully automated tracking system. Biomed Central Cell Biol 11:24. http://www.biomedcentral.com/1471-2121/11/24
22. Ichikawa T, Nakazato K, Keller PJ, Kajiura-Kobayashi H, Stelzer EHK, Mochizuki A, Nonaka S (2013) Live imaging of whole mouse embryos during gastrulation: migration analyses of epiblast and mesodermal cells. PLoS One 8(7):e64506. doi:10.1371/journal.pone. 0064506
23. Inoué S, Oldenbourg R (1998) Microtubule dynamics in mitotic spindle displayed by polarized light microscopy. Mol Biol Cell 9:1603–1607
24. Knoll M, Talbot P (1998) Cigarette smoke inhibits oocyte cumulus complex pickup by the oviduct in vitro independent of ciliary beat frequency. Reprod Toxicol 12(1):57–68
25. Knoll M, Shaoulian R, Magers T, Talbot P (1995) Ciliary beat frequency of hamster oviducts is decreased in vitro by exposure to solutions of mainstream and sidestream cigarette smoke. Biol Reprod 53(1):29–37
26. Lehnert BE, Lyer R (2002) Exposure to low-level chemicals and ionizing radiation: reactive oxygen species and cellular pathways. Hum Environ Toxicol 21:65–69
27. Li F, Zhou X, Wong STC (2010) Optimal live cell tracking for cell cycle study using time-lapse fluorescent microscopy images. Mach Learn Med Imaging 6357:124–131
28. Lin S, Talbot P (2010) Methods for culturing mouse and human embryonic stem cells. Embryonic stem cell therapy for osteodegenerative disease. Humana Press, New York, pp 31–56
29. Lin S, Talbot P (2014) Stem cells. In: Wexler P (ed) The encyclopedia of toxicology, 3rd edn. Elsevier (in press)
30. Lin S, Tran V, Talbot P (2009) Comparison of toxicity of smoke from traditional and harm-reduction cigarettes using mouse embryonic stem cells as a novel model for preimplantation development. Hum Reprod 24(2):386–397
31. Lin S, Fonteno S, Satish S, Bhanu B, Talbot P (2010) Video bioinformatics analysis of human embryonic stem cell colony growth. J Vis Exp. http://www.jove.com/index/details.stp?id=1933
32. Lin S, Fonteno Shawn, Weng Jo-Hao, Talbot P (2010) Comparison of the toxicity of smoke from conventional and harm reduction cigarettes using human embryonic stem cells. Toxicol Sci 118:202–212
33. Martin GG, Talbot P (1981) The role of follicular smooth muscle cells in hamster ovulation. J Exp Zool 216(3):469–482
34. Reichen M, Veraitchm FS, Szita N (2010) An automated and multiplexed microfluidic bioreactor platform with time-lapse imaging for cultivation of embryonic stem cells and on-line assessment of morphology and pluripotent markers. In: 14th international conference on miniaturized systems for chemistry and life sciences, Groningen, The Netherlands
35. Riveles K, Iv M, Arey J, Talbot P (2003) Pyridines in cigarette smoke inhibit hamster oviductal functioning in picomlar doses. Reprod Toxicol 17(2):191–202
36. Riveles K, Roza R, Arey J, Talbot P (2004) Pyrazine derivatives in cigarette smoke inhibit hamster oviductal functioning. Reprod Biol Endocrinol 2(1):23
37. Riveles K, Roza R, Talbot P (2005) Phenols, quinolines, indoles, benzene, and 2-cyclopenten-1-ones are oviductal toxicants in cigarette smoke. Toxicol Sci 86(1):141–151

38. Riveles K, Tran V, Roza R, Kwan D, Talbot P (2007) Smoke from traditional commercial, harm reduction and research brand cigarettes impairs oviductal functioning in hamsters (Mesocricetus auratus) in vitro. Hum Reprod 22(2):346–355
39. Schatten G (1981) The movements and fusion of the pronuclei at fertilization of the sea urchin *Lytechinus variegates:* time-lapse video microscopy. J Morphol 167:231–247
40. Talbot P (1983) Videotape analysis of hamster ovulation in vitro. J Exp Zool 225(1):141–148
41. Talbot P, Lin S (2010) Cigarette smoke's effect on fertilization and pre-implantation development: assessment using animal models, clinical data, and stem cells. J Biol Res 44:189–194
42. Talbot P, Lin S (2010) Mouse and human embryonic stem cells: can they improve human health by preventing disease? Curr Topics Med Chem 11:1638–1652. (PMID 21446909)
43. Talbot P, Geiske C, Knoll M (1999) Oocyte pickup by the mammalian oviduct. Mol Biol Cell 10(1):5–8
44. Talbot P, zur Nieden N, Lin S, Martinez I, Guan B, Bhanu B (2014) Use of video bioinformatics tools in stem cell toxicology. Handbook of Nanomedicine, Nanotoxicology and Stem Cell Use in Toxicology (in press)
45. Thannickal VJ, Fanburg BL (2000) Reactive oxygen species in cell signaling. Am J Physiol Lung Cell Mol Physiol 279:L1005–L1028
46. Tsai KL, Talbot P (1993) Video microscopic analysis of ionophore induced acrosome reactions of lobster (Homarus Americanus) sperm. Mol Reprod Dev 36(4):454–461
47. Valavanidis A, Vlachogianni T, Fiotakis K (2009) Tobacco smoke: involvement of reactive oxygen species and stable free radicals in mechanisms of oxidative damage, carcinogenesis and synergistic effects with other respirable particles. Int J Environ Res Public Health 6 (2):445–462
48. Wang Y, Moussavi F, Lorenzen P (2013) Automated embryo stage classification in time-lapse microscopy video of early human embryo development. Lect Notes Comput Sci 8150:460–467
49. Weng JH, Phandthong G, Talbot P (2014) A video bioinformatics method to quantify cell spreading and its application to cells treated with rho associated protein kinase and blebbistatin. In: Bhanu B, Talbot P (eds) Video bioinformatics. Springer, Berlin
50. Williams M, Villarreal A, Bozhilov K, Lin S, Talbot P (2013) Metal and silicate particles including nanoparticles are present in electronic cigarette cartomizer fluid and aerosol. PLoS One 8(3):e57987. doi:10.1371/journal.pone.005798
51. Wong CC, Loewke KE, Bossert NL, Behr B, De Jonge CJ, Baer TM, Reijo-Pera RA (2010) Non-invasive imaging of human embryos before embryonic genome activation predicts development to the blastocyst stage. Nat Biotechnol 28:1115–1121

Part IV
Dynamic Processes in Plant and Fungal Systems

Chapter 10
Video Bioinformatics: A New Dimension in Quantifying Plant Cell Dynamics

Nolan Ung and Natasha V. Raikhel

Abstract Microscopy of plant cells has evolved greatly within the past 50 years. Advances in live cell imaging, automation, optics, video microscopy, and the need for high content studies has stimulated the development of computational tools for manipulating, managing, and interpreting quantitative data. These tools automatically and semiautomatically determine, sizes, signal intensities, velocities, classes, and many other features of cells and subcellular structures. Quantitative methods provide data that is a basis for mathematical models and statistical analyses that lead the way to a quantitative systems outlook to cell biology. Four-dimensional video analysis provides vital data concerning the often ignored temporal dynamics within a cell. Here, we will review studies employing technology to detect regions of interest using segmentation, classify data using machine learning and track dynamics in living cells using video analysis. Many of the live cell studies presented would have been impractical without these advanced computational techniques. These examples illustrate the utility and potential for video bioinformatics to augment our knowledge of the dynamics of cells and cellular components in plants.

10.1 Introduction

Video bioinformatics is a relatively new field which can be described as the automated processing, analysis, and mining of biological spatiotemporal information from videos obtained [1]. Advancements in the field of computer vision have given biologists the ability to quantify spatial and temporal dynamics and to do so

N. Ung (✉) · N.V. Raikhel
Institute for Integrative Genome Biology, Genomics Building 4119C,
University of California, 900 University Ave, Riverside, CA 92507, USA
e-mail: nolan.m.ung@gmail.com

N.V. Raikhel
e-mail: nraikhel@ucr.edu

© Springer International Publishing Switzerland 2015
B. Bhanu and P. Talbot (eds.), *Video Bioinformatics*,
Computational Biology 22, DOI 10.1007/978-3-319-23724-4_10

in a semi automatic and automatic manner. The challenges that arise from bioimage informatics become increasingly more complicated with the addition of the time dimension. Both techniques share very similar applications and challenges including detection of regions of interest (ROIs) via segmentation, registering images, subcellular localization determination, and dealing with large amounts of image data. Here we will discuss the challenges in plant cell biology that can be addressed using automatic quantitative tools such as image and video bioinformatics and the current shortcomings that need to be improved upon as we continue to discover and describe dynamic biological phenomena at the cellular level.

Most of the image data collected to date have been interpreted subjectively, allowing for personal interpretation and a loss of objectivity [2]. In the pursuit of biological discovery, we strive for objectivity and quantitative data that we can manipulate and use to better uncover genuine biological phenomena versus artifacts or biased results. Phenotypes can be continuous and cover a large spectrum, for example, when using chemical genomics to dissect conserved cellular processes [3]. Varying concentrations of bioactive compounds or drugs can illicit proportional phenotypes [4]. Therefore, the need for quantitative image and video data is essential when interpreting data on any time scale.

Ultimately, the quantified data demonstrate the most utility when subjected to statistical analysis. Therefore, it makes sense to quantify enough data to allow for a statistically valuable sample size. This often requires large amounts of data that need. Additionally, high-throughput screens have much to gain from using quantitative metrics to screen for valuable phenotypes [5]. To meet these challenges in a practical manner, quantification needs to be automated. Automation provides decreased analysis time, and allows for reduced inter and intrauser variability. The ability to provide a consistent analysis from sample to sample provides more reliable data. Reliable data are essential to fully understand the nature of any dynamic subcellular process. Dynamic cellular phenomena such as cell division, lipid dynamics, plant defense processes, and cell wall biosynthesis, often require the measurement of various static and dynamic features [6, 7, 8]. The automated detection, tracking, and analysis of these regions of interest summarizes the major goals of video bioinformatics in a cell biological context.

Live cell imaging has become an indispensable tool for discovery throughout the basic and applied sciences. This relatively recent technique has allowed for real-time observation and quantification of dynamic biological processes on the scale of nanometers to meters and milliseconds to days [9]. The advent of green fluorescent protein (GFP) has ignited a live cell imaging revolution and has subsequently enabled the capturing of in vivo spatial and temporal dynamics [10]. Because of their versatility, GFP and its derivatives have become ubiquitous in molecular and cell biology generating large quantities of image and video data. Many of the technical advancements in bioimaging have come from a prolific collaboration between the biological sciences and engineering. The cooperation of these two disciplines has produced indispensable tools to cell biology such as the laser scanning confocal microscope [11], spinning disk confocal microscope [12], mulptiphoton microscope [13], variable-angle epifluorescence microscope (VAEM) [14], and STORM [15] to name a few. All of these imaging modalities produce large quantities of complex multidimensional data. Scientists need to work together with engineers to dissect, manage, manipulate, and ultimately make sense of the image data collected. Practitioners of both disciplines, while still working to improve the acquisition hardware, are also working together to manage and analyze the large amounts of quantifiable image data.

The traditional method of quantifying image data is to manually draw regions of interest containing the biologically relevant information. This manual measurement is the most popular method of image quantification. Software tools including ImageJ and spin-offs of ImageJ such as Fiji are free [16]. Subcellular phenotyping is time consuming and impractical when performing high-throughput screens, which are necessary for most cell biologists. This data load is only increased when analyzing videos. A recent push toward automation has favored the use of automated microscopes, and robots that perform automated high-throughput sample preparation [17]. This has lead to the development and implementation of and automated semiautomated tools that require modest to little user input [18, 19]. Automated methods can be more consistent and faster since the user does not have to provide information. However, this lack of user input can also lead to reduced flexibility and robustness. On the other hand, semiautomated methods are flexible and possibly more robust due to user input, but can often be slower because the user has to provide prior information to the software. As this analysis becomes more user friendly and practical, the ability to apply a single software tool to multiple biological problems including multidimensional data, will be favored by the biologist thereby most likely favoring the semiautomated methods.

Bioimage informatics has experienced a recent surge in popularity due to the advent of automated microscopes and the subsequent burst of image data. Engineers had to develop methods to manage and interpret large amounts of image data generated by these automated systems. Bioimage informatics relies on several engineering disciplines including computer vision, machine learning, image-processing image analysis and pattern recognition [20]. The application of theses methods aids biologists is rapid detection, quantification, and classification of biological phenomena. Bioimage informatics is generally concerned with two-dimensional data, in the x and y planes, though it is possible to deal with three-dimensional, X, Y, and Z, and four-dimensional data X, Y, Z, and frequency domain [21]. Using these dimensions, data can be accurately extracted when computational techniques are properly applied.

Here we discuss the application of three fields of computer vision as they pertain to plant cell biology including, segmentation, machine learning, and video analysis, while highlighting the recent advances that were possible due to the collaboration of biologists and engineers.

10.2   Segmentation: Detecting Regions of Interest

Ultimately, biologists want to be able to extract data from acquired multidimensional images. However, the biologist needs to be able to identify those subregions within the image that hold the most data and that are therefore more important. As expert biologists, we can accurately identify the interesting regions of an image intuitively. Segmentation is the process of partitioning the regions of an image into segments [22]. Before we can extract data, we must first detect the objects or regions that are biologically meaningful. Biological images are acquired with various modalities and therefore one segmentation method is not going to be effective for all cases Specialized methods must be applied to each case. Much progress has been made in the domain of confocal microscopy. Bright fluorophores allow for high-contrast images that facilitate robust segmentation. In the realm of plant cell biology, many organelles and protein localization sites resemble bright spots or blobs. This is due to the light diffraction limit which limits the resolution of light microscopy at 250 nm, making small objects appear as fuzzy blobs [23]. Quantifying the number or size of these bright blobs is often done manually and can take several days. Simple segmentation can greatly improve this process which can then lead to feature extraction in both static and dynamic datasets.

Static 2D images are by far the most popular type of microscopy data to analyze because of their relatively short acquisition and analysis time. The majority of subcellular imaging is focused on the localization of proteins of interest. Using fluorescent markers fused to proteins of interest and dyes, cell biologists can understand the proteins that are involved with biological processes by monitoring the abundance, size shape, and localization within organelles. Organelles are of interest to cell biologist because of their diverse and extremely important roles in plant development, homeostasis, and stress responses. Automatic tools are being developed and used to quantify protein localization and spatial features of discrete compartments [24]. Organelles often manifest as punctate dots when imaged using fluorescent confocal laser scanning microscopy. These dots are then quantified per cell area and features extracted such as area, intensity, and number of compartments [19]. Salomon et al. used such a tool to quantify the response of various endomembrane compartment to bacterial infection, cold stress, and dark treatment [25]. Crucial information can also be garnered from the cells themselves. Cell borders can be detected when labeled and size as well as shape information analyzed automatically [25]. This information can then be used to track cell growth and development. Segmentation is the first crucial step to extracting quantitative information from live cell imaging data.

Cells exist in four dimensions, X, Y, Z, and time. If cell biologists want the full complement of information from imaging data, we have to consider all four of these dimensions. Collecting and processing 3D data is computationally more expensive and more difficult to manage but can yield a greater understanding of spatial information. Most confocal microscopes can easily collect data in the Z direction and 3D reconstructions are relatively easy to do now with the capable software. Most of the images captured of dividing plant cell are two-dimensional leaving out the critical third dimension. Miart et al. used 3D constructions of a growing cell plate to understand the role of cellulose synthase complexes in cell plate formation by analyzing time lapse video data of fluorescently labeled cellulose synthase complexes [26]. Although these analyses did not take advantage of automated quantification, the visualization of the 3D cell plate greatly contributed to the understanding of how cellulose synthase complexes (CESAs) are involved in cell plate formation [26]. Quantifying temporal dynamics in a study such as this would lend insight into how fast this process happens and perhaps how the population of CESA complexes shifts from a homeostatic role to an actively dividing role.

Four-dimensional data including 3D movies of cellular phenomena, will become more popular as the tools to analyze this data become more sophisticated and more user friendly. Automated 4D analysis tools are already being used by cell biologists to analyze trichome development [9]. This system extracts the leaf surface, segmenting the mid-plane of the young leaf and detects the developing trichomes using a Hough transform which can detect circles [27]. One 3D image is registered to the next 3D image in the time series to maintain consistency and to track and compare its growth over time [28]. These tools will need to be adapted from analyzing gross morphology to tracking moving cellular structures over time.

10.3 Machine Learning Allows Automatic Classification of Cellular Components

Machine learning is a subdiscipline of artificial intelligence that is focused on the development of algorithms that can learn from given data [29]. These methods often require the use of training data and user input to learn how the data should be classified. Training the algorithm allows it to correctly identify the class to which each sample belongs. A simple example is the spam filter on most email accounts that can discern between those messages that are spam and those that are important.

A logical application of machine learning in cell biology was determining the subcellular localization of fluorescent markers based on extracted features. Traditionally, cell biologists have to colocalize their protein of interest and markers of known localization to determine where the protein is located. Biologists could simply analyze a confocal micrograph with a machine learning program and receive the location of their protein of interest. An additional advantage to the machine learning methods over traditional cellular methods, other than reduced time is that these methods provide statistics as to how likely the determined localization is to be true [30]. Though there seems to be reasonable progress in determining subcellular localization using machine learning, the biological community has yet to adopt the methodology. Prediction of subcellular localization will streamline experimental design and support traditional colocalization assays.

Machine learning is a powerful tool for gene discovery and organelle dynamics. It can help uncover relationships that we otherwise could not. Because organelle dynamics can be complex and variable, it is valuable to simplify dynamics. Using Baysian networks, Collinet et al. found that endosome number, size, concentration of cargo, and position are mediated by genetic regulation and not random [31]. Furthermore, they used this method to discover novel components regulating endocytosis by clustering endocytic phenotypes caused by screening siRNA libraries [31]. Statistical analysis was similarly used to summarize and classify organelle movement in Arabidopsis stomata (Higaki 2012). The result is an atlas of organelle movement in stomata that can be compared to various conditions. Organelle movement patterns were compared between open and closed stomata revealing differences in ER position in response to stomatal opening. These new findings emphasize the need for statistical methods to manage complex data and present this data in forms we can easily understand and manipulate.

Though we are interested in cell autonomous processes, cells do not exist in a vacuum. We are also interested in how a cell influences the development and function of its neighboring cells. To address this challenge, segmentation coupled with machine learning was used to jointly detect and classify cell types in whole tissues. three-dimensional images of propidium iodide stained roots were used to automatically find cell files in longitudinal and transverse sections using watershed segmentation and a support vector machine to classify cell types [32]. An alternative approach used histological sections of Arabidopsis hypocotyls to differentiate tissue layers and predict the location of phloem bundle cells [33]. The true utility of these tools will be realized when they are used to compare wild-type cell profiles with mutants, possibly being used in large content screening.

10.4 Quantifying Temporal Dynamics Adds a New Dimension of Data

Once an object has been detected and classified, it is often very important to follow its movement through time and space. This extremely important problem of tracking has been tackled by many engineers developing the field of computer vision. A multitude of tools are available for tracking cells and organelles, most of these being manual and semi automated [19]. Tracking organelles are difficult because rarely do they have a straight forward movement model. It is because of the diversity and variability of tracking problems that semiautomated methods are the most widely used. Common problems include object moving out of the plane of focus when using 2D images. It is because of this issue that 3D movies are such valuable data sets [34]. Therefore, automatically tracking object in a 3D image set is an invaluable tool [35]. Other challenges include maintaining identity when two objects fuse or break off from one another, and maintaining multiple tracks at the same time. A perfect tracking algorithm would overcome all of these problems, while maintaining minimal user input and accurate segmentation.

The purpose of quantifying movement and movement patterns is to gain useful biological insight such as diffusion rates, types of motion including Brownian motion, non-Brownian motion, confined motion, directed motion, or anomalous diffusion [36]. Ung et al. correlated multiple dynamic features which suggested that when tobacco pollen tubes where treated with specific bioactive compounds the contained Golgi bodies increased in size and this increase in size was correlated to an increase in signal intensity and a decrease in straightness suggesting that these were possibly multiple fused Golgi and that this fusion disrupted movement [19]. Similar correlations were made by collinet et al. when examining endocytosis in mammalian cells [31]. Indeed these data are consistent with the hypothesis proposed by Hamilton et al. (2007) including conservation of surface area, measurement of volume, flux across a membrane, the role of pressure and tension and vesicle fusion. These biological details would not be obtainable without quantitative video analysis.

Although each challenge was presented separately, they are by no means mutually exclusive. The vast majority of image analysis problems require identification of regions of interest before they can be quantified, tracked, or classified.

10.5   Opportunities for Innovation in Cellular Video Bioinformatics

As video bioinformatics tools become increasingly accurate and biologist friendly, they will be more widely used in biological studies. The future of video analysis is moving toward automatic quantification of cellular dynamics in four dimensions (3D time lapse images). The amount of data that can be extracted from 3D movies will increase with the availability and ease of use of software. Biologists will be able to quantify difference in movement possibly identifying underlying principals of movement and other components essential to cellular dynamics. As these video analysis tools become more fully automated, it will be more practical to screen for factors that influence dynamics. In this manner, biologists will be able to directly screen for changes in cellular dynamics.

Creating the tools to quantify cellular dynamics is futile unless biologists use them to produce data. The pipeline from engineer to the biological community needs to be stronger. This could be enhanced by taking advantage of open source repositories of image analysis tools. A small pool of these repositories currently available and will grow in popularity as the need for these programs becomes greater [37]. As we take advantages of quantitative methods we will produce large

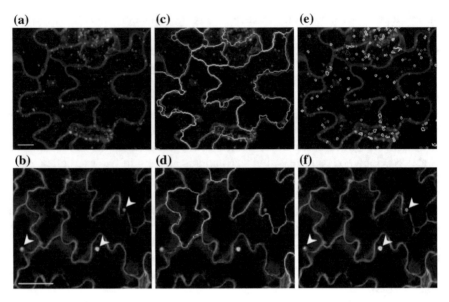

Fig. 10.1 Pattern recognition of membrane compartments in leaf epidermal tissue at cellular and subcellular resolution. Merged confocal laser microscopic images show Arabidopsis leaf epidermal cells. *Top section* GFP-2xFYVE plants were imaged at 403 magnification (scale bar = 20 mm) and images analyzed by the endomembrane script. *Bottom section* GFP-PEN1 plants were imaged at 203 magnification (scale bar = 50 mm) and images analyzed by the plasma membrane microdomain script. **a** and **b**, Merged pseudo images. **c** and **d**, Recognition of epidermal cells is shown by *colored lines*. **e**, Recognition of GFP-2xFYVE-labeled endosomal compartments is shown by *colored circles*. **f**, Recognition of (**b**). Graminis induced GFP-PEN1 accumulation beneath attempted fungal entry sites (indicated by *arrowheads*) is shown by *colored circles*. Color coding is random, different colors indicate individual cells compartments [25]

amounts of data that has the potential to fuel mathematical models or other future studies [38]. Many mathematical models require know numerical parameters to be of use.

As live cell imaging modalities and acquisition methods become more advanced including super resolution methods and as biological systems change, our analysis methods will to have to adapt. In the future, these efforts will be spearheaded by a handful of interdisciplinary scientists that will be trained in biological principals, experimental design, computer programming, and image analysis's tool design. Future biologists will have to be well-versed in computer programming basics and be able to design tools that are applicable to their specific research topic, while having a basic understanding of the appropriate algorithms all while being able to communicate with engineers. Engineers on the other hand, will have to understand biological limitations, know which features are useful, experimental design, and acquisition methods (Figs. 10.1, 10.2 and 10.3).

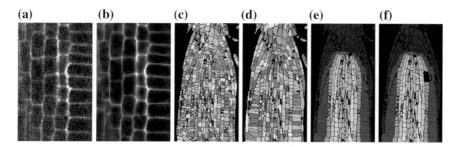

Fig. 10.2 Cell wall stained by Propidium iodide: **a** Raw data of a test image. **b** After anisotropic diffusion. **c** Watershed segmentation (from local minima) shown in random color. **d** Final segmentation result. **e** Classification of the segments. **f** Manually generated ground-truth for segmentation and classification (modified from Liu et al. [32])

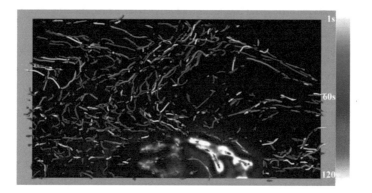

Fig. 10.3 Representation obtained by automatic tracking method of the same tracking result from a cell tagged with GFP-Rab6. **a** 3D trajectories of the detected objects. The color indicates the time window when an object was detected. The z projection of the first 3D image of the time series is shown below the trajectories to visualize the cell morphology (modified from Racine et al. [35])

Acknowledgment This work was supported in part by the National Science Foundation (NSF) Integrative Graduate Education and Research Traineeship (IGERT) in Video Bioinformatics (DGE-0903667) and NSF Graduate Research Fellowship Program (GRFP). Nolan Ung is an IGERT and GRFP Fellow.

References

1. Bhanu B (2009) IGERT Program. In: The UC Riverside integrated graduate education research and training program in video bioinformatics. Univeristy of California, Riverside, 26 Oct 2009. Web. 24 Mar 2014
2. Peterson RC, Wolffsohn JS (2007) Sensitivity and reliability of objective image analysis compared to subjective grading of bulbar hyperaemia. Br J Ophthalmol 91(11):1464–1466

3. Drakakaki G, Robert S, Szatmari AM, Brown MQ, Nagawa S, Van Damme D, Leonard M, Yang Z, Girke T, Schmid SL, Russinova E, Friml J, Raikhel NV, Hicks GR (2011) Clusters of bioactive compounds target dynamic endomembrane networks in vivo. Proc Natl Acad Sci USA 108(43):17850–17855

4. Robert S, Chary SN, Drakakaki G, Li S, Yang Z, Raikhel NV, Hicks GR (2008) Endosidin1 defines a compartment involved in endocytosis of the brassinosteroid receptor BRI1 and the auxin transporters PIN2 and AUX1. Proc Natl Acad Sci USA 105(24):8464–8469

5. Beck M, Zhou J, Faulkner C, MacLean D, Robatzek S (2012) Spatio-temporal cellular dynamics of the Arabidopsis flagellin receptor reveal activation status-dependent endosomal sorting. Plant Cell 24(10):4205–4219

6. Tataw OM, Liu M, Roy-Chowdhurry A, Yadav RK, Reddy GV (2010) Pattern analysis of stem cell growth dynamics in the shoot apex of arabidopsis. In: 17th IEEE international conference on image processing (ICIP), pp 3617–3620

7. Liu J, Elmore JM, Lin ZJ, Coaker G (2011) A receptor-like cytoplasmic kinase phosphorylates the host target RIN4, leading to the activation of a plant innate immune receptor. Cell Host Microbe 9(2):137–146

8. Sampathkumar A, Gutierrez R, McFarlane HE, Bringmann M, Lindeboom J, Emons AM, Samuels L, Ketelaar T, Ehrhardt DW, Persson S (2013) Patterning and lifetime of plasma membrane-localized cellulose synthase is dependent on actin organization in Arabidopsis interphase cells. Plant Physiol 162(2):675–688

9. Domozych DS (2012) The quest for four-dimensional imaging in plant cell biology: it's just a matter of time. Ann Bot 110(2):461–474

10. Brandizzi F, Fricker M, Hawes C (2002) A greener world: the revolution in plant bioimaging. Nat Rev Mol Cell Biol 3(7):520–530

11. Rajadhyaksha M, Anderson R, Webb RH (1999) Video-rate confocal scanning laser microscope for imaging human tissues; In Vivo. Appl Opt 38(10):2105–2115

12. Nakano A (2002) Spinning-disk confocal microscopy–a cutting-edge tool for imaging of membrane traffic. Cell Struct Funct 27(5):349–355

13. Meyer AJ, Fricker MD (2000) Direct measurement of glutathione in epidermal cells of intact Arabidopsis roots by two-photon laser scanning microscopy. J Microsc 198(3):174–181

14. Konopka CA, Bednarek SY (2008) Variable-angle epifluorescence microscopy: a new way to look at protein dynamics in the plant cell cortex. Plant J 53(1):186–196

15. Rust MJ, Bates M, Zhuang X (2006) Sub-diffraction-limit imaging by stochastic optical reconstruction microscopy (STORM). Nat Methods 3(10):793–796

16. Schindelin J, Arganda-Carreras I, Frise E, Kaynig V, Longair M, Pietzsch T, Preibisch S, Rueden C, Saalfeld S, Schmid B, Tinevez JY, White DJ, Hartenstein V, Eliceiri K, Tomancak P, Cardona A (2012) Fiji: an open-source platform for biological-image analysis. Nat Methods 9(7):676–682

17. Hicks GR, Raikhel NV (2009) Opportunities and challenges in plant chemical biology. Nat Chem Biol 5(5):268–272

18. Kuehn M, Hausner M, Bungartz HJ, Wagner M, Wilderer PA, Wuertz S (1998) Automated confocal laser scanning microscopy and semiautomated image processing for analysis of biofilms. Appl Environ Microbiol 64(11):4115–4127

19. Ung N, Brown MQ, Hicks GR, Raikhel NV (2012) An approach to quantify endomembrane dynamics in pollen utilizing bioactive chemicals. Mol Plant

20. Eils R, Athale C (2003) Computational imaging in cell biology. J Cell Biol 161(3):477–481

21. Cheung G, Cousin MA (2011) Quantitative analysis of synaptic vesicle pool replenishment in cultured cerebellar granule neurons using FM dyes. J Vis Exp (57)

22. Shapiro LG, Stockman GC (2001) Computer Vision. New Jersey, Prentice-Hall, pp 279–325. ISBN 0-13-030796-3

23. de Lange F, Cambi A, Huijbens R, de Bakker B, Rensen W, Garcia-Parajo M, van Hulst N, Figdor CG (2001) Cell biology beyond the diffraction limit: near-field scanning optical microscopy. J Cell Sci 114(23):4153–4160

24. Sethuraman V, Taylor S, Pridmore T, French A, Wells D (2009) Segmentation and tracking of confocal images of Arabidopsis thaliana root cells using automatically-initialized Network Snakes. In: 3rd international conference on bioinformatics and biomedical engineering, 2009. ICBBE 2009
25. Salomon S, Grunewald D, Stüber K, Schaaf S, MacLean D, Schulze-Lefert P, Robatzek S (2010) High-throughput confocal imaging of intact live tissue enables quantification of membrane trafficking in arabidopsis. Plant physiol 154(3):1096
26. Miart F, Desprez T, Biot E, Morin H, Belcram K, Höfte H, Gonneau M, Vernhettes S (2013) Spatio-temporal analysis of cellulose synthesis during cell plate formation in Arabidopsis. Plant J
27. Illingworth J, Kittler J (1987) The adaptive Hough transform. IEEE Trans Pattern Anal Mach Intell 5:690–698
28. Bensch R, Ronneberger O, Greese B, Fleck C, Wester K, Hulskamp M, Burkhardt H (2009) Image analysis of arabidopsis trichome patterning in 4D confocal datasets. In: IEEE international symposium on biomedical imaging: from nano to macro, 2009. ISBI'09, pp 742–745
29. Goldberg DE, Holland JH (1988) Genetic algorithms and machine learning. Mach Learn 3 (2):95–99
30. Hua S, Sun Z (2001) Support vector machine approach for protein subcellular localization prediction. Bioinformatics 17(8):721–728
31. Collinet C, Stöter M, Bradshaw CR, Samusik N, Rink JC, Kenski D, Habermann B, Buchholz F, Henschel R, Mueller MS, Nagel WE, Fava E, Kalaidzidis Y, Zerial M (2010) Systems survey of endocytosis by multiparametric image analysis. Nature 464(7286):243–249
32. Liu K, Schmidt T, Blein T, Durr J, Palme K, Ronneberger O (2013) Joint 3d cell segmentation and classification in the arabidopsis root using energy minimization and shape priors. In: IEEE 10th international symposium on biomedical imaging (ISBI), pp 422–425
33. Sankar M, Nieminen K, Ragni L, Xenarios I, Hardtke CS (2014) Automated quantitative histology reveals vascular morphodynamics during Arabidopsis hypocotyl secondary growth. eLife 3
34. Carlsson K, Danielsson P-E, Liljeborg A, Majlöf L, Lenz R, Åslund N (1985) Three-dimensional microscopy using a confocal laser scanning microscope. Opt Lett 10 (2):53–55
35. Racine V, Sachse M, Salamero J, Fraisier V, Trubuil A, Sibarita JB (2007) Visualization and quantification of vesicle trafficking on a three-dimensional cytoskeleton network in living cells. J Microsc 225(Pt 3):214–228
36. Saxton MJ, Jacobson K (1997) Single-particle tracking: applications to membrane dynamics. Annu Rev Biophys Biomol Struct 26(1):373–399
37. Swedlow JR, Eliceiri KW (2009) Open source bioimage informatics for cell biology. Trends Cell Biol 19(11):656–660
38. Phillips R, Milo R (2009) A feeling for the numbers in biology. Proc Natl Acad Sci USA 106 (51):21465–21471

Chapter 11
Understanding Growth of Pollen Tube in Video

Asongu L. Tambo, Bir Bhanu, Nan Luo and Zhenbiao Yang

Abstract Pollen tubes are tube-like structures that are an important part of the sexual reproductive cycle in plants. They deliver sperm to the ovary of the plant where fertilization occurs. The growth of the pollen tube is the result of complex biochemical interactions in the cytoplasm (protein–protein, ions, and cellular bodies) that lead to shape deformation. Current tube growth models are focused on capturing these internal dynamics and using them to show changes in tube length/area/volume. The complex nature of these models makes it difficult for them to be used to verify tube shapes seen in experimental videos. This chapter presents a method of verifying the shape of growing pollen tubes obtained from experimental videos using video bioinformatics techniques. The proposed method is based on a simplification of the underlying internal biological processes and their impact on cell morphogenesis. Experiments are conducted using videos of growing pollen tubes to show the model's performance.

A.L. Tambo (✉) · B. Bhanu
Center for Research in Intelligent Systems, University of California, Chung Hall 216, Riverside, CA 92521, USA
e-mail: atamb001@student.ucr.edu

B. Bhanu
e-mail: bhanu@ee.ucr.edu

N. Luo · Z. Yang
Department of Botany and Plant Sciences, University of California, 2150 Batchelor Hall, Riverside, CA 92521, USA
e-mail: nan.luo@email.ucr.edu

Z. Yang
e-mail: zhenbiao.yang@ucr.edu

© Springer International Publishing Switzerland 2015
B. Bhanu and P. Talbot (eds.), *Video Bioinformatics*,
Computational Biology 22, DOI 10.1007/978-3-319-23724-4_11

11.1 Introduction

Biological systems are complex naturally occurring systems. Attempts to understand their behavior often involve laboratory experimentation to identify key substances (proteins, ions, etc.), parts of the organisms, and whatever relationships exist between them. These relationships involve, but are not limited to, signaling pathways between system parts and how these parts respond to and affect changes in their environment. To simplify this task, scientists search for relatively simple organisms/systems, termed 'model organisms', which are representative of more complex systems. Experiments are conducted on model organism in hopes of gleaning knowledge about the behavior of more complex systems. Pollen tubes are a good model system, not only because they are essential for plant reproduction, but also because they grow through localized cell wall expansion (apical growth) rather than uniform cell wall expansion.

In recent years, growth models have been presented to address the complexity of apical growth [1–5]. Each model explains polar growth in pollen tubes by capturing different cellular processes in the form of differential equations that describe the various interactions that lead to growth: protein–protein interactions, ion dynamics, turgor pressure, cell wall resistance, etc. Depending on the readers' background, any of the above models can prove challenging to understand. The models in [1, 2] focus on transport phenomena while [3, 5] focus on the physics of the cell wall and how it is deformed by turgor pressure. Using equations, they all explain what is known about different aspects of the growth process. Owing to the complex nature of the problem, the above models suffer from one major limitation: verification of the model with experiments, i.e., using the model to explain/predict the dynamics of variables in experimental videos. Such a task would involve making accurate measurements from the images, parameter estimation to determine variable values, and/or statistics of unobservable variables. The model presented in this chapter is based on aspects of the tip growth process that are unobservable as well as observable phenomenon whose behavior can only be approximated.

The rest of this chapter is organized as follows: Sect. 11.2 gives a detailed outline and comparison of three of the above models, and the motivation for a simplistic, intuitive, video-based model. Section 11.3 details the technical approach of the model with relevant equations and assumptions. Section 11.4 covers experimental results and discussion of these results, and Sect. 11.5 concludes the chapter.

11.2 Related Work and Motivation

11.2.1 Related Work

Available models of pollen tube growth are predictive in nature. They are mainly used to predict what the effect of changing the values of one or more variables will have on the growth behavior of the pollen tube. As Fig. 11.1 shows, cell growth is

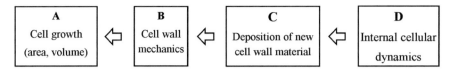

Fig. 11.1 Separation of the growth process into key blocks. Cell growth requires the deposition of new cell wall material. This deposition is controlled by internal cellular dynamics, which are by far the most complex part of the process

the end result of a sequence of reactions that get more complex as one digs deeper. In order for the cell to grow, there must be available material. The availability of cell wall material is controlled by internal cellular dynamics (protein–protein interactions, ion concentrations and reaction rates, etc.). As a result of this complexity, current models approach the problem from different angles and make assumptions about the behavior/values of other un-modeled parameters. In this section, we shall examine some of these models and their assumptions.

In [1] the authors present a model for cell elongation that can be said to cover parts A, B and C of Fig. 11.1. Growth is represented as an increase in cell tip volume due to the expansion/stretching of a thin film that quickly hardens to form new cell wall. As a result of positive changes in osmotic and turgor pressure between the cell and its surroundings, water enters the cell and induces a stress–strain relationship at the tip. This causes stretching of deposited cell wall material. To prevent the thinning of the cell wall, this stretching is balanced by a material deposition system. The interplay between stretching and deposition determines and maintains the thickness of the cell wall.

Similarly, the authors of [3] present a model that focuses on parts A and B of Fig. 11.1. This model represents growth as a result of internal stresses (in two principal directions) that deform the cell wall and lead to various cell shapes. The cell wall is represented as a viscous fluid shell. Cell wall thickness is maintained by local secretion of cell wall material. This secretion rate is maximal at the apex of the cell and decreases in the direction of the shank. With simulations, the authors show that cell radius and velocity can be determined as a function of turgor pressure.

In [5] a growth model is presented based solely on the rheology of the cell wall. Cell growth is modeled as the stretching of a thin viscoplastic material under turgor pressure. The stress–strain relationship is modeled after Lockhart's equation [6] for irreversible plant cell elongation. Like in [3], this model focuses on the physics of cell wall deformation and requires material deposition to maintain wall thickness. Simulations show that various shapes can be observed using this model.

11.2.2 Motivation

The models presented above take different approaches to solving the problem of tip growth in pollen tubes. They all take into account certain aspects of the process that are universally agreed upon: turgor pressure is necessary for growth, and deposition of material is needed to maintain cell wall thickness. Each model can be used to predict how cell growth varies under certain conditions prior to conducting an experiment. This is a useful tool in understanding the effects of variables in a complex system.

The main limitation of these methods is that they are not used to determine parameter values for a given experiment, i.e., they are neither used to determine variable behavior/values during the course of an experiment nor to check whether the behavior of un-modeled variables agrees with model assumptions. In the next section, we develop a model that is capable of accounting for changes observed in experimental videos of growing pollen tubes. The model leverages underlying biology in the development and unification of appropriate functions that explain tip growth.

11.3   Technical Approach

As indicated above, pollen tube growth involves a complex series of interactions between cell organelles and their environments. Modeling cell growth is an attempt to express the interactions between several variables (ion and protein concentrations, turgor pressure, wall stress/strain, rate of exocytosis/endocytosis, cell length/volume, etc.) over time with a system of mathematical equations, usually differential equations. A proper solution to such a system would require experimental data of how these variables are changing. Some of the readily available variables include cell length/volume and ion/protein concentrations. Those variables whose behavior cannot be readily observed are therefore inferred.

This section presents a model to account for cell growth using a minimal number of variables while inferring the behavior of other components. Section 3.1 covers the behavior of the protein ROP1, which is key in pollen tube growth. Section 3.2 covers tube elongation (growth) as a resource-consumption process whereby material deposited at the cell wall via exocytosis is used for growth. Although this is commonly known, the above models do not explicitly take advantage of it.

11.3.1   Dynamics of ROP1-GTP Protein Determines Region and Start of Exocytosis

It is accepted among the pollen tube community that pollen tube growth requires the deposition of material to maintain the thickness of the cell wall. If the rate of cell expansion is greater than the rate of cell wall reconstruction, then the thickness of the cell wall reduces. This leads to bursting at the cell tip. On the other hand, if the rate of cell wall reconstruction is greater, then the cell wall thickens and inhibits growth. So the availability of cell wall material is crucial in maintaining cell integrity. This material is deposited by vesicles that have travelled from within the cell to the site of exocytosis.

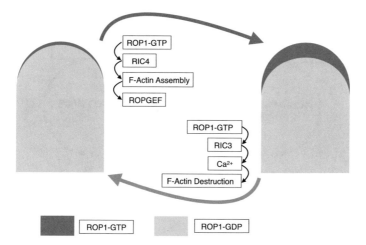

Fig. 11.2 The dynamics of ROP1 in both the active form (-GTP) and the inactive form (-GDP). In positive feedback (*left*-to-*right*), ROP1-GTP promotes RIC4, which promotes the assembly of the F-Actin network that conveys ROPGEF to the sites of ROP1-GTP. ROPGEF converts ROP1-GDP to ROP1-GTP. In negative feedback (*right*-to-*left*), ROP1-GTP promotes RIC3, which promotes Ca^{2+} accumulation. Ca^{2+} destroys the F-Actin network and stops the transportation of ROPGEF. With diminished localized concentration of ROPGEF, global conversion of ROP1-GTP into ROP1-GDP prevails

The authors of [7, 8] have shown that the localization and concentration of the active form of the protein ROP1 (ROP1-GTP) indicates the region of exocytosis. As shown in Fig. 11.2, the localization of active ROP1 promotes the synthesis of proteins whose dynamics contribute to both positive and negative feedback of ROP1-GTP. Positive feedback of ROP1-GTP leads to an increase in localized concentrations of ROP1-GTP. In images, this is represented by an increase in localized brightness. It is considered that the assembly of the F-Actin network contributes to vesicle accumulation at the site of exocytosis. Negative feedback of ROP1-GTP is believed to contribute to an increase in the concentration of apical Ca^{2+}. Ca^{2+} may also accumulate at the tip due to stretch-activated pores during cell growth [9]. Accumulation of Ca^{2+} destroys the F-Actin network and fosters the conversion of ROP1-GTP into ROP1-GDP. When this occurs, there is a drop in brightness of the fluorescing pollen tube. We consider this event to be the signal that starts exocytosis. During exocytosis, accumulated vesicles fuse with the cell membrane, adding to the surface area of the cell membrane, and releasing their contents to the cell wall. This idea is supported by the authors of [10], who show that exocytosis in pollen tubes leads to an increase in tube growth rate.

In this study, we consider that the dynamics of apical Ca^{2+} cannot be observed, but its influence can be seen in the accumulation/decay profile of ROP1-GTP. We represent the dynamics of ROP1 as a 2-state (accumulation and decay states) Markov process. We also consider that the transition from accumulation to decay is the signal for the start of exocytosis as this coincides with the destruction of the F-actin network (Fig. 11.3).

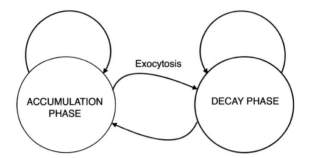

Fig. 11.3 Representation of ROP1-GTP activity as a 2-state Markov process. The transition from accumulation-to-decay is assumed to be the signal for the start of exocytosis, which leads to growth

11.3.2 Growth as a Resource-Consumption Process

Geometrically, the pollen tube can be divided into two main parts: the shank and the tip. The shank is the cylindrical part of the tube while the tip is the hemispherical part. As the tube grows, the tip moves forward leaving behind a wall that forms the shank. Let $X = \left\{ \begin{bmatrix} x_i \\ y_i \end{bmatrix}, i = 1, \ldots, N \right\}$ be the set of points along the tip of the tube, and assuming that the cell is elongating vertically. Then there exist an affine transformation that will project the tip points at time $(t - 1)$ to the next shape at time t, i.e.,

$$X(t) = \begin{bmatrix} A(t-1) & T \\ \vec{0} & 1 \end{bmatrix} X(t-1), \tag{11.1}$$

where $A(t)$ is a (2×2) matrix of rotation, scale, and shear parameters, T is a 2-dimensional normally distributed random variable that represents system noise, and $X(.)$ is a ($2 \times N$) matrix of tip points. The matrix A can be decomposed into its component parts using QR-decompositions into a product of a rotation matrix (R) and an upper-triangular matrix U:

$$A = R * U \tag{11.2}$$

The matrix U can be further decomposed into a product of scale and shear matrices. Put together, A can be written as

$$A(t) = R \begin{bmatrix} 1 & c(t) \\ 0 & 1 \end{bmatrix} \begin{bmatrix} a(t) & 0 \\ 0 & b(t) \end{bmatrix}, \tag{11.3}$$

where $a(t)$, $b(t)$ represent scaling factors that control expansion (x-direction) and elongation (y-direction), and $c(t)$ is a shear parameter that controls the turning of the tip. R is a (2×2) rotation matrix whose impact is a result of system noise. The elongation and expansion of the cell are governed by

$$\begin{bmatrix} x_f \\ y_f \end{bmatrix} = \begin{bmatrix} a & 0 \\ 0 & b \end{bmatrix} \begin{bmatrix} x_i \\ y_i \end{bmatrix} \rightarrow \begin{cases} a - 1 = \dot{\epsilon}_x = \dot{\phi}\sigma_\theta \\ b - 1 = \dot{\epsilon}_y = \dot{\phi}\sigma_s \end{cases}, \qquad (11.4)$$

where ϕ is the extensibility of the cell wall, ϵ_i, σ_i are the strain and stress, respectively, in the ith-direction as depicted in Fig. 11.3. Similarly,

$$\frac{x_f - x_i}{y_i} = c \qquad (11.5)$$

The stress/strain relationship used above was proposed by Lockhart [6]. A similar equation is used in [5, 11]. We consider that the cell is made extensible due to the stretching of weakened cellulose fibers in the cell wall caused by exocytosis. We also assume that deposited material is used to restore the fibers to their original thickness, thus maintaining the thickness of the cell wall. To represent this resource-consumption process, we use the Gompertz function, which is one of several specific cases of a logistic function. Logistic functions have been used extensively in studies of population growth [12, 13]. The choice of the Gompertz function limits the logistic function to 2 parameters. Thus, we consider that cell wall extensibility is governed by

$$\dot{\phi}(t) = -r\phi(t) \ln\left(\frac{K}{\phi(t)}\right), \qquad (11.6)$$

where $\phi(t)$ denotes the extensibility of the cell wall over time, $K \geq 0$ is the maximum attainable extensibility of the cell wall, and r is the extension rate of the cell wall. The growth rate depends on a number of factors that include the rate of exocytosis and the rate of cell wall reconstruction. Since these quantities cannot be measured in experiments, we assume that the rate of cell stretching is proportional to the rate of exocytosis (material deposition). This ensures that the thickness of the cell wall is maintained.

Since we assume that the thickness of the cell wall is kept constant, the change in length of the cell is related to the volume of material deposited by vesicles during exocytosis. A measure of the number of vesicles that were involved in exocytosis is given in [14] by considering that tip growth adds cylindrical rings of the same thickness as the cell wall at the tip/shank junction. To determine this ratio we use the change in the circumference of the points along the tip. The rate of vesicle deposition becomes

$$\frac{d}{dt}N(t) = \frac{3}{8}\frac{R_t \delta}{r_v^3}\frac{d}{dt}h(t), \qquad (11.7)$$

where R_t is the radius of the tip/shank junction, δ is the thickness of the cell wall, h is the change in the circumference of the tip, and r_v is the radius of a vesicle.

11.3.3 Video Analysis

The above equations are used to determine the dynamics of the described variables. So far, the prescribed method works best with videos of growing pollen tubes showing fluorescent markers for ROP1. Given one such image, the contour of the cell is segmented from the background. The fluorescence along the cell contour is measured and used to segment the cell into tip and shank regions. The average fluorescence of the tip, $f_{avg}(t)$, (Table 11.1, line 4) is also recorded. Cell shape analysis begins when a maxima turning point is detected in $f_{avg}(t)$. Let this maxima turning point be at time $t_\alpha > 0$. The line separating the tip and the shank at the time of the maxima turning point is used to separate future shapes into tip-shank regions. Given that t_α is the start of the growth process, we implement an initialization phase (Table 11.1, line 6bi) requiring a delay of three frames, i.e., the cell tip is detected from two consecutive images in a set of three images and used to determine the

Table 11.1 Pseudo-code for video analysis of growing pollen tubes

---

Given:
Fluorescent video of growing pollen tube experiment
N: fluorescence smoothing window (buffer size)

For i = 1 to video_length:
  1. *Obtain the ith image.*
  2. *S = $\{(x, y)\}$ ←contour of segmented pollen tube*
  3. *tip = $\{(x, y) \in S\}$ ← uniform region of relatively high fluorescence along contour*
  4. *$f_{avg}(i)$ ← average brightness/fluorescence around tip*
  5. *T ← maxima turning points in smoothed $f(i)$ using N*
  6. *If isNotEmpty(T)*
        a. *if isNewPoint(T(end))*
              i. *predicted_tip ←tip*
        b. *else*
              i. *Initialization phase for $a(t), b(t), c(t)$*
              ii. *$tip_{predicted} \leftarrow A(t) \times tip_{previous}$*
              iii. *if $error(tip_{predicted}, tip) > e_{threshold}$*
                    1. *update $tip_{predicted}$*
                    2. *update $a(t), b(t), c(t)$*
                    3. *end*
              iv. *$tip_{previous} \leftarrow tip_{predicted}$*
  7. *End.*
End

---

values of $a(t)$, $b(t)$, $c(t)$. Using these values, predictions for the transformation parameters for the next time are made as

$$a(t+1) = \frac{1}{3} \sum_{i=\max(0,t-2)}^{t} a(i) \qquad (11.8)$$

Equation (11.8) applies for $b(t + 1)$ and $c(t + 1)$. These predictions are used to transform/grow the cell tip.

If the error between the predicted cell tip and the detected tip is greater than a prescribed threshold, the predicted shape and the predicted values for $a(t)$, $b(t)$, and $c(t)$ are corrected. This correction involves computing the transformation that takes the predicted shape to the measured shape. The values for scale and shear parameters extracted from this new transformation are used to update their respective current values. This process continues until a new maxima is detected in $f_{avg}(t)$, at which point, the method resets (Table 11.1, line 6a).

11.4 Experiments

In this section, we present the results of using the above method to analyze a video of a growing pollen tube. The pollen tube grows from pollen grains taken from the plant *Arabidopsis thaliana*. This plant is considered a *model* plant for pollen tube study because it grows quickly and its entire genome is known. Since the genome is known, mutants can be easily made and studied to understand the effect of proteins/ions on the growth process.

11.4.1 Materials and Methods

Arabidopsis RIP1 (AT1G17140, also named ICR1) was fused with GFP on the C-terminus and used as a marker of active ROP1. The RIP1-GFP fragment was fused into a binary vector pLat52::NOS, which contains a pollen tube-specific promoter, to generate the binary construct pLat52::RIP1-GFP construct. pLat52:: RIP1-GFP was introduced to *Arabidopsis* wild-type Col0 background using Agrobacterium mediated flower dip method.

Arabidopsis thaliana plants were grown at 22 °C in growth rooms under a light regime of 16 h of light and 8 h of dark. Pollens were germinated on a solid medium (18 % sucrose, 0.01 % boric acid, 1 mM $CaCl_2$, 1 mM $Ca(NO_3)_2$, 1 mM $MgSO_4$, pH 6.4 and 0.5 % agar) for 3 h before observation under microscope. GFP-ROP1, GFP-RIC1, GFP-RIC4, or RLK-GFP expressing pollen tubes were observed under a Leica SP2 confocal microscope (488 nm excitation, 500–540 nm emission). Median plane of pollen tube tip was taken for time-lapse images.

11.4.2 Results

The diagrams below show the output of the above method in video analysis of growing pollen tubes. Figure 11.4 shows model output on two video sequences. Fluorescence analysis detected 4 growth cycles in the first video and 3 cycles in the second video. The first video (first row) contains 220 images with a time interval of about 1 s. To reduce computational time, the images were downsized by 50 % to a size of (200 × 232) pixels. A buffer size of 3 (parameter N in Table 11.1) was used in determining the maxima turning points in the curve of ROP1-GTP. Each pixel has a length of 58.13 nm. The average radius of the pollen tube is 1.84 μm. The second video (second row) contains 300 images with a time interval of 1 s. Images were resized to (256 × 256) pixels. Due to the fast acquisition time for both videos, the analysis was performed on every 5th image to ensure observable growth between two time points. Each image of Fig. 11.4 shows an overlay of the cell tip predicted by the model (green), the cell contour as detected by image segmentation (red), and the initial cell tip position (yellow) that marked the start of the growth cycle. In the first two rows, there is very little elongation or change in cell tip area. The last two rows show the most elongation of the cell.

Figure 11.5 shows a Receiver Operating Characteristic (ROC) plot of the average accuracy and standard deviation (error bars) of the predicted shape as the pixel acceptance threshold increases from [0–10] pixels for the first experiment

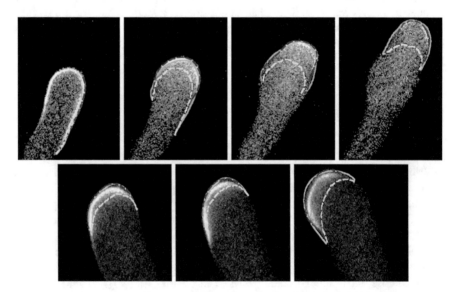

Fig. 11.4 Images showing the results of tip tracking for two experimental videos (*rows*) of growing pollen tubes. The number of images in each *row* denotes the number of detected growth cycles. Each image shows the initial tip shape (*yellow*), the models estimate of the final tip shape (*green*) and the observed tip shape (*red*). The first experiment shows straight growth while the second shows both straight and turning tip behavior. Please see Supplemental Materials for a video

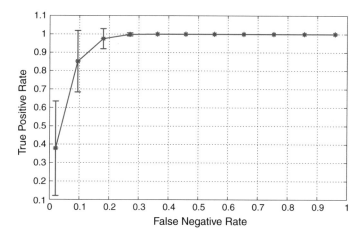

Fig. 11.5 ROC plot of average accuracy of predicted shape as pixel acceptance threshold increases from 0 to 10 pixels (0–0.58 μm). Error bars indicate standard deviation. Statistics are obtained for approximately 64 predicted shapes from experiment 1 in Fig. 11.4. Over 90 % accuracy is achieved within 0.1 μm (5.43 % of cell radius)

shown in Fig. 11.4. The statistics are obtained on 64 predicted shapes, i.e., each point in Fig. 11.5 shows the average accuracy across 64 trials when the pixel acceptance threshold is n-pixels wide. Points on the predicted tip shape are considered accurate if they are within a distance of n-pixels of an observed point. It is expected that as n increases, the number of accurate points will increase as well as the number of misclassified points. The figure shows that over 90 % accuracy is achieved at a distance of 3-pixels (0.1 μm or 5.43 % of cell radius).

Figure 11.6 shows the change in length over time of the growing pollen tube computed from the changing length of the deforming growth region (black) and the fitted curve (red) when using the Gompertz equation for cell wall weakening.

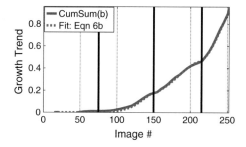

Fig. 11.6 Elongation trend in the pollen tube shown in Fig. 11.4, row 1. The *blue curve* is the cumulative sum of the measured affine parameter for elongation (**b**) and the *red curve* (*dotted*) shows the fit of Eq. (11.6) to the data for each growth cycle (between *vertical bars*). Curve agreement indicates that the Gompertz function is suitable for explaining observed cell wall dynamics

There are four growth segments representing row 1 of Fig. 11.4. The accuracy of the fitted curve indicates that the Gompertz equation is a good choice for determining cell wall dynamics.

11.5 Conclusions

In the above sections, we presented an intuitive model for tracking the tip of a pollen tube during growth. Instead of explicitly accounting for every known variable in the growth process, our approach focused on those variables that can be observed/measured in experimental videos of growing pollen tubes. In the development, we suggest suitable functions for propagating the shape of the cell over time. These functions are based on an understanding and simplification of the growth process. Given an initial tip shape, an affine transformation is performed to obtain subsequent tip shapes. The parameters of the transformation matrix rely on the biological processes that affect cell wall extensibility. Since the growth process is cyclical, the model searches for cyclical patterns in the fluorescence intensity signal and reinitializes the shape process when a new cycle is detected. The results show that our model can detect appropriate growth cycles within experimental videos and track cell tip changes within each cycle. This ability is important in quantifying the effects of various agents on the growth process.

Acknowledgment This work was supported in part by the National Science Foundation Integrative Graduate Education and Research Traineeship (IGERT) in Video Bioinformatics (DGE-0903667). Asongu Tambo is an IGERT Fellow.

References

1. Hill AE, Shachar-hill B, Skepper JN, Powell J, Shachar-hill Y (2012) An osmotic model of the growing pollen tube. PLoS One 7(5):e36585
2. Liu J, Piette BMAG, Deeks MJ, Franklin-Tong VE, Hussey PJ (2010) A compartmental model analysis of integrative and self-regulatory ion dynamics in pollen tube growth. PLoS One 5 (10):e13157
3. Campàs O, Mahadevan L (2009) Shape and dynamics of tip-growing cells. Curr Biol 19 (24):2102–2107
4. Kroeger JH, Geitmann A, Grant M (2008) Model for calcium dependent oscillatory growth in pollen tubes. J Theor Biol 253(2):363–374
5. Dumais J, Shaw SL, Steele CR, Long SR, Ray PM (2006) An anisotropic-viscoplastic model of plant cell morphogenesis by tip growth. Int J Dev Biol 50(2–3):209–222
6. Lockhart JA (1965) An analysis of irreversible plant cell elongation. J Theor Biol 8(2):264–275
7. Yang Z (2008) Cell polarity signaling in arabidopsis. Annu Rev Cell Dev Biol 24:551–575
8. Qin Y, Yang Z (2011) Rapid tip growth: insights from pollen tubes. Semin Cell Dev Biol 22 (8):816–824

 9. Steinhorst L, Kudla J (2013) Calcium: a central regulator of pollen germination and tube growth. Biochim Biophys Acta 1833(7):1573–1581
10. McKenna ST, Kunkel JG, Bosch M, Rounds CM, Vidali L, Winship LJ, Hepler PK (2009) Exocytosis precedes and predicts the increase in growth in oscillating pollen tubes. Plant Cell 21(10):3026–3040
11. Kroeger JH, Zerzour R, Geitmann A (2011) Regulator or driving force? The role of turgor pressure in oscillatory plant cell growth. PLoS One 6(4):e18549
12. Law R, Murrell D, Dieckmann U (2003) Population growth in space and time: spatial logistic equations. Ecology 84(1):252–262
13. Tsoularis A, Wallace J (2002) Analysis of logistic growth models. Math Biosci 179(1):21–55
14. Bove J, Vaillancourt B, Kroeger J, Hepler PK, Wiseman PW, Geitmann A (2008) Magnitude and direction of vesicle dynamics in growing pollen tubes using spatiotemporal image correlation spectroscopy and fluorescence recovery after photobleaching. Plant Physiol 147(4): 1646–1658

Chapter 12
Automatic Image Analysis Pipeline for Studying Growth in Arabidopsis

Katya Mkrtchyan, Anirban Chakraborty, Min Liu and Amit Roy-Chowdhury

Abstract The need for high-throughput quantification of cell growth and cell division in a multilayer, multicellular tissue necessitates the development of an automated image analysis pipeline that is capable of processing high volumes of *live imaging* microscopy data. In this work, we present such an image processing and analysis pipeline that combines cell image registration, segmentation, tracking, and cell resolution 3D reconstruction for confocal microscopy-based time-lapse volumetric image stacks. The first component of the pipeline is an automated landmark-based registration method that uses a *local graph*-based approach to select a number of landmark points from the images and establishes correspondence between them. Once the registration is acquired, the cell segmentation and tracking problem is jointly solved using an adaptive segmentation and tracking module of the pipeline, where the tracking output acts as an indicator of the quality of segmentation and in turn the segmentation can be improved to obtain better tracking results. In the last module of our pipeline, an adaptive geometric tessellation-based 3D reconstruction algorithm is described, where complete 3D structures of individual cells in the tissue are estimated from sparse sets of 2D cell slices, as obtained from the previous components of the pipeline. Through experiments on Arabidopsis

K. Mkrtchyan (✉)
Department of Computer Science, University of California, Riverside, CA, USA
e-mail: mkrtchyk@cs.ucr.edu

A. Chakraborty · M. Liu · A. Roy-Chowdhury
Department of Electrical Engineering, University of California, Riverside, CA, USA
e-mail: anirban.chakraborty@email.ucr.edu

M. Liu
e-mail: mliu009@ucr.edu

A. Roy-Chowdhury
e-mail: amitrc@ee.ucr.edu

© Springer International Publishing Switzerland 2015
B. Bhanu and P. Talbot (eds.), *Video Bioinformatics*,
Computational Biology 22, DOI 10.1007/978-3-319-23724-4_12

shoot apical meristems, we show that each component in the proposed pipeline provides highly accurate results and is robust to 'Z-sparsity' in imaging and low SNR at parts of the collected image stacks.

12.1 Introduction

The causal relationship between cell growth patterns and gene expression dynamics has been a major topic of interest in developmental biology. However, most of the studies in this domain have attempted to describe the interrelation between the gene regulatory network, cell growth, and deformation qualitatively. A proper quantitative analysis of the cell growth patterns in both the plant and the animal tissues has remained mostly elusive so far. Information such as rates and patterns of cell expansion plays a critical role in explaining cell growth and deformation dynamics and thereby can be extremely useful in understanding morphogenesis. The need for quantifying these biological parameters (such as cell volume, cell growth rate, cell shape, mean time between cell divisions, etc.) and observing their time evolution is, therefore, of utmost importance to biologists.

For complex multilayered, multicellular plant and animal tissues, the most popular method to capture individual cell structures and to estimate the aforementioned parameters for growing cells is the Confocal Microscopy-based Live Cell Imaging. Confocal Laser Scanning Microscopy (CLSM) enables us to visually inspect the inner parts of the multilayered tissues. With this technique, optical cross sections of the cells in the tissue are taken over with multiple observational time points to generate spatio-temporal image stacks. For high-throughput analysis of these large volumes of image data, development of fully automated image analysis pipelines are becoming necessities, thereby giving rise to many new automated visual analysis challenges.

The image analysis pipeline for gathering the cell growth and division statistics comprises of four main parts: image registration, cell segmentation, cell tracking, and 3D reconstruction. All four components of the pipeline encounter significant challenges that depend on the dataset characteristics, which make the problem a nontrivial image processing problem.

In spite of the extreme usefulness of CLSM-based live cell imaging for analyzing such tissue structures, there are number of technical challenges associated with this imaging technique that make the problem of automated cell growth estimation nontrivial. To keep the cells alive and growing, we have to limit the laser radiation exposure to the specimen, i.e., if dense samples in one time point are collected, it is highly unlikely that we will be able to get time-lapse images as the specimen will not continue to grow in time due to high radiation exposure. Therefore, the number of slices in which a cell is imaged is often low (2–4 slices per cell). Again, the fluorescent signal fades as we image the deeper layers of the tissue, thereby bringing in the problem of low SNR in parts of the confocal image stack.

Please note that in some cases, a two photon excitation microscopy or light sheet microscopy can be better choices for live cell imaging for more efficient light detection and less photo-bleaching effect. But, a large number of data sets exist that are imaged using CLSM or exhibit the characteristic of our data, and our method can be useful in analyzing them. We have found that two photon excitations are more toxic to SAM (Shoot Apical Meristem) cells than the single photon CLSM and since the SAM is surrounded by several developing flower buds, the side ward excitation may not be possible. Also, by designing an image-analysis method that is capable of handling the worse quality data, we can ensure that the same or better accuracy can be achieved on a dataset having superior image quality and resolution.

In this work, we have looked at the problem of image registration, cell segmentation and tracking, and 3D reconstruction of a tightly packed multilayer tissue from its Z-sparse confocal image slices. As a special example, in this work, we apply our proposed pipeline in analyzing large volumes of image data for Shoot Apical Meristem (SAM) of Arabidopsis Thaliana. As hinted in [1], this pipeline can be applied on root meristem as well. SAM, also referred to as the stem cell niche, is a important part of a plant body plan because it supplies cells for all the above ground plant parts such as leaves, branches, and stem. A typical Arabidopsis SAM is a densely packed multilayered cell cluster consisting of about five hundred cells where the cell layers are clonally distinct from one another. Using CLSM technique, time lapse 3D image stacks are captured which are essentially 4D video data. Our objective in this work is to extract quantifiable parameters describing growth from this large volume of video data. Therefore, it is one of the most important video bioinformatics problems from the plan sciences perspective.

12.2 Method

As discussed above, the main components of the pipeline are image registration, segmentation of the tissue into individual cellular regions, spatio-temporal cell tracking, and cell resolution 3D reconstruction. The flow of the pipeline is shown in Fig. 12.1. We are going to discuss the individual components of the pipeline in the next subsections.

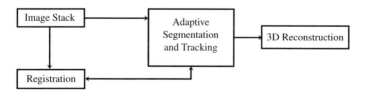

Fig. 12.1 The general workflow in the image analysis pipeline

12.2.1 Registration

12.2.1.1 Motivation and Related Work

In practice, the live cell imaging of a plant comprises of several steps, where the plant has to be physically moved between different places. For normal growth of the plant, it has to be kept in a place having specific physical conditions (temperature of 24 °C). The plant is moved and placed under microscope at the imaging/observational time points, before it is placed back in the aforementioned place once again. For 72 hours overall, this process is repeated for every 3 hours. Because of this process of replacement of the plant under the microscope and also since the plant keeps growing during these 72 h, various shifts can occur between two Z-stacks of images taken in consecutive time points, though images in any Z-stack are automatically registered. So, to get the accurate and longer cell lineages and cell division statistics, image stacks should be aligned.

Recently, there has been some work done on SAM cells [2], where cells are segmented by watershed algorithm [3] and tracked by local graph matching method. The method in [2] was constrained to focus on datasets that are approximately registered. Therefore, registration is of utmost importance to be able to work with varied datasets. Popular registration method based on maximization of the mutual information [4, 5], fails to provide accurate registration as it uses the pixel intensities to acquire the registration. Pixel intensities in the Arabidopsis SAM images are not discriminative features. The landmark-based methods are more suitable to register such images. A recent paper [6] uses SAM images acquired from multiple angles to automate tracking and modeling. For pair of images to be registered, the user identified correspondences by pairing a few anchor points (referred as landmark points in this work). In this section, we present a fully automated landmark-based registration method that can find out correspondences between two images and utilize these correspondences to yield a better registration result. In the registration results from Sect. 12.2.1.3, we show that landmark-based registration is more suitable for noisy and sparse confocal images, than registration based on maximization of the mutual information.

The most common landmark-based registration algorithm is the Iterative Closest Point (ICP) algorithm [7], where a set of landmark point pair correspondences are constructed between two images and then the images are aligned so as to minimize the mean square error between the correspondences. The ICP algorithm is sensitive to initialization; it provides a good estimate of the correct correspondence when the images are approximately aligned with each other. There are different additions to the basic ICP algorithm, e.g., Iterative Closest Point using Invariant Features (ICPIF) [8], which uses features like eccentricity and curvature to overcome the issue. But in Arabidopsis SAM, because densely packed cells have similar features, the eccentricity, curvature, and other common features such as shape, color, etc., are not discriminative enough to be used for registration. Thus, available landmark-based registration approaches may not be able to properly align the SAM images. This is

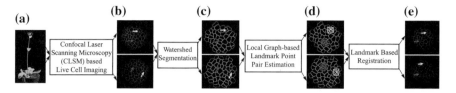

Fig. 12.2 Registration methodology in sparse confocal image stacks—**a** SAM located at the tip Arabidopsis shoot, **b** raw images taken at consecutive time instances, **c** segmented images after applying watershed segmentation, **d** estimation of the corresponding landmark point pairs, and **e** bottom image is registered to the top image (the *same color arrows* represent the same cell)

why we need to develop a novel feature that can be used to register SAM images. The proposed landmark estimation method uses features of the local neighborhood areas to find corresponding landmark pairs for the image registration [9].

In this section, we present an automatic landmark-based registration method for aligning Arabidopsis SAM image stacks. The flow of the proposed registration method is described in Fig. 12.2. In the rest of this subsection, we call the image that we wish to transform as the input image, and the reference image is the image against which we want to register the input.

12.2.1.2   Detailed Registration Method

Landmark Identification There is accumulated random shift, rotation, or scaling between the images taken at different time points. The performance of tracking is affected by the registration. The quality of the image registration result depends on the accuracy of the choice of landmark points. Common features such as shape, color, etc., cannot be used to choose corresponding landmark pairs. Motivated by the idea presented in [2], we use the relative positions and ordered orientation of the neighboring cells as unique features. To exploit these properties, we represent these local neighborhood structures as graphs and select the best candidate landmark points that have the minimum distance between the local graphs built around them.

Local graphs as features Graphical abstraction is created on the basos of collection of cells. Vertices in the graph are the centers of the cells and neighboring vertices are connected by an edge. Neighborhood set $N(C)$ of a cell C contains the set of cells that share a boundary with C. Thus, every graph consists of a cell C and a set of clockwise ordered neighboring cells (Fig. 12.3a, d). The ordering of the cells in $N(C)$ is important because under nonreflective similarity transformation, the absolute positions of the neighboring cells could change but the cyclic order of the cells remains invariant.

Landmark point pair estimation from local graphs Cell divisions happen throughout the 72 h intervals but at the consecutive images, taken every 3 h apart, only several cell divisions are present. Ideally, in the areas where there is no cell division, the local graph topology should not change (segmentation errors will circumvent this in practice). We exploit these conditions to find the corresponding

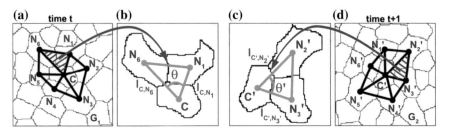

Fig. 12.3 a, d The local graphs G_1 and G_2 at time t and $t+1$ with the central cells C and C', respectively, and clockwise ordered neighboring cell vertices $N_1, \ldots, N_6$ and $N'_1, \ldots, N'_6$, **b, c** two enlarged triangle subgraphs with indicated features N_{i_1}, N_{i_2}—i_1th and i_2th neighboring cells of C, N'_{j_1}, N'_{j_2}—j_1th and j_2th neighboring cells of C', $\theta_{N_{i_1}, C, N_{i_2}}(t)$—angle between $\overline{N_{i_1}C}$ and $\overline{N_{i_2}C}$, $\theta_{N'_{j_1}, C', N'_{j_2}}(t)$—angle between $\overline{N'_{j_1}C'}$ and $\overline{N'_{j_2}C'}$, $l_{C,N_{i_1}}(t), l_{C,N_{i_2}}(t)$—neighbor edge lengths, $l_{C',N'_{j_1}}(t+1), l_{C',N'_{j_1}}(t+1)$—edge lengths, $A_{N_{i_1}}(t), A_{N_{i_2}}(t)$—areas of the cells N_{i_1}, N_{i_2} $A_{N'_{j_1}}(t+1)$, $A_{N'_{j_2}}(t+1)$—areas of the cells N'_{j_1}, N'_{j_2}

landmark pairs in two images. Let, $G_1^{(t)}$ and $G_2^{(t+1)}$ be two local graphs constructed around the cells C and C' in consecutive temporal slices (Fig. 12.3). For each triangle subgraph of the local graph $G(t)$, we define feature vector of five components where the components are the angles between the edges (Fig. 12.3 $\theta_{N_{i_1}, C, N_{i_2}}(t)$), lengths of the neighboring edges (Fig. 12.3 $l_{C,N_{i_1}}(t), l_{C,N_{i_2}}(t)$), and the areas of the neighboring cells (Fig. 12.3 $A_{N_{i_1}}(t), A_{N_{i_2}}(t)$). Having these features defined on the subgraphs, we compute distance between two triangle subgraphs by considering summation of the square of the normalized differences of the corresponding components of feature vectors. To ensure that our landmark estimation method takes care of the rotation of local area, we consider all cyclic permutations of the clockwise ordered neighbor set of the cell C' from the input image. Then, we define distance between the local graphs for the fixed permutation k of the neighboring set of the cell $C', D(G_1, G_2^k)$, as the summation of the distance of triangle subgraphs. Finally, the distance $D^{\star}(G_1, G_2)$ between two graphs G_1 and G_2 corresponding to cells (C, C') for all permutations k is

$$D^{\star}(G_1, G_2) = D(G_1, G_2^{k^{\star}}), \tag{12.1}$$

where $k^{\star} = \arg_k \min D(G_1, G_2^k), k \in \{0, 1, \ldots, (m-1)\}$. This guarantees that our landmark estimation method is invariant of the rotation in the local area.

For all cell pairs C_i, C'_j and corresponding graphs G_i, G_j from two consecutive images, we compute the distance $D^{\star}(G_i, G_j)$. Then, the cell pairs are ranked according to their distances $D^{\star}$ and the top q cell pairs are chosen as landmark point pairs. The choice of q is described later.

Image Registration Once we have the landmark point pairs corresponding to the reference and input images, we find the spatial transformation between them.

Finding the nonreflective similarity transformation between two images is a problem of solving a set of two linear equations. As mentioned before, for better accuracy of transformation parameters, the top q landmark point pairs are used in a least-square parameter estimation framework. The choice of q depends on the quality of the input and base image as choosing more landmark point pairs, generally increases the risk of having more false positive landmark point pairs. In our experiments, we choose four, five, or six landmark pairs depending on the dataset image quality.

12.2.1.3 Registration Results

We have tested our proposed automatic landmark-based registration method, combined with the watershed segmentation [3] and local graph matching-based tracking [2]. We compared tracking results of the proposed method with results obtained without registration, with semi-automated registration (the landmark pairs are chosen manually, the transformation is obtained automatically) and with MIRIT software [4].

Figure 12.4a–e shows cell tracking results from two consecutive images (30th and 36th hour). The results with MIRIT registration and without registration show incorrect cell tracks. Whereas, the proposed method and semi-automated registration correctly registered two images with 100 % correct tracking results. Detailed results for the same dataset are shown in Fig. 12.4f, g. In Fig. 12.4f, we can see that from 33 and 27 cells, present in the images at time points five (30th hour) to six (36th hour), respectively, none are tracked by the tracker run on the images registered with the MIRIT software and not registered images (as in Fig. 12.4a–e). The same result is seen for the tracking results in images at time points 6–7. But the tracking results obtained with proposed and semi-automated methods provided results close to manual results. Figure 12.4g shows lengths of the cell lineages calculated with the proposed method, semi-automated registration, MIRIT registration, and without registration. We can see that in tracking without registration and after registration with MIRIT software, there are no cells that have lineage lengths greater then four as opposed to the case with the proposed and semi-automated registration, where cells have lineages for the entire 72 h. The reason for such results is that there is a big shift between two images from consecutive time points, in the middle time points. Without proper registration, the tracking algorithm is not able to provide correct cell correspondence results, which interrupts the lineage of the cells. The result in Fig. 12.4g can also be related to Fig. 12.4f. Since no cells have been tracked in frames five to six, and overall there are eleven frames, then no cell can have a lineage life with the length greater than or equal to five. Table 12.1 shows the number of cell divisions in 72 h. We can see that the semi-automated and the proposed registration provide results that are close to the manual results as opposed to without registration and MIRIT software.

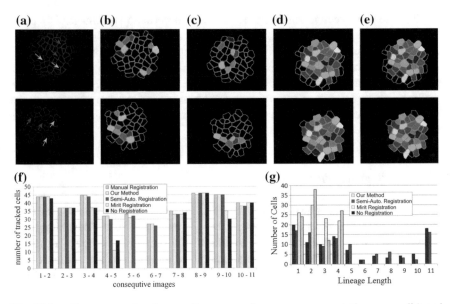

Fig. 12.4 a Raw consecutive images (the *same color arrows* represent the same cells) and tracking results obtained **b** without registration **c** with MIRIT registration, **d** with manual registration, and **e** with proposed automatic registration. The *same colors* represent the same cell. **f** Number of tracked cells across two consecutive images. **g** Length of cell lineages

Table 12.1 Total number of cell divisions/ground-truth

Data	Our method	Manual	MIRIT [4]	No registration
1	28/34	30/34	23/34	25/34
2	17/21	17/21	11/21	12/21

12.2.2 Segmentation and Tracking

12.2.2.1 Motivation and Related Work

The manual tracking of cells through successive cell divisions and gene expression patterns is beginning to yield new insights into the process of stem cell homeostasis [10]. However, manual tracking is laborious and impossible as larger and larger amounts of microscope imagery are collected worldwide. More importantly, manual analysis will not provide quantitative information on cell behaviors, besides cell-cycle length and the number of cell divisions within a time period. There are significant challenges in automating the segmentation and tracking process. The SAM cells in a cluster appear similar with few distinguishing features, cell division event changes the relative locations of the cells, and the live images are inherently noisy.

There has been some work on automated tracking and segmentation of cells in time-lapse images, both plants and animals. One of the well-known approaches for segmenting and tracking cells is based on level sets [11–13]. However, the level set method is not suitable for tracking of SAM cells because the cells are in close contact with each other and share similar physical features. The Softassign method uses the information on point location to simultaneously solve both the problem of global correspondence as well as the problem of affine transformation between two time instants iteratively [14–16]. A recent paper [6] uses SAM images acquired from multiple angles to automate tracking and modeling. Since SAMs are imaged from multiple angles, it imposes a limitation on the temporal resolution. This precludes a finer understanding of spatial-temporal dynamics through dynamical modeling. In an earlier study, we have used level set segmentation and local graph matching method to find correspondence of cells across time points by using live imagery of plasma membrane labeled SAMs [2] imaged at 3 h intervals. However, this study did not make an attempt to integrate segmentation and tracking so as to minimize the segmentation and tracking errors, which are major concerns in noisy live imagery. Here, we have combined the local graph matching-based tracking methodology from [2] with the watershed segmentation in an adaptive framework in which tracking output is integrated with the segmentation (Fig. 12.5) [17]. In our recent works [18, 19], we provided new approaches for the cell tracking problem.

The local graph matching-based tracking methodology from [2] was combined with the watershed segmentation in an adaptive framework in which tracking output is integrated with the segmentation. The system diagram is also shown (Fig. 12.5).

12.2.2.2 Detailed Segmentation and Tracking Method

We used watershed transformation [20, 21] to segment cell boundaries. Watershed treats the input image as a continuous field of basins (low-intensity pixel regions), barriers (high-intensity pixel regions), and outputs the barriers that represent cell boundaries. It has been used to segment cells of Arabidopsis thaliana root meristem [22]. It outperforms the level set method in two aspects. On the one hand, it reaches more accurate cell boundaries. On the other hand, it is faster, which provides the opportunity to implement it in an adaptive framework efficiently. However, the main drawback is that it results in both over-segmentation and under-segmentation of cells, especially those from deeper layers of SAMs that are noisy. So, prior to

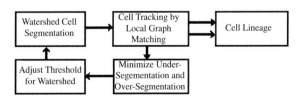

Fig. 12.5 The overall diagram of the adaptive cell segmentation and tracking scheme

applying the watershed algorithm, the raw confocal microscopy images undergo *H*-minima transformation in which all the pixels below a certain threshold percentage *h* are discarded [23]. The *H*-minima operator was used to suppress shallow minima, namely those whose depth is lower than or equal to the given *h*-value. The watershed segmentation after the *H*-minima operator with a proper threshold can produce much better segmentation results than level set segmentation.

Since the *H*-minima threshold value *h* plays a crucial role in the watershed algorithm, especially when the input images are noisy, it is extremely important to choose an appropriate threshold value such that only the correct cell boundaries are detected. Generally, a higher value of the threshold parameter *h* performs under-segmentation of the images and, conversely, a lower value over-segments the images. The numbers of over-segmented cells in a noisy image slice increases as we choose lower values for the *H*-minima threshold *h* and, on the other hand, a larger value of *h* produces more under-segmented cells. Since the cell size is fairly uniform for most cells of the SAM, the watershed should ideally produce a segmented image that contains similarly sized cells. Thus, a good starting threshold could be the value of *h* such that variance of cell areas in the segmented image is minimal. This optimal value of h is what we are trying to obtain, as will be explained later.

The tracker performance depends heavily on the quality of the segmentation output. However, due to a low Signal-to Noise Ratio (SNR) in the live cell imaging dataset, the cells are often over- or under-segmented. Therefore, the segmentation and tracking have to be carried out in an integrated and adaptive fashion, where the tracking output for a particular slice acts as an indicator of the quality of segmentation and the segmentation can be improved so as to obtain the best tracking result.

Design of Integrated Optimization Function Due to the rapid deterioration of image quality in deeper layers of the *Z*-stack, the existing segmentation algorithms tend to under-segment or over-segment image regions (especially in the central part of the image slices). Even a manual segmentation of cells is not always guaranteed to be accurate if each slice in the deeper layers is considered separately due to low SNR. In fact, in such cases, we consider the neighboring slices, which can provide additional contextual information to perform segmentation of the noisy slice in a way that provides the best correspondence for all the segmented cells within the neighborhood. The automated method of integrated segmentation and tracking proposed here, involves correcting faulty segmentation of cells by integrating their spatial and temporal correspondences with the immediate neighbors as a feedback from the tracking to the segmentation module. In the next few paragraphs, we formalize this framework as a spatial and temporal optimization problem and elaborate the proposed iterative solution strategy that yields the best segmentation and tracking results for all the cell slices in the 4D image stack.

The advantage of using watershed segmentation is that it can accurately find the cell boundaries, while its main drawback is over-segmentation and under-segmentation, which can be reduced by choosing the proper *H*-minima threshold. Due to the over-segmentation errors in regions of low SNR, the watershed algorithm often tends to generate spurious edges through the cells. In the cases in which a cell is imaged at multiple slices along the Z-stack and is over-segmented in one of the slices, the tracker can identify this over-segmentation error as a spurious 'spatial cell-division' event. Clearly, this is not legitimate and is a result of faulty segmentation. Additionally, cell-merging in the temporal direction (again an impossible event) can arise from under-segmentation, where the watershed algorithm fails to detect a legitimate edge between two neighboring cells. The intuition behind the proposed method in this paper is to reduce the over-segmentation errors by minimizing the spurious spatial cell divisions and reduce the under-segmentation errors by minimizing the number of merged cells. Specifically, for frame S_k^t the kth image slice at time point t, we are going to minimize the number of spurious 'cell divisions' between it and its spatial neighbor S_{k-1}^t and the number of spurious cell-merging events in S_k^t from its temporal predecessor S_k^{t-1}, as shown in Fig. 12.6. (Although it may be possible to identify that such spurious events have happened through simple rules disallowing a cell division in the Z-direction or a merging in the forward temporal direction, correcting for them is a lot harder, as the structure of the collection of cells needs to be maintained. Our approach will allow detection of not only such spurious events, but also their correction.)

The optimization goal here is to minimize the number of spurious cell divisions $\left(N_{(k-1,k)}^t\right)$ for the frame S_k^t from its upper slice S_{k-1}^t and the number of cell-merging events $N_k^{(t,t-1)}$ in S_k^t from its previous slice. The error caused by over-segmentation, $\left(N_{(k-1,k)}^t\right)$ monotonically decreases with the increment in threshold h, whereas the error due to under-segmentation, $\left(N_k^{(t,t-1)}\right)$ monotonically increases with h. Hence, the optimal segmentation result can be obtained through finding the value of h for

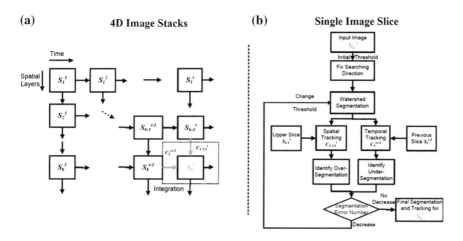

Fig. 12.6 Optimization Scheme. **a** Schematic showing how to integrate the spatial and temporal trackers for 4D image stacks. **b** Adaptive segmentation and tracking scheme for a certain image slices S_k^t (the kth slice at the t time point)

which the summation $(N_k^{(t,t-1)} + N_{(k-1,k)}^t)$ attains a minimum. The cost function $(N_k^{(t,t-1)} + N_{(k-1,k)}^t)$ is essentially an indicator of the overall error in segmentation (combining both over and under-segmentation) and can be optimized by varying the H-minima threshold h for S_k^t. Formally, the optimal value h_k^t is found as a solution to the following optimization problem:

$$h_k^t = \min_h (N_k^{(t,t-1)}(h) + N_{(k-1,k)h}^t)$$

With the variation of h_k^t (by either increasing or decreasing), the cost function decreases to a minimum (ideally 0). The threshold h_k^t for which the cost function attains this minimum is the optimum value of the threshold for H-minima transformation.

12.2.2.3 Segmentation and Tracking Results

We have tested our proposed adaptive cell segmentation and lineage construction algorithm on two SAM datasets. Datasets consist of 3D image stacks taken at 3 h intervals for a total of 72 h (24 data points). Each 3-D stack consists of 30 slices in one stack, so the size of the 4D image stack is $888 \times 888 \times 30 \times 24$ pixels. After image registration, we used the local graph matching-based algorithm to track cells, the detailed information about which can be found in [2]. We demonstrate that, by integrating this within the proposed adaptive scheme, we are able to obtain significantly better results. The adaptive segmentation and tracking method is run on every two consecutive images in both the spatial and temporal directions. The thresholds of H-minima for images in the given 4D image stack are determined sequentially along the direction of the arrows shown in Fig. 12.6a. In the segmentation, we normalized the image intensities in the range of [0 1] and set the searching range for the optimal H-minima threshold h in [0.005 0.09]. The step size used in the search was ± 0.005, and the sign depends on the search direction ($h_k^t > h_{k(\text{init})}^t$ or $h_k^t < h_{k(\text{init})}^t$). We manually verified the accuracy of the cell lineages obtained by the proposed algorithm.

Figure 12.7 is a typical example of the segmentation result and tracking result using our proposed adaptive method, with seven simultaneously dividing cells being detected. In Fig. 12.8, the segmentation and tracking results in selected 3D image stacks along three time instances (6, 9, and 15 h) are shown. The tracker is able to compute both spatial and temporal correspondences with high accuracy as a result of improved segmentation.

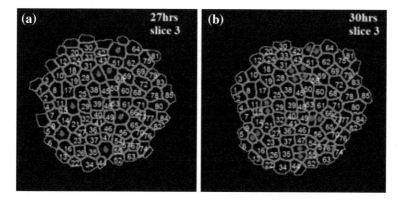

Fig. 12.7 The segmentation and tracking results using adaptive method

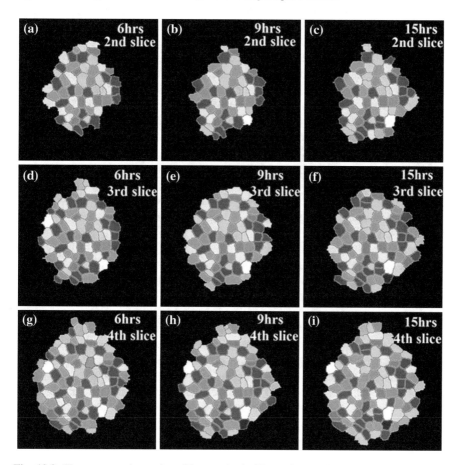

Fig. 12.8 The segmentation and tracking results in 3D stacks at selected time instances. The segmented cells shown in *same color* across consecutive slices (second, third, and fourth slices) represent same cells

12.2.3 3D Reconstruction

12.2.3.1 Motivation and Related Work

There are several methods of shape and size estimations for individual cells such as impedance method [24] and light microscopy methods [25]. Methods such as [26] are used to study changes in cell sizes found in cell monolayers. In live plant tissues, a number of works focused on the surface reconstruction [27, 28]. But we are looking at a much more challenging problem, where the subject of study is a dense cluster of cells. Plant meristem is one example of such cell clusters where hundreds of small cells are densely packed into a multilayer structure. In such cases, now-a-days, the most popular practice is to use Confocal Laser Scanning Microscopy (CLSM) to image cell or nucleus slices at a high spatial resolution and then reconstruct the 3D volume of the cells from those serial optical slices which have been shown to be reasonably accurate [6, 29, 30].

A recent method [6] accurately reconstructs the Shoot Apical Meristem of Arabidopsis. This method uses a dataset containing fine slice images acquired from three different angles, each at a Z-resolution of 1 μm. They have reported 24 h as the time resolution in imaging. But, for analyzing the growth dynamics of cell clusters where the time gap between successive cell divisions is in the range of 30–36 h, we need a much higher time resolution in imaging, in order to capture the exact growth dynamics. To obtain longer cell lineages at high time resolution, we may have to sacrifice the spatial or depth resolution and hence the number of image slices in which a cell is present which can be really small. With such a limited amount of image data, the existing 3D reconstruction/segmentation techniques cannot yield a good estimate of cell shape. In the present work, we have addressed this problem of reconstructing plant cells in a tissue when the number of image slices per cell is limited.

There is a basic difference between the segmentation problem at hand and a classical 3D segmentation scheme. A classical method solves the segmentation problem using the pixel intensities and cannot work when no intensity information is provided for a majority of 3D pixels in the image. In such situations, the most intuitive way to perform the segmentation is to first segment sections in the image with known intensity information using a classical segmentation scheme, and then to extrapolate between these sparse segments using a known geometric model or a function which could be generic or data-specific. In [31], we have shown that 3D shapes of individual cells in the SAM can be approximated by a deformed truncated ellipsoidal model. In a recent work [1], we explain how [31] can be described as a special case of the Mahalanobis distance-based Voronoi tessellation under specific conditions and in that respect, the present method is a generalization of the work in [31]. In our work, we have shown that a quadratic Voronoi tessellation is an accurate choice for such a geometric model for segmenting a tissue with anisotropically growing tightly packed cells starting with a sparse set of 2D Watershed segmented slices per cell.

The Shoot Apical Meristem is a multilayer and multicellular structure where the cells are tightly packed together with hardly any void in between. Motivated by this physical structure of SAM, we propose our novel cell resolution 3D reconstruction in a geometric tessellation framework. A tessellation is a partition of a space into closed geometric regions with no overlap or gap among these regions. In case of the SAM tissue, each cell is represented by such a closed region and any point in the 3D SAM structure must be the part of one and only one cell. In fact, there are some recent works in the literature as [32] which predicted that the 3D structures of Arabidopsis SAM cells could be represented by convex polyhedrons forming 3D 'Voronoi' tessellation pattern.

A Voronoi tessellation is one of the simplest form of partitioning of the metric space, where the boundaries between two adjacent partitions are equidistant from a point inside each of these regions, also known as the 'sites'. In [32, 33], these sites are the approximate locations of the center of the cell nuclei about which the tissue is tessellated into individual cells. However, this work used a dataset where both the plasma membrane as well as the nucleus of each cell is marked with fluorescent protein, whereas, in our case, only the plasma membrane is visible under the confocal microscope.

In [1], we presented and evaluated a fully automated cell resolution 3D reconstruction framework for reconstructing the Arabidopsis SAM where the number of confocal image slices per cell is limited. The framework comprises of different modules such as cell segmentation, spatial cell tracking, SAM surface reconstruction, and finally a 3D tessellation module. We proposed a quadratic distance-based anisotropic Voronoi tessellation, where the distance metric for each cell is estimated from the segmented and tracked sparse data points for the cell. This method is applicable to the densely packed multicellular tissues and can be used to reconstruct tissues without voids between cells with sufficiently high accuracy. Note that, for the proposed 3D reconstruction module, we start with a handful of data points on each segmented cell which are pre-clustered through the cell tracking method (i.e., an incomplete segmentation of the 3D) whereas, the final output of our algorithm is a complete tessellation of the entire 3D structure of the SAM, where each cell is represented by a dense point cloud. These point clouds for individual cells can be visualized by 3D convex polyhedrons that approximate the shape of the cells.

12.2.3.2 Detailed 3D Reconstruction Method

Generation of Dense Point Cloud to Be Partitioned Into Cells: Global Shape of SAM At this stage, we estimate the 3D structure of the SAM by fitting a smooth surface to its segmented contours. The surface fitting is done in two steps. In step one, the SAM boundary in every image slice is extracted using the 'Level Set' method. A level set is a collection of points over which a function takes on a constant value. We initialize a level set at the boundary of the image slice for each SAM cross section, which behaves like an active contour and gradually shrinks toward the boundary of the SAM. Let the set of points on the segmented SAM contours be $P^{SAM}(\{x^{SAM}, y^{SAM}, z^{SAM}\})$.

In the second step, we fit a surface on the segmented points P^{SAM}. Assuming that the surface can be represented in the form $z = f(x, y)$ (where the function f is unknown), our objective is to predict z at every point (x, y) on a densely sampled rectangular grid of points bounded by $\left[x_{min}^{SAM}, y_{min}^{SAM}, x_{max}^{SAM}, y_{max}^{SAM}\right]$. As the segmented set of data points are extremely sparse, this prediction is done using a linear interpolation on a local set of points on the grid around the point (x, y). As the value (z) for the point (x, y) is approximated by a linear combination of the values at a few neighboring points on the grid, the interpolation problem can be posed as a linear least-square estimation problem. We also impose a smoothness constraint in this estimation by forcing the first partial derivatives of the surface evaluated at neighboring points to be as close as possible.

Once the SAM surface (S^{SAM}) is constructed, we uniformly sample a dense set of 3D points (P_{dense}) such that every point in P_{dense} must lie inside S^{SAM}. Thus, $P_{dense} = \{\mathbf{x}_1, \mathbf{x}_2, \ldots \mathbf{x}_N\}$ and the required output from the proposed algorithm is a clustering of these dense data points into C cells/clusters such that $P_{dense} = \{\hat{P}_{dense}^{(1)}, \hat{P}_{dense}^{(2)}, \ldots \hat{P}_{dense}^{(C)}\}$ starting from the sparse set of segmented and tracked points $P_{sparse} = \{P_{sparse}^{(1)}, P_{sparse}^{(2)}, \ldots P_{sparse}^{(C)}\}$ obtained from the confocal slice images of individual cells.

An Adaptive Quadratic Voronoi Tessellation (AQVT) for Nonuniform Cell Sizes and Cell Growth Anisotropy: In a tissue like SAM, cells do not grow uniformly along all three axes (X, Y, Z). In fact, most of the cells show a specific direction of growth. Again, neighboring cells in SAM, especially in the central region (CZ), are not likely to grow along the same direction. Thus, even if a tessellation is initially an affine Voronoi diagram, it is not likely to remain so after a few stages of growth. Such cases of nonuniform cell sizes and anisotropic growth can be captured in a more generalized nonaffine Voronoi tessellation called the 'Anisotropic Voronoi Diagrams.' In the most general form of such diagram for point sites, the distance metric has a quadratic form with an additive weight [34].

Following similar notations used in previous paragraphs, for a set of anisotropic sites $\mathscr{S} = \{\mathbf{s}_1, \mathbf{s}_2, \ldots \mathbf{s}_n\}$ in $\mathbb{R}^d$, the anisotropic Voronoi region for a site $\mathbf{s}_i$ is given as

$$V_A(\mathbf{s}_i) = \left\{\mathbf{x} \in \mathbb{R}^d | d_A(\mathbf{x}, \mathbf{s}_i) \leq d_A(\mathbf{x}, \mathbf{s}_j) \quad \forall j \in \{1, 2, \ldots n\}\right\}, \tag{12.2}$$

where

$$d_A(\mathbf{x}, \mathbf{s}_i) = (\mathbf{x} - \mathbf{s}_i)^T \Sigma_i (\mathbf{x} - \mathbf{s}_i) - \omega_i \tag{12.3}$$

Σ_i is a $d \times d$ positive definite symmetric matrix associated with the site $\mathbf{s}_i$ and $\omega_i \in \mathbb{R}$. Thus, each of the anisotropic Voronoi regions is parameterized by the triplet $(\mathbf{s}_i, \Sigma_i, \omega_i)$. Further assuming $\omega_i = \omega_j \ \forall i, j \in \{1, 2, \ldots n\}$, the distance function becomes

$$d_Q(\mathbf{x}, \mathbf{s}_i) = (\mathbf{x} - \mathbf{s}_i)^T \Sigma_i (\mathbf{x} - \mathbf{s}_i) \tag{12.4}$$

As the bisectors of such a Voronoi diagram are quadratic hypersurfaces, these diagrams are called 'Quadratic Voronoi Diagrams', wherein every Voronoi cell i is parameterized by $(\mathbf{s}_i, \Sigma_i)$ pairs.

$$V_Q(\mathbf{s}_i) = \left\{ \mathbf{x} \in \mathbb{R}^d \,|\, d_Q(\mathbf{x}, \mathbf{s}_i) \leq d_Q(\mathbf{x}, \mathbf{s}_j) \quad \forall j \in \{1, 2, \ldots n\} \right\} \tag{12.5}$$

From Eq. 12.4, it can be observed that Σ_i is essentially a weighting factor that nonuniformly weights distances in every Voronoi regions along every dimension. When all the Voronoi regions are equally and uniformly weighted along every axis, $\Sigma_i = I_{dxd} \ \forall i = 1, 2, \ldots n$ and the resulting diagram for point sites becomes an Euclidean distance-based Voronoi diagram.

Estimating the Distance Metric from Sparse Data: Minimum Volume Enclosing Ellipsoid Now, the problem at hand is to estimate the parameter pair for each cell/quadratic Voronoi regions from the sparse data points, as obtained from the segmented and tracked slices, that belongs to the boundary of each cell. Given the extreme sparsity of the data, there is no available method that would provide Σ_is for each region. We, in this work, propose an alternative way of estimating $(\mathbf{s}_i, \Sigma_i)$ pairs directly from the sparse data points. The motivation of this estimation strategy can be found in [1].

After registration, segmentation and identification of a cell in multiple slices in the 3D stack, we can obtain (x, y, z) co-ordinates of the set of points on the perimeter of the segmented cell slices. Let this set of points on the cth cell be $P^{(c)}_{\text{sparse}} = \{p_1, p_2, \ldots p_k\} \in \mathbb{R}^3$. We estimate the minimum volume ellipsoid which encloses all these k points in $\mathbb{R}^3$ and we denote that with $\mathscr{E}$. An ellipsoid in its center form is represented by

$$\mathscr{E}(s, \Sigma) = \{p \in \mathbb{R}^3 \,|\, (p - s)^T \Sigma (p - s) \leq 1\}, \tag{12.6}$$

where $s \in \mathbb{R}^3$ is the center of the ellipsoid $\mathscr{E}$ and $\Sigma \in \mathbb{R}^{3 \times 3}$. Since all the points in $P^{(c)}_{\text{sparse}}$ must reside inside $\mathscr{E}$, we have

$$(p_i - s)^T \Sigma (p_i - s) \leq 1 \quad \text{for } i = 1, 2, \ldots k \tag{12.7}$$

and the volume of this ellipsoid is

$$\text{Vol}(\mathscr{E}) = \frac{4}{3} \pi \{\det(\Sigma)\}^{-\frac{1}{2}} \tag{12.8}$$

Therefore, the problem of finding the Minimum Volume Enclosing Ellipsoid (MVEE) for the set of points $P_{\text{sparse}}^{(c)}$ can be posed as

$$
\begin{aligned}
&\min_{\Sigma, s} \; -\log \det(\Sigma) \\
&\text{s.t. } (p_i - s)^T \, \Sigma(p_i - s) \le 1 \text{ for } i = 1, 2, \ldots k \\
&\qquad \Sigma \succ 0
\end{aligned}
\tag{12.9}
$$

To efficiently solve Problem 12.9, we convert the primal problem into its dual problem since the dual is easier to solve. A detailed analysis on the problem formulation and its solution can be found in [35, 36]. Solving this problem individually for each sparse point set $P_{\text{sparse}}^{(1)}, P_{\text{sparse}}^{(2)}, \ldots P_{\text{sparse}}^{(C)}$, the parameters of the quadratic distance metrics are estimated as $\{(\hat{s}_1, \hat{\Sigma}_1), (\hat{s}_2, \hat{\Sigma}_2), \ldots (\hat{s}_C, \hat{\Sigma}_C)\}$.

As we are estimating the parameters of quadratic distance metric associated with every individual Voronoi cell separately and then using the distance metrics, thus obtained, to tessellate a dense point cloud, we choose to call the resulting Voronoi tessellation as the 'Adaptive Quadratic Voronoi Tessellation' (AQVT).

3D Tessellation Based on the Estimated Parameters of AQVT: The Final Cell Shapes As soon as the parameters of the quadratic distance metrics are estimated from the previous step, the dense point cloud P_{dense} can be partitioned into different Voronoi regions based on Eq. (12.8), i.e., the dense point cloud belonging to cell c is given as

$$
\hat{P}_{\text{dense}}^{(c)} = \left\{ \mathbf{x} \in P_{\text{dense}} \,\middle|\, (\mathbf{x} - \hat{\mathbf{s}}_c)^T \hat{\Sigma}_c (\mathbf{x} - \hat{\mathbf{s}}_c) \le (\mathbf{x} - \hat{\mathbf{s}}_j)^T \hat{\Sigma}_j (\mathbf{x} - \hat{\mathbf{s}}_j) \quad \forall j \in \{1, 2, \ldots C\} \right\}
\tag{12.10}
$$

For visualization purpose of the cell resolution 3D reconstruction results, we fit convex polyhedrons to $\hat{P}_{\text{dense}}^{(1)}, \hat{P}_{\text{dense}}^{(2)}, \ldots \hat{P}_{\text{dense}}^{(C)}$ to represent each cell (Fig. 12.9).

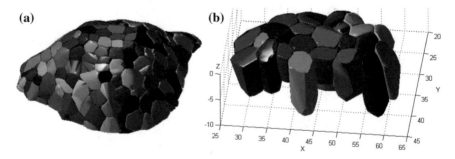

Fig. 12.9 Visualization of the AQVT-based 3D reconstruction of SAM cell cluster. **a** Visualization of the 3D reconstructed structure of a cluster of around 220 closely packed cells using convex polyhedron approximations of the densely clustered data points for each cell, as obtained from the proposed 3D reconstruction scheme, **b** a subset of cells from the same tissue

12.2.3.3   3D Reconstruction Results

Figure 12.6a shows a cell resolution reconstruction of the cell cluster in SAM using AQVT. Note that for 3D visualization purpose of the 3D structure only, we have represented each cell as a convex polyhedron fitted to the dense point cloud clustered to the cells, as obtained from our 3D reconstruction/3D segmentation scheme. For better understanding of the 3D structures of individual cells, we have shown the reconstructed shapes of a smaller cluster of cells in Fig. 12.6b.

Validation on 3D SAM Data There is hardly any biological experiment which can directly validate the estimated growth statistics for individual cells in a sparsely sampled multilayered cluster. In fact, the absence of a method to estimate growth statistics directly using noncomputational methods in a live imaging developmental biology framework is the motivation for the proposed work and we needed to design a method for computationally validating our 3D reconstruction technique. Once the 3D reconstruction is achieved, we can computationally re-slice the reconstructed shape along any arbitrary viewing plane by simply collecting the subset of reconstructed 3D point cloud that lies on the plane (Fig. 12.10).

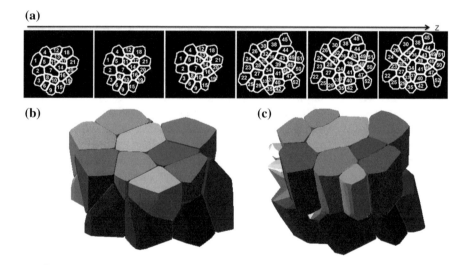

Fig. 12.10 Reconstruction of a cluster of cells using Euclidean distance-based Voronoi tessellation and the proposed AQVT for comparison of the 3D reconstruction accuracy. **a** Segmented and tracked cell slices for a cluster of 52 cells from the L1 and L2 layers of SAM. A dense confocal image stack is subsampled at a z-resolution of 1.35 μm to mimic the 'z-sparsity' observed in a typical Live-Imaging scenario. The slices belonging to the same cell are marked with the same number to show the tracking results. **b** 3D reconstructed structure for a subset of these cells when reconstructed using the Euclidean distance-based Voronoi Tessellation. **c** The AQVT-based reconstruction result for the same cell cluster

To show the validation of our proposed method, we have chosen a single time point dataset that is relatively densely sampled along Z (0.225 μm between successive slices). Then, we resampled this dense stack at a resolution of 1.35 μm to generate a sparser subset of slices that mimic the sparsity generally encountered in a live imaging scenario. The sparsely sampled slices for a cluster of cells spanning two layers (L1 and L2) in the SAM are shown in Fig. 12.7. The aforementioned tracking method [2] is used to obtain correspondences between slices of the same cells. Different slices of the same cells imaged at different depths in Z are shown using the same number in Fig. 12.7a. Next, we reconstructed the cell cluster first by the standard Voronoi tessellation using the Euclidean distance metric and then using our proposed method (AQVT) with a quadratic distance metric adapted for each of these cells. The reconstruction results for a subset of the cells for each of these methods are shown in Fig. 12.7b, c, respectively, for a direct comparison. It can be observed that not only our proposed method accurately reconstructed the cell shapes but also it has captured the multilayer architecture of these SAM cells more closely in comparison to its Voronoi counterpart with the Euclidean distance metric.

12.3 Conclusion

In this chapter, we have presented a fully automated image analysis pipeline to process large volumes of *live imaging* data of tightly packed multilayer biological tissues. We have provided the necessary details of each of the multiple components of such a pipeline, viz., image registration, cell segmentation, spatio-temporal cell tracking, and cell resolution 3D reconstruction of the tissue. We have shown experimental results on such datasets obtained through confocal microscopy on the shoot meristem of a model plant Arabidopsis thaliana. We have shown that our proposed method is capable of processing data consisting hundreds of cells of stereotypical shapes and sizes imaged through multiple hours of observations. Our method is shown to be robust to low SNRs at parts of the images and can be used to generate large volume of cell growth and cell division statistics in a high-throughput manner. These important statistics, as obtained from this pipeline, would be useful for biologists to gain quantitative insight into the broader problem of morphogenesis.

Acknowledgment We gratefully acknowledge Prof. Venugopala Reddy from Plant Biology at the University of California, Riverside for providing us the datasets on which results are shown. This work was supported in part by the National Science Foundation Integrative Graduate Education and Research Traineeship (IGERT) in Video Bioinformatics (DGE-0903667). Katya Mkrtchyan is an IGERT Fellow.

References

1. Chakraborty A, Perales M, Reddy GV, Roy-Chowdhury AK (2013) Adaptive geometric tessellation for 3D reconstruction of anisotropically developing cells in multilayer tissues from sparse volumetric microscopy images. PLoS ONE
2. Liu M, Yadav RK, Roy-Chowdhury A, Reddy GV (2010) Automated tracking of stem cell lineages of Arabidopsis shoot apex using local graph matching. Plant J
3. Vincent L, Soille P (1991) Watersheds in digital spaces: an efficient algorithm based on immersion simulations. IEEE Trans Pattern Anal Mach Intell
4. Maes F, Collignon A, Vandermeulen D, Marchal G, Suetens P (1997) Multimodality image registration by maximization of mutual information. IEEE Trans Med Imaging
5. Viola P, Wells WM (1997) Alignment by maximization of mutual information. Int J Comput Vis
6. Fernandez R, Das R, Mirabet V, Moscardi E, Traas J, Verdeil J, Malandain G, Godin C (2010) Imaging plant growth in 4D: robust tissue reconstruction and lineaging at cell resolution. Nat Methods
7. Besl P, McKay N (1992) A method for registration of 3-D shapes. IEEE Trans Pattern Anal Mach Intell
8. Sharp GC, Lee SW, Wehe DK (2002) Invariant features and the registration of rigid bodies. IEEE Trans Pattern Anal Mach Intell
9. Mkrtchyan K, Chakraborty A, Roy-Chowdhury A (2013) Automated registration of live imaging stacks of Arabidopsis. In: International symposium on biomedical imaging
10. Reddy GV, Meyerowitz EM (2005) Stem-cell homeostasis and growth dynamics can be uncoupled in the Arabidopsis shoot apex. Science
11. Chan T, Vese L (2001) Active contours without edges. IEEE Trans Image Process
12. Li K, Kanade T (2007) Cell population tracking and lineage construction using multiple-model dynamics filters and spatiotemporal optimization. Microscop Image Anal Appl Biol
13. Cunha AL, Roeder AHK, Meyerowitz EM (2010) Segmenting the sepal and shoot apical meristem of Arabidopsis thaliana. Annu Int Conf IEEE Eng Med Biol Soc
14. Chui H (2000) A new algorithm for non-rigid point matching. IEEE Comput Soc Conf Comput Vis Pattern Recogn
15. Gor V, Elowitz M, Bacarian T, Mjolsness E (2005) Tracking cell signals in fluorescent images. In: IEEE workshop on computer vision methods for bioinformatics
16. Rangarajan A, Chui H, Bookstein FL (2005) The soft assign procrustes matching algorithm. Inf Process Med Imag
17. Liu M, Chakraborty A, Singh D, Gopi M, Yadav R, Reddy GV, Roy-Chowdhury A (2011) Adaptive cell segmentation and tracking for volumetric confocal microscopy images of a developing plant meristem. Mol Plant
18. Chakraborty A, Roy-Chowdhury A (2014) Context aware spatio-temporal cell tracking in densely packed multilayer tissues. Med Image Anal
19. Chakraborty A, Roy-Chowdhury A (2014) A conditional random field model for tracking in densely packed cell structures. IEEE Int Conf Image Process
20. Beucher S, Lantuejoul C (1979) Use of watersheds in contour detection. In: International workshop on image processing: realtime edge and motion detection/estimation
21. Najman L, Schmitt M (1994) Watershed of a continuous function. Signal Process
22. Marcuzzo M, Quelhas P, Campilho A, Mendonca AM, Campilho AC (2008) Automatic cell segmentation from confocal microscopy images of the Arabidopsis root. In: IEEE international symposium on biomedical imaging
23. Soille P (2003) Morphological image analysis: principles and applications, 2nd edn. Springer, New York
24. Nakahari T, Murakami M, Yoshida H, Miyamoto M, Sohma Y, Imai Y (1990) Decrease in rat submandibular acinar cell volume during ACh stimulation. Am J Physiol

25. Farinas J, Kneen M, Moore M, Verkman AS (1997) Plasma membrane water permeability of cultured cells and epithelia measured by light microscopy with spatial filtering. J General Physiol
26. Kawahara K, Onodera M, Fukuda Y (1994) A simple method for continuous measurement of cell height during a volume change in a single A6 cell. Jpn J Physiol
27. Kwiatkowska D, Routier-Kierzkowska A (2009) Morphogenesis at the inflorescence shoot apex of Anagallis arvensis: surface geometry and growth in comparison with the vegetative shoot. J Exp Botany
28. Tataw O, Liu M, Yadav R, Reddy V, Roy-Chowdhury A (2010) Pattern analysis of stem cell growth dynamics in the shoot apex of Arabidopsis. IEEE Int Conf Image Process
29. Zhu Q, Tekola P, Baak JP, Belikin JA (1994) Measurement by confocal laser scanning microscopy of the volume of epidermal nuclei in thick skin sections. Anal Quant Cytol Histol
30. Errington RJ, Fricker MD, Wood JL, Hall AC, White NS (1997) Four-dimensional imaging of living chondrocytes in cartilage using confocal microscopy: a pragmatic approach. Am J Physiol
31. Chakraborty A, Yadav RK, Reddy GV, Roy-Chowdhury A (2011) Cell resolution 3D reconstruction of developing multilayer tissues from sparsely sampled volumetric microscopy images. IEEE Int Conf Bioinform Biomed
32. Mjolsness E (2006) The growth and development of some recent plant models: a viewpoint. J Plant Growth Regul (Springer)
33. Gor V, Shapiro BE, Jönsson H, Heisler M, Reddy GV, Meyerowitz EM, Mjolsness E (2005) A software architecture for developmental modelling in plants: the computable plant project. Bioinf Genome Regul Struct
34. Boissonnat J, Wormser C, Yvinec M (2006) Curved Voronoi diagrams, effective computational geometry for curves and surfaces. Mathematics and visualization. Springer
35. Khachiyan LG (1996) Rounding of polytopes in the real number model of computation. Math Methods Oper Res
36. Kumar P, Yildirim EA (2005) Minimum-volume enclosing ellipsoids and core sets. J Opt Theory Appl

Chapter 13
Quantitative Analyses Using Video Bioinformatics and Image Analysis Tools During Growth and Development in the Multicellular Fungus *Neurospora crassa*

Ilva E. Cabrera, Asongu L. Tambo, Alberto C. Cruz,
Benjamin X. Guan, Bir Bhanu and Katherine A. Borkovich

Abstract *Neurospora crassa* (*Neurospora*) is a nonpathogenic multicellular fungus. *Neurospora* has many attributes that make it an ideal model organism for cell biology and genetic studies, including a sequenced genome, a predominantly haploid life cycle and the availability of knock-out mutants for the ∼ 10,000 genes. *Neurospora* grows by polar extension of tube-like structures called hyphae. *Neurospora* has a complex life cycle, with two asexual sporulation pathways and a sexual cycle that produces meiotic progeny. This study analyzes stages during the formation of a colony, from asexual spore to mature hyphae with the use of video bioinformatics and

Electronic Supplementary Material Supplementary material is available in the online version of this chapter at 10.1007/978-3-319-23724-4_13. Videos can also be accessed at http://www.springerimages.com/videos/978-3-319-23723-7

I.E. Cabrera · K.A. Borkovich (✉)
Department of Plant Pathology and Microbiology, University of California,
900 University Avenue, Riverside, CA 92521, USA
e-mail: Katherine.Borkovich@ucr.edu

I.E. Cabrera
e-mail: icabr001@ucr.edu

A.L. Tambo · A.C. Cruz · B.X. Guan · B. Bhanu
Center for Research in Intelligent Systems, University of California,
Chung Hall 216, Riverside, CA 92521, USA
e-mail: atamb001@student.ucr.edu

A.C. Cruz
e-mail: albert.cureg.cruz@gmail.com

B.X. Guan
e-mail: xguan001@ucr.edu

B. Bhanu
e-mail: bhanu@ee.ucr.edu

© Springer International Publishing Switzerland 2015
B. Bhanu and P. Talbot (eds.), *Video Bioinformatics*,
Computational Biology 22, DOI 10.1007/978-3-319-23724-4_13

image analysis tools. We are the first to analyze the asexual spore size, hyphal compartment size and hyphal growth rate in an automated manner, using video and image analysis algorithms. Quantitative results were obtained for all three phenotypic assays. This novel approach employs phenotypic parameters that can be utilized for streamlined analysis of thousands of mutants. This software, to be made publicly available in the future, eliminates subjectivity, and allows high-throughput analysis in a time saving manner.

13.1 *Neurospora* Life Cycle and Genomics

Neurospora is a eukaryotic filamentous fungus belonging to the phylum Ascomycota. Filamentous fungi include a large group of plant and animal pathogens that affect crop production and human health [1, 2]. Due to the feasibility of genetic manipulations, a predominantly haploid life cycle and a relatively fast growth rate, genetic studies in *Neurospora* have further advanced the understanding of multiple pathways shared with both important pathogens and higher eukaryotes [3, 4]. Access to a sequenced genome and knock-out mutants for almost all of the 10,000 genes [5, 6] are among the reasons that *Neurospora* has been designated as an NIH model system for filamentous fungi (http://www.nih.gov/science/models/).

Neurospora grows by polar extension, branching, and fusion of tube-like structures termed hyphae. All the three mechanisms of growth (extension, branching, and fusion) involve polarization of the hyphal tip [7]. A typical *Neurospora* hypha is 4–12 µm in diameter and contains multinucleate cellular compartments, separated by incomplete cell walls (septa) [8]. As the cell divides, it lays down the septum during the final stage of mitosis (cytokinesis). The septal wall contains a pore that allows for cytoplasmic communication and movement of organelles between the cells. Fusion and branching of hyphae results in formation of a colony, consisting of interconnected multinucleated hyphae [9].

Neurospora uses three different developmental pathways to produce three different types of spores (Fig. 13.1). The major asexual sporulation developmental pathway is known as macroconidiation. Macroconidia, also referred to as conidia, are multinucleate, containing 3–5 nuclei per cell [10]. These conidia develop in chains at the end of aerial hyphae and are easily dispersed into the environment upon maturation. Spores produced during the second asexual sporulation pathway, microconidia, contain only one nucleus and develop from basal vegetative hyphae [11]. The third type of spore is produced during sexual reproduction. Sexual reproduction in *Neurospora* involves the fusion of cells from the two mating types, *mat A* and *mat a*. Nitrogen starvation induces the production of female reproductive structures termed protoperithecia. The presence of an asexual spore (male) of opposite mating type in the vicinity will cause a hypha from the protoperithecium to grow toward it and fuse, with uptake of the male nucleus. Fertilization and meiosis occur and the female structure then enlarges to form a fruiting body (perithecium)

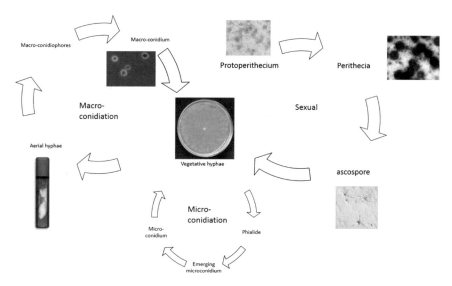

Fig. 13.1 *N. crassa* life cycle. *Neurospora* grows by extension, branching, and fusion of hyphae to form a colony. *Neurospora* has two asexual sporulation pathways (macroconidiation and microconidiation) and a sexual cycle

containing the progeny (ascospores) [12]. Ascospores are ejected from the mature perithecium into the environment, and are able to germinate to form a colony upon heat activation [13].

Neurospora has made a large impact in the fungal genetics community, and will continue to do so, due to the feasibility of genetic manipulations [3, 14, 15]. The *Neurospora* genomic sequence became available in 2003, revealing a 43 Mb genome containing approximately 10,000 genes [5]. The discovery of *mus*-51 and *mus*-52, two genes homologous to *KU*70 and *KU*80 in humans, which are responsible for non-homologous end joining, has facilitated the production of gene deletion mutants [16]. Transformation of *Neurospora* strains from these two genetic backgrounds (Δ*mus*-51 and Δ*mus*-52) ensures 100 % homologous recombination of gene replacement cassettes containing a hygromycin resistance gene in place of each target gene [17].

13.2   Hyphal Growth in *Neurospora*

Growth is an essential process required for colonization of dead and decaying plant material by *Neurospora*. In fungal animal and plant pathogens, tip-based (polarized) growth is used for invasion of the host and defects in hyphal morphogenesis compromise the virulence of pathogens [18, 19].

In filamentous fungi, the complex hypothesized to be responsible for hyphal polarized growth is the Spitzenkorper. The Spitzenkorper is the vesicle organization site, from which vesicles containing components needed for cell wall degradation and synthesis are shuttled [20]. Cell-wall-synthesizing activity occurs at the hyphal apex to promote hyphal extension [21]. In addition, the position of the Spitzenkorper correlates with the local direction of growth [22].

The cytoskeleton is also important for polarized growth. The absence of actin filaments blocks polarized growth and prevents septum formation [23, 48]. Microtubules, another component of the cytoskeleton, have also been proposed to play a role during polarized growth. The absence of microtubules leads to loss of directed extension of hyphae and irregular growth patterns [24, 25].

13.2.1 Quantifying Hyphal Compartment Size

Cell compartment properties are quantified by examining morphology. Conventionally, data is gathered via manual analysis by multiple experts who would repeat the same measurements multiple times for the same image in order to increase the accuracy of the results. A method is needed that does not require a user to perform manual segmentation. Such fully automatic methods improve the objectivity of results and significantly improve the throughput of experiments. For comparison, the time required to manually segment the pavement cells of *Arabidopsis thaliana* in a single image is measured in 10^2–10^3 s. The time required for automatic methods is measured in 10^0–10^{-1} s [26].

The system overview is as follows: (A) First, the region of interest (ROI), the region of the image belonging to the network, is identified with a Poisson-based ROI extraction procedure. This method detects the compartments. (B) Edges are extracted with Gabor filters and a non-classical receptive field inhibitor. This method detects the walls that separate the compartments, reducing the ROI into separate regions corresponding to the compartments. (C) Morphological operators refine the results by separating erroneously joined compartments and filling in holes. (D) Connected components identify each cell and, finally, (E) properties of each compartment, such as area, orientation, length, and width, are extracted.

13.2.1.1 Cell Growth and Microscopy

This study used wild-type *N. crassa* strain FGSC 2489 [17]. Conidia were inoculated in the center of a 100 × 15 mm petri plate containing 12 ml of Vogel's minimal medium (VM) with 1 % agar as a solidifying agent [27]. Cultures were incubated for 20–22 h at 30 °C in the dark. Vegetative hyphae were stained using a 1:1 ratio of Calcoflour-white in VM liquid medium [28]. The "inverted agar block method" described in [29] was used to visualize the sample using an Olympus IX71 inverted microscope (Olympus America, Center Valley, PA) with a 40× phase objective. Several images were collected for the strain. Manual analysis of the length and diameter were also performed, (data not shown). Statistical analysis (t-test) was used to compare methods and performed using Excel [30].

13.2.1.2   Poisson-Based ROI Extraction

State-of-the-art methods model the pixel intensity value with a probabilistic framework, and assume a normal distribution in computation. One study [31] utilized a probabilistic method to correct intensity loss in confocal microscopy images, assuming a normal distribution. In Pound et al. [32], a normal distribution is applied as a point spread function and for segmentation. In a third approach, [33], a semi-supervised learning algorithm segments cells, assuming a normal distribution. We propose modeling the intensity as a Poisson distribution. When capturing the images in a confocal microscope, a pixel measures the arrival of photons at that point in the sensor. Because the Poisson distribution models the probability of a given number of events occurring, it is better suited for the segmentation model. The model in Gopinath et al. [32] was used as groundwork for the proposed Poisson-based segmentation model.

13.2.1.3   Non-classical Receptive Field Inhibition Based Edge Detection

Grigorescue et al. [34] proposed a method for edge detection that models the human visual system. This method is based on the Gabor filter, modeling of gradients in the V1 cortex of the human visual system. It expands on Gabor filters by modeling the process by which the human visual system can extract meaningful edges in noisy data. It is called non-classical receptive field inhibition. This method was exploited for microtubule detection in *A. thaliana* pavement cells and was found to reduce false alarm rate over state-of-the-art methods by a factor of 2.14 [26].

13.2.1.4   Results

The automated method allows the user to select the cell or cells of interest in a rapid, non-biased manner. Once the cells have been selected, quantitative data for length, width (diameter), area, and perimeter are stored as a CSV file (Table 13.1). Statistical analysis confirmed that there was no statistical difference between manual analysis and the automated method (Data not shown). The automated method detects the walls that separate the hyphal compartments, reducing the ROI into separate compartments. Figure 13.2 shows the primary image as colored compartments to easily recognize the different cells for the user to select for analysis.

Table 13.1 Results from the cell compartment size automated program

Measurement	Length (μm)	Width (μm)	Area (μm^2)	Perimeter (μm)
Mean	79.85	10.65	625.88	174.01
Standard error	5.32	0.62	79.69	12.06
Standard deviation	26.58	3.11	398.46	60.30
Minimum	25.71	7.085	148.96	60.31
Maximum	142.26	16.91	1557.90	303.61

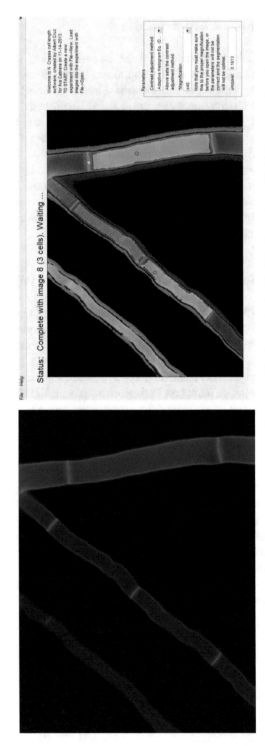

Fig. 13.2 Automated analysis of hyphal cell compartments. (*Left*) Image of hyphae stained with Calcoflour-white. (*Right*) Demonstration of automatic analysis software results before selection of *Neurospora* regions

13.2.2 Measuring Hyphal Growth Rate

Neurospora is one of the most rapidly growing filamentous fungi, extending at a rate of 5 mm/h on solid medium [15, 35]. Capturing hyphal tip growth using video microscopy will enable us to quantitate the growth rate of any strain in real time. Many mutants possess a growth defect and this method will provide an unbiased and accurate growth rate measurement.

13.2.2.1 Hyphal Growth and Video Microscopy

The *N. crassa* strain and culture methods are identical to those described in Sect. 13.2.1.1. The "inverted agar block method" described in [29] was used to visualize the sample using an Olympus IX71 inverted microscope (Olympus America, Center Valley, PA) with a 100× phase objective. Frames were captured at a speed of one frame every second for 5 min (300 s).

13.2.2.2 Video Analysis

These operations are performed using MATLAB software [36]. For each image in the video sequence, image segmentation is performed to distinguish the hypha (foreground) from the media (background). This process is achieved using an edge-based active contour model as outlined in [37]. The algorithm is provided with an initial contour that it collapses using gradient descent to find the shape of the hypha. The result of this process is a binary image where the foreground pixels have a value of 1 and the background pixels are 0. The extracted shape of the hypha is further collapsed inward using gradient descent and the distance transform image to find the internal line of symmetry of the hypha. The tip of the line is projected to intersect with the contour of the shape to form the complete line of symmetry. This line is used to compute the length of the hypha via Euclidean distance. Video 1 represents this method on one frame of the growing hypha.

The results with wild type (Fig. 13.3) suggest that overall hyphal growth rate is linear, but that the velocity is oscillating over short time periods. These results agree with those from a prior video microscopy study reporting that *Neurospora* grows in a pulsatile manner [38]. Previous results involving manual calculation reported the growth rate for wild type vegetative hyphae as 0.201 μm/s [38], which is very close to our finding of 0.207 μm/s.

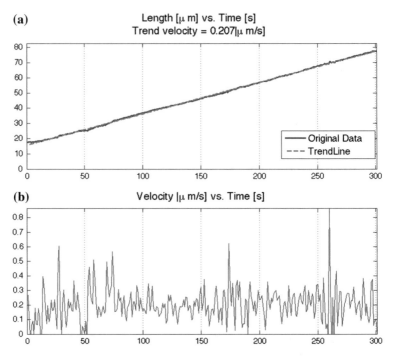

Fig. 13.3 Determination of hyphal growth rate. **a** Length of the midline of the hypha over time. The *green dashed line* is the smoothed measurements from the *solid blue line*. The smoothed measurements were obtained with a moving average. **b** Rate of change in length of the hypha over time (velocity). The results show that hyphal growth is oscillating over time

13.3 Asexual Sporulation

As mentioned above, asexual spores (conidia) are essential for dispersal of *Neurospora*. Conidia formation is initiated by light, nutrient deprivation and/or exposure to oxygen [39]. These conditions lead to growth of aerial hyphae that rise perpendicular to the basal hyphae [40]. Aerial hyphae begin to bud at their tips to form macroconidia, each containing 2–5 nuclei [41]. Most macroconidia are 5–9 μm in diameter [39]. Mature conidia are held together by septa, and can be easily separated by natural forces such as wind, which will aid their dispersal in the environment [41].

13.3.1 Quantifying Conidia Area

In this study, we propose an automated algorithm for conidia cell region detection. The cell region is the foreground, F, and the rest of noncell regions are background,

B. The cell has three basic characteristics: (1) A strong halo surrounded the cell; (2) a dark oval or circular ring circumscribes the cell body; (3) the cell center has similar intensity as, or brighter intensity than, the background.

13.3.1.1 Conidia Propagation and Image Capture

This study used the *N. crassa* strain described in Sect. 13.2.1.1 conidia from a 5–7-day-old slant cultures were collected using 1 ml of sterile water. The suspension was diluted 1:50 and visualized using an Olympus IX71 inverted microscope (Olympus America, Center Valley, PA) with a 40× phase objective. Several images were collected for analysis.

13.3.1.2 Algorithm

The algorithm exploits three characteristics for cell region detection. The algorithm uses a Gabor filter to enhance the effect of halo for cell region detection [42]. The effect of image property with/without Gabor filter can be seen in Fig. 13.4. After filtering, a simple thresholding by Otsu's method is sufficient for segmenting out cell regions from noncell regions [43]. As shown in Fig. 13.5, the halo intensity distribution is located at the right of the histogram, while the background distribution is on the left. As the result, halo stands out as part of detected cell regions by Otsu's. However, there is an overlapped region between foreground and background intensity distribution (Fig. 13.5). As a result, a morphological method is used to obtain the entire cell region [44, 45, 46]. Since the cell body without halo is the only interest, oversegmentation is inevitable in the algorithm, which is shown in Fig. 13.4. The concept of region shrinking is used after Otsu's method to reduce over-segmentation [47]. The region shrinking is applied to each detected cell region. Figure 13.4 shows the final result of the algorithm. The equations for the algorithm are shown below:ą

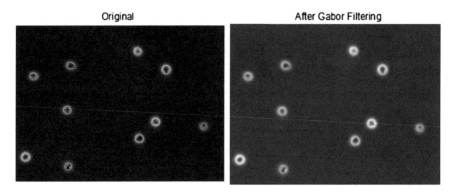

Fig. 13.4 Conidia images. Original image (*left*) and after being subjected to Gabor filtering (*right*)

Fig. 13.5 Halo intensity histogram. The halo intensity distribution is located at the *right* of the histogram, while the background distribution is on the *left*. As the result, the halo stands out as part of detected cell regions by Otsu's

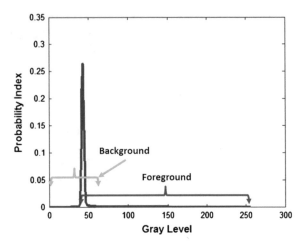

$$W_k = e^{-\frac{\left(\mu_{e_k} - \mu_{\text{halo}}\right)^2}{\sigma_{\text{halo}}^2}} \tag{13.1}$$

$$SNR_k = \frac{\mu_{\text{cell}}(k)}{\sigma_{\text{cell}}(k)} \tag{13.2}$$

$$M(k) = W_k \times SNR_k \tag{13.3}$$

$$k_{\text{opt}} = \max_k (M(k)) \tag{13.4}$$

W_k	is the weight for edge at the k_{th} iteration
e_k	is a vector that contain edge intensity values at the k_{th} iteration
μ_{e_k}	is the mean intensity value of e_k
μ_{halo}	is the mean intensity value of halo
σ_{halo}^2	is the intensity variance of halo
$\mu_{\text{cell}}(k)$	is the mean intensity value of cell region at the k_{th} iteration
$\sigma_{\text{cell}}(k)$	is the intensity standard deviation of cell region at the k_{th} iteration
SNR_k	is the signal to noise ratio at the k_{th} iteration
$M(k)$	is the weighted SNR at the k_{th} iteration
$F_j(k_{\text{opt}})$	is the j_{th} cell region at the k_{opt} iteration

Algorithm: Cell detection

Input:
I: Phase contrast image.
k_{limit}: Iteration limit for morphological erosion.
Output:
F: Foreground (the cell regions).
B: Background (the non-cell regions).
==

Procedure Detection_Algo(I);
{
1. Initialize F and B with a zero matrix with size of image, I.
2. I_{Gabor} is a result of Gabor Filtering of image, I (Mehrotra, Namuduri et al. 1992).
3. BW is a binary result by Otsu's method on I_{Gabor} (Otsu 1979).
4. Apply morphological filtering on BW (Soille 1999).
5. Apply connected component on BW (Haralick and Shapiro 1991).
6. BW_{Label} is a connected component labeling of BW.
7. For (j = 1 to total number of components in BW_{Label}){
a. L is the j_{th} component of BW_{Label}.
b. For (k = 1 to k_{limit}){
i. Obtain L_{erode} by apply morphological erosion on L (van den Boomgard and van Balen 1992).
ii. Obtain edge locations, E from i (Li, Huang et al. 2006).
iii. Obtain e_i from image, I, for all edge locations in E.
iv. Calculate μ_{e_i} from e_i.
v. Calculate W_i, SNR_i and $M(i)$ with equation (1), (2) and (3).
vi. $L \leftarrow L_{erode}$.
vii. $F_j(k) \leftarrow L_{erode}$.
}
c. Calculate k_{opt} using equation (4).
d. $F \leftarrow F \cup F_j(k_{opt})$.
8. }
9. $B \leftarrow 1 - F$.
};
==

13.4 Conclusion

The use of algorithm analysis on biological organisms has revolutionized the way scientists analyze their data. Here, collaborations between biologists and engineers have enhanced the analysis process with video and image analyses algorithms. These methods have saved tremendous amounts of time and money. In addition, data analysis becomes unbiased by reducing human error, thus ensuring the authenticity of the results. In this study, we have analyzed *Neurospora* during growth and development. This includes analysis of asexual spores and mature basal hyphae. Quantifiable data was extracted using images, while growth rate was measured using video bioinformatics. These methods may be applied to different *Neurospora* gene deletion strains, thus revealing defects not captured by other phenotypic assays. In addition, other fungi or any organism with similar features can be analyzed using these programs. For example, there are other fungi that produce spores and quantifying the area of these spores is feasible, as long as phase contrast microscopy is used. Growth rate analysis on other polarized tip-growing organisms can also be quantified, thus expanding the use of these programs.

Acknowledgements We thank Jhon Gonzalez for his insightful conversations and comments on this work. We also thank Alexander Carrillo and Caleb Hubbard for assistance in determining the optimal conditions for capture of microscopic images. I.E.C., A.L.T, A.C.C., and B.X.G. were supported by NSF IGERT Video Bioinformatics Grant DGE 0903667.

References

1. Agrios GN (1997) Plant Pathology, 4th edn. Academic Press, San Diego
2. Latge JP (1999) *Aspergillus fumigatus* and aspergillosis. Clin Microbiol Rev 12:310–350
3. Perkins DD (1992) *Neurospora*: the organism behind the molecular revolution. Genetics 130:687–701
4. Davis RH, Perkins DD (2002) *Neurospora*: a model of model microbes. Nat Rev Genet 3:397–403
5. Galagan JE, Calvo SE, Borkovich KA, Selker EU, Read ND, Jaffe D, FitzHugh W, Ma LJ, Smirnov S, Purcell S, Rehman B, Elkins T, Engels R, Wang S, Nielsen CB, Butler J, Endrizzi M, Qui D, Ianakiev P, Bell-Pedersen D, Nelson MA, Werner-Washburne M, Selitrennikoff CP, Kinsey JA, Braun EL, Zelter A, Schulte U, Kothe GO, Jedd G, Mewes W, Staben C, Marcotte E, Greenberg D, Roy A, Foley K, Naylor J, Stange-Thomann N, Barrett R, Gnerre S, Kamal M, Kamvysselis M, Mauceli E, Bielke C, Rudd S, Frishman D, Krystofova S, Rasmussen C, Metzenberg RL, Perkins DD, Kroken S, Cogoni C, Macino G, Catcheside D, Li W, Pratt RJ, Osmani SA, DeSouza CP, Glass L, Orbach MJ, Berglund JA, Voelker R, Yarden O, Plamann M, Seiler S, Dunlap J, Radford A, Aramayo R, Natvig DO, Alex LA, Mannhaupt G, Ebbole DJ, Freitag M, Paulsen I, Sachs MS, Lander ES, Nusbaum C, Birren B (2003) The genome sequence of the filamentous fungus *Neurospora crassa*. Nature 422:859–868
6. Borkovich KA, Alex LA, Yarden O, Freitag M, Turner GE, Read ND, Seiler S, Bell-Pedersen D, Paietta J, Plesofsky N, Plamann M, Goodrich-Tanrikulu M, Schulte U, Mannhaupt G, Nargang FE, Radford A, Selitrennikoff C, Galagan JE, Dunlap JC, Loros JJ, Catcheside D, Inoue H, Aramayo R, Polymenis M, Selker EU, Sachs MS, Marzluf GA, Paulsen I, Davis R, Ebbole DJ, Zelter A, Kalkman ER, O'Rourke R, Bowring F, Yeadon J, Ishii C, Suzuki K, Sakai W, Pratt R (2004) Lessons from the genome sequence of *Neurospora crassa*: tracing the path from genomic blueprint to multicellular organism. Microbiol Mol Biol Rev 68:1–108
7. Riquelme M, Yarden O, Bartnicki-Garcia S, Bowman B, Castro-Longoria E, Free SJ, Fleissner A, Freitag M, Lew RR, Mourino-Perez R, Plamann M, Rasmussen C, Richthammer C, Roberson RW, Sanchez-Leon E, Seiler S, Watters MK (2011) Architecture and development of the *Neurospora crassa* hypha—a model cell for polarized growth. Fungal Biol 115:446–474
8. Bruno KS, Aramayo R, Minke PF, Metzenberg RL, Plamann M (1996) Loss of growth polarity and mislocalization of septa in a *Neurospora* mutant altered in the regulatory subunit of cAMP-dependent protein kinase. EMBO J 15:5772–5782
9. Fleissner A, Simonin AR, Glass NL (2008) Cell fusion in the filamentous fungus, *Neurospora crassa*. Methods Mol Biol 475:21–38
10. Kolmark G, Westergaard M (1949) Induced back-mutations in a specific gene of *Neurospora crassa*. Hereditas 35:490–506
11. Barratt RW, Garnjobst L (1949) Genetics of a colonial microconidiating mutant strain of *Neurospora crassa*. Genetics 34:351–369

12. Perkins DD (1996) Details for collection of asci as unordered groups of eight projected ascospores. Neurospora Newsl. 9:11
13. Lindegren CC (1932) The genetics of *Neurospora* I. The inheritance of response to heat-treatment. Bull Torrey Bot Club:85–102
14. Horowitz NH (1991) Fifty years ago: the *Neurospora* revolution. Genetics 127:631–635
15. Perkins DD, Davis RH (2000) *Neurospora* at the millennium. Fungal Genet Biol 31:153–167
16. Ninomiya Y, Suzuki K, Ishii C, Inoue H (2004) Highly efficient gene replacements in *Neurospora* strains deficient for nonhomologous end-joining. Proc Natl Acad Sci USA 101:12248–12253
17. Colot HV, Park G, Turner GE, Ringelberg C, Crew CM, Litvinkova L, Weiss RL, Borkovich KA, Dunlap JC (2006) A high-throughput gene knockout procedure for *Neurospora* reveals functions for multiple transcription factors. Proc Natl Acad Sci USA 103:10352–10357
18. Nichols CB, Fraser JA, Heitman J (2004) PAK kinases Ste20 and Pak1 govern cell polarity at different stages of mating in *Cryptococcus neoformans*. Mol Biol Cell 15:4476–4489
19. Castillo-Lluva S, Alvarez-Tabares I, Weber I, Steinberg G, Perez-Martin J (2007) Sustained cell polarity and virulence in the phytopathogenic fungus *Ustilago maydis* depends on an essential cyclin-dependent kinase from the Cdk5/Pho85 family. J Cell Sci 120:1584–1595
20. Harris SD, Read ND, Roberson RW, Shaw B, Seiler S, Plamann M, Momany M (2005) Polarisome meets spitzenkorper: microscopy, genetics, and genomics converge. Eukaryot Cell 4:225–229
21. Bartnicki-Garcia S, Bartnicki DD, Gierz G, Lopez-Franco R, Bracker CE (1995) Evidence that Spitzenkorper behavior determines the shape of a fungal hypha: a test of the hyphoid model. Exp Mycol 19:153–159
22. Reynaga-Pena CG, Gierz G, Bartnicki-Garcia S (1997) Analysis of the role of the Spitzenkorper in fungal morphogenesis by computer simulation of apical branching in *Aspergillus niger*. Proc Natl Acad Sci USA 94:9096–9101
23. Riquelme M, Reynaga-Pena CG, Gierz G, Bartnicki-Garcia S (1998) What determines growth direction in fungal hyphae? Fungal Genet Biol 24:101–109
24. Konzack S, Rischitor PE, Enke C, Fischer R (2005) The role of the kinesin motor KipA in microtubule organization and polarized growth of *Aspergillus nidulans*. Mol Biol Cell 16:497–506
25. Takeshita N, Higashitsuji Y, Konzack S, Fischer R (2008) Apical sterol-rich membranes are essential for localizing cell end markers that determine growth directionality in the filamentous fungus *Aspergillus nidulans*. Mol Biol Cell 19:339–351
26. Harlow GJ, Cruz AC, Li S, Bhanu B, Yang Z (2013) Pillars of plant cell polarity: 3-D automated microtubule order & asymmetric cell pattern analysis. NSF IGERT Video and Poster Competition, Washington, DC
27. Vogel HJ (1964) Distribution of lysine pathways among fungi: evolutionary implications. Am Nat 98:435–446
28. Elorza MV, Rico H, Sentandreu R (1983) Calcofluor white alters the assembly of chitin fibrils in *Saccharomyces cerevisiae* and *Candida albicans* cells. J Gen Microbiol 129:1577–1582
29. Hickey PC, Swift SR, Roca MG, Read ND (2004) Live-cell Imaging of filamentous fungi using vital fluorescent dyes and confocal microscopy. Meth Microbiol 34:63–87
30. (2013) Excel. Microsoft, Redmond, Washington
31. Gopinath S, Wen Q, Thakoor N, Luby-Phelps K, Gao JX (2008) A statistical approach for intensity loss compensation of confocal microscopy images. J Microsc 230:143–159
32. Pound MP, French AP, Wells DM, Bennett MJ, Pridmore TP (2012) CellSeT: novel software to extract and analyze structured networks of plant cells from confocal images. Plant Cell 24:1353–1361
33. Su H, Yin Z, Huh S, Kanade T (2013) Cell segmentation in phase contrast microscopy images via semi-supervised classification over optics-related features. Med Image Anal 17:746–765
34. Grigorescu C, Petkov N, Westenberg MA (2003) Contour detection based on nonclassical receptive field inhibition. IEEE Trans Image Process 12:729–739

35. Ryan FJ, Beadle GW, Tatum EL (1943) The tube method of measuring the growth rate of *Neurospora*. Amer J Bot 30:784–799
36. (2010) MATLAB software .7.10.0.584 Natick, MA
37. Li C, Xu C, Gui C, Fox MD (2010) Distance regularized level set evolution and its application to image segmentation. IEEE Trans Image Process 19:3243–3254
38. Lopez-Franco R, Bartnicki-Garcia S, Bracker CE (1994) Pulsed growth of fungal hyphal tips. Proc Natl Acad Sci USA 91:12228–12232
39. Springer ML, Yanofsky C (1989) A morphological and genetic analysis of conidiophore development in *Neurospora crassa*. Genes Dev 3:559–571
40. Davis RH, deSerres FJ (1970) Genetic and microbiological research techniques for *Neurospora crassa*. Methods Enzymol 71A:79–143
41. Davis R (2000) Neurospora: contributions of a model organism. Oxford University Press, New York
42. Mehrotra KN, Namuduri KR, Ranganathan N (1992) Gabor filter-based edge detection. Pattern Recognit 25:1479–1492
43. Otsu N (1979) A threshold selection method from gray-level histograms. IEEE Trans Syst Man Cybern SMC-9: 62–66
44. Haralick RM, Shapiro LG (1991) Computer and robot vision. Addison-Wesley Longman Publishing Co. Inc., Boston
45. van den Boomgard R, van Balen R (1992) Methods for fast morphological image transforms using bitmapped binary images. CVGIP: Graph Models Image Process 54:252–258
46. Soille P (1999) Morphological image analysis: principles and applications. Springer, Berlin
47. Li ZH, Huang FG, Liu YM (2006) A method of motion segmentation based on region shrinking. In: Proceedings of intelligent data engineering and automated learning—Ideal 2006, vol 4224, pp 275–282
48. Torralba S, Raudaskoski M, Pedregosa AM, Laborda F (1998) Effect of cytochalasin A on apical growth, actin cytoskeleton organization and enzyme secretion in *Aspergillus nidulans*. Microbiology 144(Pt 1):45–53

Part V
Dynamics of Intracellular Molecules

Chapter 14
Quantification of the Dynamics of DNA Repair to Ionizing Radiation via Colocalization of 53BP1 and γH2AX

Torsten Groesser, Gerald V. Fontenay, Ju Han, Hang Chang, Janice Pluth and Bahram Parvin

Abstract Cellular response to stress can be manifested and visualized by measuring induced DNA damage. However, cellular systems can repair the damage through a variety of DNA repair pathways. It is important to characterize the dynamics of DNA repair in a variety of model systems. Such a characterization is another example of the video bioinformatics through harvesting and fixing of a large sample size at different time points. This chapter provides background and motivation for quantifying the dynamics of DNA damage induction and repair in cycling and stationary cells. These model systems indicate that the repair kinetics have a similar profile for gamma radiation; however, following iron ion exposure residual unrepaired damage is noted at longer times when assayed in stationary cells. Repair kinetics are visualized by immunofluorescence staining of phosphorylated histone gamma-H2AX and the DNA repair protein 53BP1. The kinetics are then quantified using cell-based segmentation, which provides a context for repair measurements and colocalization analysis. For enhanced robustness, cell-based segmentation and protein localization leverage geometric methods. Subsequently, cellular profiles are stored in a database, where colocalization analysis takes place through specially design database queries.

T. Groesser · G.V. Fontenay · J. Han · H. Chang · J. Pluth · B. Parvin (✉)
Lawrence Berkeley National Laboratory, Berkeley, CA, USA
e-mail: B_Parvin@lbl.gov

T. Groesser
e-mail: TGroesser@lbl.gov

G.V. Fontenay
e-mail: GVFontenay@lbl.gov

J. Han
e-mail: JHan@lbl.gov

H. Chang
e-mail: HChang@lbl.gov

© Springer International Publishing Switzerland 2015
B. Bhanu and P. Talbot (eds.), *Video Bioinformatics*,
Computational Biology 22, DOI 10.1007/978-3-319-23724-4_14

14.1   Introduction

Immunofluorescence staining of proteins involved in DNA repair or histone modifications such as phosphorylation is a widely used technique to study DNA damage and repair, in those cases where the repair mechanism is a dynamic process. In recent years, quantitative analysis of DNA damage and repair has become a necessity, given the large volumes of multifactorial data. In order to manage experimental factors, images, and the analysis data, the BioSig imaging bioinformatics system [1] has been significantly extended, and one of its views is shown in Fig. 14.1. Each row represents a subset of samples under identical experimental conditions. The experimental factors are (i) radiation dosage (e.g., 0 cGy, 200 cGy), (ii) cycling and non-cycling cells (e.g., with or without EGF), and harvest time (e.g., 48, 72 h post radiation). The phenotypic responses, in terms of the number of colocalizations (e.g., spatial overlap between two different fluorescent labels with different emission wavelength, where each label corresponds to a different DNA damage markers) on each nucleus, provide an index for the repair dynamics.

The phosphorylation of histone H2AX on serine 139 (γH2AX) is one of the most studied histone modifications, as it occurs at the sites of double strand breaks (DSB) and provides a good biomarker for DSB kinetics. During the DNA repair process γH2AX demarcates large chromatin domains on both sides of the DNA damage [2].

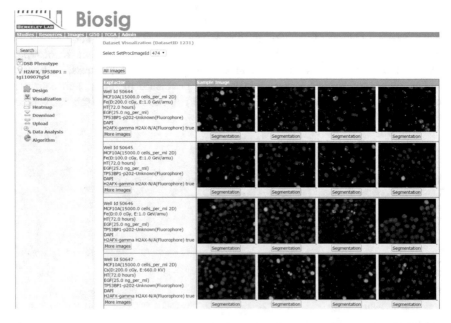

Fig. 14.1 Multifactorial experimental design and associated images can be visualized through the web. Each row corresponds to thumbnail views of samples prepared and imaged with the same set of experimental factors

Using specific antibodies, this highly amplified response can then be visualized as distinctive γH2AX foci. Another widely studied DNA repair protein is tumor suppressor p53-binding protein 1 (53BP1), which colocalizes with γH2AX at sites of DNA damage. We will refer to these sites as "foci" in the remainder of the chapter. It is possible to detect the histone modification or the protein accumulation in the cell nucleus through specific primary antibodies, and to visualize them with fluorescence-labeled secondary antibodies. The distinct small foci can then be used as a surrogate marker for the induction and repair of DNA damage. Immediately after ionizing radiation damage foci numbers are shown to correlate well with induced DNA double-strand breaks (dsb), therefore it is assumed that foci measurement allows one to study the induction and repair of DNA DSBs. In this chapter, an integrated computational and experimental assay is developed, to investigate the dynamics of the DNA repair subject to damage induced by heavy charged particles. Understanding the repair dynamics is crucial for radiation risk assessment and protection. From a computational perspective, sites of DNA damage are detected on a cell-by-cell basis, and registered with a database for subsequent colocalization analysis. In this case, colocalization refers to detecting two different proteins or phosphorylations occuring in the same spatial position. The problem and challenges of correctly defining the number and co-localization of foci on a per cell basis in an unbiased way are illustrated by an example shown in Fig. 14.2. From an image analysis perspective, the difficulties can be reduced by segmentation of nuclear and phosphorylation regions (e.g., foci) within each cell.

Background Investigating the colocalization of two different fluorescence signals can give insights into the biological interaction of the labeled targets, and thus is crucial to understanding their role in biological processes [3]. In addition, it can be used to understand DNA repair in a more complex way than just measuring foci

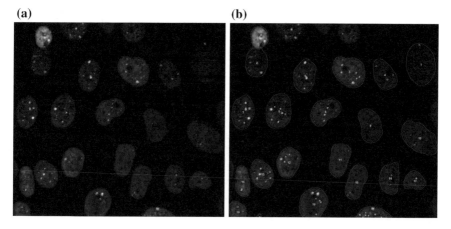

Fig. 14.2   a An example of γH2AX and 53BP1 co-localization on a cell-by-cell basis. *Blue* is the DAPI (nuclear) stained channel. *Green* corresponds to γH2AX, and red corresponds to 53BP1. **b** Each nucleus and the corresponding sites of foci are segmented

numbers of a single surrogate marker. Some of the data sets that are used here for measuring co-localization of γH2AX and 53BP1 were published earlier [4]. DNA damage was induced by iron ion exposure in cycling and non-cycling MCF10A cells, a human mammary epithelial cell (HMEC) line. DNA repair studies with heavy charged particles can give insights into the repair of complex DNA damage, and are crucial for radiation risk assessment and protection.

From a computational perspective, the main issues for detection of foci and co-localization [5] are (i) detection of overlapping nuclei, which provide context for quantitative analysis on a cell-by-cell basis; (ii) heterogeneity in the fluorescent signals in each channel; and (iii) methods for co-localization studies.

14.2   Materials and Methods

Cell culture conditions, radiation exposure as well as the immunostaining protocol and image acquisition were reported in detail in [4]. In short, cycling or stationary human mammary epithelial cells MCF10A were exposed to 1 or 2 Gy of 968 MeV/amu iron ions (LET = 151 keV/μm). Immunostaining for γH2AX and 53BP1 foci formation was performed on cell monolayers at different times after exposure. Cells were counterstained with 4′,6-diamidino-2-phenylindole (DAPI), air-dried, and mounted with Vectashield before image acquisition. Image acquisition was performed using a Zeiss Axiovert epifluorescence microscope with a Zeiss plan-apochromat 40× dry lens and a scientific grade 12-bit charge-coupled device camera. All images within the same data set were captured with the same exposure time so that intensities were within the 12-bit linear range and could be compared between specimens. Images were taken in 11 focal planes with 0.7 mm steps over a range of 7 mm total, to capture foci in different focal plans. A single 2D image was then constructed via maximum projection of all 3D slices. A total of 216 images were collected, and at least 100 cells per treatment group have been analyzed for each independent experiment.

14.3   Quantitative Analysis

(i) **Delineating nuclei** Nuclear segmentation, from DAPI-stained samples, has been widely utilized by the community, and remains an active area of research [6–8]. The main challenges are heterogeneous fluorescent signals due to changes in cell cycle, nuclei touching in a clump of cells, and variation in the background signal. The primary barrier has been designating individual nuclei when nuclei are touching, and a number of approaches have been proposed to enable analysis on a cell-by-cell basis. These include seed detection followed by watershed or evolving fronts [9], gradient edge flow [10, 11], and geometric reasoning [12]. In the current system, nuclear segmentation is based on separating foreground and background, followed by geometric reasoning for validation and separating touching nuclei [13, 14]. The foreground-background separation is based on a combination of gradient and zero-crossing filters, which delineate foreground regions consisting of both individual nuclei and clumps of nuclei. This method is preferred over classical thresholding techniques since gradient-based methods are less sensitive to variations in foreground and background intensities. Having delineated the foreground regions, the system utilizes geometric reasoning to separate touching nuclei, with the main assumption being that normal nuclei have a convex morphology. As a result, ambiguities associated with the delineation of overlapping nuclei can be resolved by detecting concavities and partitioning them through geometric reasoning. The process, shown in Fig. 14.3, consists of the following steps:

a. Detection of Points of Maximum Curvature: The contours of the nuclear mask were extracted, and the curvature along the contour is computed by using $k = x'y'' - y'x''(x'2 + y'2)3/2$, where x and y are coordinates of the boundary points. The derivatives were then computed by convolving the boundary with derivatives of the Gaussian filters. An example of detected points of maximum curvature is shown in Fig. 14.3.

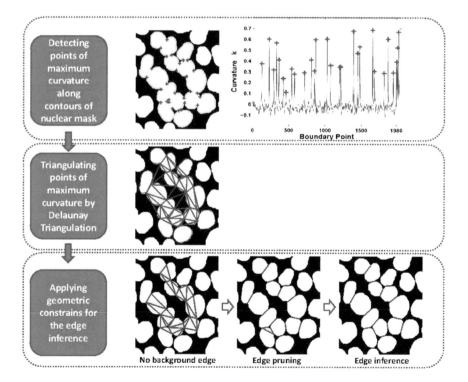

Fig. 14.3 Steps in delineating clumps of nuclei

b. Delaunay Triangulation (DT) of Points of Maximum Curvature for Hypothesis Generation and Edge Removal: DT was applied to all points of maximum curvature to hypothesize all possible groupings. The main advantage of DT is that the edges are non-intersecting, and the Euclidean minimum spanning tree is a sub-graph of DT. This hypothesis space was further refined by removing edges based on certain rules, e.g., no background intersection.

c. Geometric reasoning: Properties of both the hypothesis graph (e.g., degree of vertex), and the shape of the object (e.g., convexity) were integrated for edge inference.

The protocol is similar to the method presented in [12]; however, Delaunay Triangulation offers a significant performance improvement. For more details, see [14, 15].

(ii) **Detection and segmentation of foci** Nuclear segmentation enables quantitative foci analysis on a cell-by-cell basis. Here, foci detection refers to the phosphorylation sites of γH2AX or 53BP1, as shown in Fig. 14.2a. The foci morphology is typically round and punctate; however, the intensity and the signal amplitude can vary. Typically, spot detection is based on thresholding; however, thresholding is sensitive to the background intensity and presence of foci in the same proximity. More importantly, foci may vary slightly in size and shape. The current protocol is based on iterative radial voting [16], which has been validated [12] to be quite effective in spot detection and which remains robust to variations in the background and foreground intensity. This is an important feature since the signal intensity tends to vary as a result of the dynamic of the DNA repair mechanism. The iterative radial voting utilizes the local spatial gradient to detect "blob-shaped" objects and is tolerant to variations in shape and object size. It is essentially a form of clustering that gradually improves foci detection iteratively. Foci detection is followed by segmentation to quantify foci profiling (e.g., intensity, morphology), with an example shown in Fig. 14.4.

Fig. 14.4 Detection of foci on a cell-by-cell basis for two nuclei

(a)   **(b)**

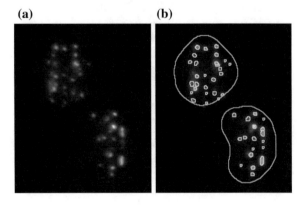

(iii) **Colocalization analysis** The database schema captures foci profile, on a cell-by-cell basis, which includes morphological and intensity-based indices. One of the novel aspects of our analysis is that co-localization analysis is performed at the database level. Each foci (e.g., γ-H2AX, 53BP1) has an assigned bounding box, and the ratio of the overlap between bounding boxes, computed from different channels, reflects co-localization being true or false. The overlap function takes advantage of the geometric data types and optimized geometric operators native to the PostgreSQL relational database management system [17]. The functionality leverages the *box data type* and the *intersection operator* in order to store the amount of intersection for each possible pair of foci within any single nucleus. With these stored tables, database functions are developed to quantify co-localization of foci, in different imaging channels, by defining the minimum percentage of overlap that should qualify as overlap or intersection. Each query has three parameters: percentage of overlap, the source channel, and the target channel. Subsequently, the system exports co-localization data per radiation dose on a cell-by-cell basis. One of the advantages of the co-localization studies through the database queries is that the query can be repeated, with a different parameter setting for the amount of overlap between computed foci from different imaging channels, to improve sensitivity and robustness in co-localization studies.

14.4   Results

In this section, we summarize quantitative results of co-localization experiments that were outlined earlier [4], which have not yet been reported. Figure 14.5a–f shows a representative set of phenotypic signatures for the site of DNA damage and repair. Accordingly, the response is heterogeneous, which requires a large dataset for appropriate statistical interpretation, and the repair kinetic is complete within the first 15 h following the radiation exposure. Figure 14.6a, b show the amount of co-localization over time for 53BP1 (source channel) with γH2AX (target channel) for 50 % (0.5), 75 % (0.75), and 95 % (0.95) foci overlap in stationary or cycling MCF10A cells. Results represent the average amount of co-localization from 2–3 independent experiments; for better visualization, error bars were not included. The amount of colocalization (e.g., the overlap between segmented foci regions in 53BP1 and γH2AX) drops for a stricter requirement of overlap for treated and control cells. Higher co-localization can be detected in exposed cells compared to the unexposed cells. In addition, a drop in co-localization over time is also noted. Unexposed cells show a more constant level of co-localization over time with only minor fluctuations, (20–40 % for stationary cells and 10–30 % for cycling cells), depending on the amount of overlap. In Fig. 14.6c the amount of co-localization for 50 % foci overlap in cycling cells is plotted over time, with error bars that represent the standard deviation of three independent experiments. It is clear that the highest

(a)   (b)   (c)

(d)   (e)   (f)

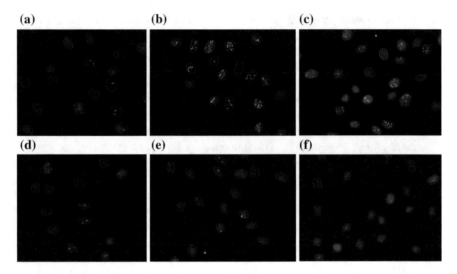

Fig. 14.5 Representative phenotypic signatures of the DNA repair sites, as a function of time, indicates that the repair mechanism is typically complete within the first 15 h following the radiation exposure

co-localization in the irradiated cells occurs between 2 and 6 h after exposure, and that there is a constant drop over time. Cells exposed to 2 Gy iron ions seem to have a slightly higher amount of co-localization than those exposed to 1 Gy. We did not observe 100 % co-localization in any of our samples at any time. Maximum levels were around 85 % of co-localization in cells, 6 h after a 2 Gy exposure (for 50 % overlap).

14.5   Discussion and Conclusion

An integrated approach for co-localization studies has been presented, where the spot counting method has been validated in contrast to human read outs in our earlier papers [18]. There is extensive literature on the co-localization of 53BP1 with γH2AX. Some studies have suggested nearly complete co-localization while others have reported either partial or no co-localization [5]. Our unbiased quantitative analyses of 53BP1 and γH2AX foci co-localization in MCF10A cell line, show a dose and post-irradiation time-dependency. The fraction of foci that co-localize drops over time, after reaching a maximum at about 2–6 h. In addition, radiation-induced foci in exposed cells show a higher level of co-localization as compared to unirradiated cells. Our results are in good agreement with co-localization data published by [19], in VH-10 fibroblasts and HeLa cells, where they also reported a lack of co-localization in a fraction of 53BP1 and γH2AX foci, despite generally good co-localization for the majority of the foci. Other authors

Fig. 14.6 a Amount of colocalization of 53BP1 (source channel) and γH2AX (target channel) foci in 2D cultures of stationary MCF10A cells after iron ion exposure over time. The amount of foci overlap is given in parenthesis. Data points represent the average of three independent experiments. **b** Amount of colocalization of 53BP1 (search channel) and γH2AX (target channel) foci in 2D cultures of cycling MCF10A cells after iron ion exposure over time. Amount of foci overlap is given in parenthesis. Data points represent the average of two independent experiments. **c** Amount of co-localization of 53BP1 (source channel) and γ-H2AX (target channel) foci in 2D cultures of stationary MCF10A cells after iron ion exposure over time for 50 % (0.5) foci overlap. Error bars represent the standard deviation of three independent experiments

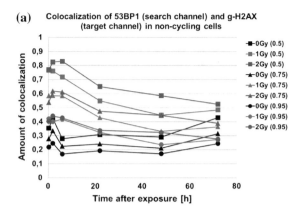

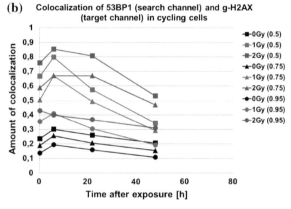

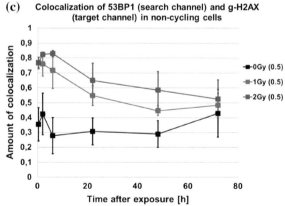

have reported complete co-localization of 53BP1 and γH2AX up to 24 h after iron ion exposure [20]. The variation in the extent of co-localization reported in the literature is most likely due to the use of different cell lines and radiation qualities, as well as the way co-localization was determined. In summary, the quantitative approach for foci detection and co-localization outlined here removes several degrees of freedom (e.g., cell line, dosage, fixation and staining) leading to additional uncertainties, enables unbiased analysis for large scale data, and integrates sensitivity analysis for cellular profiling.

Acknowledgment Funding: *National Institute of Health [grant R01 CA140663] and carried out at Lawrence Berkeley National Laboratory under Contract No. DE-AC02-05CH11231.*

References

1. Parvin B, Yang Q, Fontenay G, Barcellos-Hoff M (2003) BioSig: an imaging bioinformatics system for phenotypic studies. IEEE Trans Syst Man Cybernetic B33(5):814–824
2. Rossetto D, Avvakumov N, Côté J (2012) Histone phosphorylation: a chromatin modification involved in diverse nuclear events. Epigenetics 7(10):1098–1108
3. Bolte S, Cordelieres F (2006) A guided tour into subcellular colocalization analysis in light microscopy. Microscopy 224:213–232
4. Groesser T, Chang H, Fontenay G, Chen J, Costes S, Barcellos-Hoff M, Parvin B, Rydberg B (2011) Persistent of gamma-H2AX and 53BP1 fori in proliferating and non-proliferating human mammary epithelial cells after exposure to gamma-rays or iron ions. Int J Radiat Res 87(7):696–710
5. Belyaev I (2010) Radiation-induced DNA repair foci: spatio-temporal aspects of formation, application for assessment of radiosensitivity and biological dosimetry. Mutat Res 704:132–141
6. Coelho L, Shariff A, Murphy R (2009) Nuclear segmentation in microscope cell images: a hand-segmented dataset and comparison of algorithms. In: IEEE International symposium on biomedical imaging: from nano to macro, Boston, MA
7. Carpenter A, Jones T, Lamprecht M, Clarke C, Kang I, Friman O, Guertin D, Chang J, Lindquits R, Moffat J, Golland P, Sabatini D (2006) Cell profiler: image analysis software for identifying and quantifying cell phenotype. Gen Bio **7**(10)
8. Han J, Chang H, Andrarwewa K, Yaswen P, Barcellos-Hoff M, Parvin B (2010) Multidimensional profiling of cell surface proteins and nuclear markers. In: IEEE Transactions on computational biology and bioinformatics
9. Chang H, Yang Q, Parvin B (2007) Segmentation of heterogeneous blob objects through voting and level set formulation. Pattern Recogn Lett 28(13):1781–1787
10. Yang Q, Parvin B (2003) Harmonic cuts and regualrized centroid transform for localization of subcellular structures. IEEE Trans Bioeng 50(4):469–476
11. Lin G, Adiga U, Olson K, Guzowski JF, Barnes CA, Roysam B (2003) A hybrid 3-D watershed algorithm incorporating gradient cues & object models for automatic segmentation of nuclei in confocal image stacks. Cytometry Part A 56A(1):23–36
12. Raman S, Maxwell C, Barcellos-Hoff MH, Parvin B (2007) Geometric approach to segmentation and protein localization in cell-cultured assays. Microscopy 225(a)
13. Chang H, Wen Q, Parvin B (2015) Coupled segmentation of nuclear and membrane-bound macromolecules through voting and multiphase level set. Pattern Recogn 48(3):882–893
14. Chang H, Han J, AD B, Loss L, Gray J, Spellman P, Parvin B (2013) Invariant delineation of nuclear architecture in Glioblastoma Multiforme for clinical and molecular association. IEEE Trans Med Imaging 32(4):670–682
15. Wen Q, Chang H, Parvin B (2009) A Delaunay triangulation approach for segmenting a clump of nuclei. In: IEEE international synposium on biomedical imaging: from nano to macro. Boston, MA

16. Parvin B, Yang Q, Han J, Chang H, Rydberg B, Barcellos-Hoff MH (2007) Iterative voting for inference of structural saliency and characterization of subcellular events. IEEE Trans Image Proc 16(3):1781–1787
17. Urbano F, Cagnacci F (2014) Spatial database for GPS wildlife tracking tata: a practical guide to creating a data management system with PostgreSQL/PostGIS and R. Springer, New York
18. Raman S, Maxwell C, Barcellos-Hoff MH, Parvin B (2007) Geometric approach to segmentation and protein localization in cell culture assays. J Microsc 225(1):22–30
19. Markova E, Schultz N, Belyaev I (2007) Kinetics and dose-response of residual 53BP1/g - H2AX foci: co-localization, relationship with DSB repair and clonogenic survival. Int J Radiat Biology 83(5):319–329
20. Asaithamby A, Uematsu N, Chatterjee A, Story M, Burma S, Chen D (2008) Repair of HZE-particle-induced DNA double-strand breaks in normal human fibroblasts. Radiat Res 169 (4):437–446

Chapter 15
A Method to Regulate Cofilin Transport Using Optogenetics and Live Video Analysis

Atena Zahedi, Vincent On and Iryna Ethell

Abstract Alzheimer's disease (AD) is a neurodegenerative disease, where early stages of learning and memory loss are associated with a pronounced loss of synapses and dendritic spines. The actin-severing protein cofilin regulates the remodeling of dendritic spines in neurons, which are small protrusions on the surface of dendrites that receive inputs from neighboring neurons. However, the underlying mechanisms that mediate this are unclear. Previous studies have reported that phosphorylation regulates cofilin activity, but not much is known about the spatiotemporal dynamics of cofilin in synapses and spines. Here, an optogenetic method was developed to modulate the activity of cofilin, and video bioinformatics tools were used to study cofilin transport in dendritic spines and its effects on synapses. Gaining further insight into the workings of cofilin in spines can lead to potential therapies that regulate synaptic connectivity in the brain. In this chapter, a light-inducible, multichannel, live video imaging system was used to track the localization of cofilin, regulate its activity, and modulate synaptic connectivity in cultured hippocampal neurons.

A. Zahedi
Bioengineering Department, University of California Riverside,
Riverside, CA 92521, USA
e-mail: azahe001@ucr.edu

V. On
Electrical and Computer Engineering Department, University of California Riverside,
Riverside, CA 92521, USA
e-mail: von001@ucr.edu

I. Ethell (✉)
Biomedical Sciences Department, University of California Riverside,
Riverside, CA 92521, USA
e-mail: iryna.ethell@ucr.edu

© Springer International Publishing Switzerland 2015
B. Bhanu and P. Talbot (eds.), *Video Bioinformatics*,
Computational Biology 22, DOI 10.1007/978-3-319-23724-4_15

15.1 Introduction

Alzheimer's disease (AD) is a progressive neurodegenerative disorder characterized by the abnormal accumulation and aggregation of β-amyloid $(A\beta)_{1-42}$ peptides, which lead to the formation of Aβ plaques in the brain [1]. AD results in a decline of cognitive abilities and progressive dementia followed by loss of fine motor skills and severe neurological impairments that eventually lead to death [2]. However, early stages of the disease reveal a prominent loss of synapses/spines associated with memory decline [2–4]. The actin-severing protein cofilin has been implicated in the underlying mechanisms that govern synaptic loss prior to plaque formation [5, 6]. Here, a video bioinformatics approach was developed to study the spatiotemporal dynamics of cofilin in dendritic spines using photoactivatable probes and live multiphoton imaging. This approach can be used to further investigate the role of cofilin in loss of dendritic spines/synapses.

Dendritic spines are small, actin-enriched structures that house the post-synaptic sites of most excitatory synapses in the central nervous system (CNS) [7–11]. Dendritic spines modulate the formation, maintenance, and removal of synaptic connections [12–16]. The size of dendritic spines directly correlates with the number of neurotransmitter receptors in the post-synaptic density (PSD) [10, 11, 17–20]. Dendritic spines can rapidly undergo cytoskeletal changes that lead to their morphological enlargement, reduction, or deletion [13, 14, 19, 21]. These structural modifications correlate with plasticity at the level of synapses, and their long-term changes are important in cognitive functions such as learning and memory [12–16, 19, 22]. These activity-dependent modifications of spines are regulated by changes in the organization of filamentous actin (F-actin) [12–16], which is accomplished by actin-binding proteins, such as cofilin [6, 21, 23–25].

Pathological changes in actin organization can lead to synaptic loss and may underlie the cognitive decline seen in AD [6]. Brain tissues from human and transgenic AD mouse models show the presence of intracellular actin aggregates called cofilin-actin rods [6, 21]. These actin rods can cause abnormal protein inclusions, blockade of transport, neuronal swelling, and death [6, 26]. The formation of these rods is generally triggered by cofilin dephosphorylation and represents early signs of cytoskeletal alterations [6, 16, 21, 23–25]. Aβ-induced actin rods are associated with a decrease in spine density [27] and signaling events that resemble long-term depression (LTD) [28, 29]. Moreover, previous studies have shown that elevation of Aβ enhances LTD [30] and activation of N-methyl-D-aspartate receptor (NMDAR)-dependent LTD results in translocation of cofilin to spines where it causes synaptic loss [31].

Cofilin is an actin-severing protein that can bind and disassemble filamentous (F)-actin, creating new barbed-ends (Fig. 15.1) [6, 16, 21, 23–25, 32–36]. This process increases the level of G-actin monomers [32, 33, 37] and can lead to actin remodeling and destabilization/loss of synapses [38–41]. Very low levels of cofilin lead to stability of the actin filaments, slightly higher concentrations lead to actin depolymerization, and even higher levels result in release of inorganic phosphate (Pi) without severing

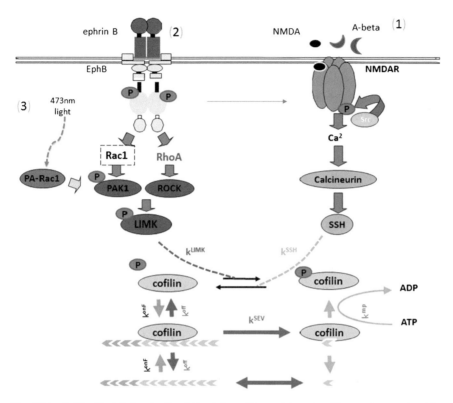

Fig. 15.1 Cofilin Modulation in Dendritic Spines. The activity of cofilin can be regulated by altering its phosphorylation state: (*1*) NMDAR-mediated Ca^{2+} influx, which activates calcineurin and dephosphorylates cofilin by SSH; (*2*) Rac1-PAK1-LIMK signaling cascade downstream of Ephrin/EphB receptors, which phosphorylates cofilin; and (*3*) activation via a light-controllable PA-Rac1. Pathway (*1*, *green*) can lead to F-actin severing and loss and/or reorganization of dendritic spines, or formation of cofilin-actin aggregates called rods. Pathways (*2*, *red*) and (*3*) are proposed to lead to polymerization of F-actin and stabilization of spines

and the formation of actin-cofilin rods [23–25, 42]. Another important consideration is the localization of cofilin in specific subcellular compartments such as dendritic spines, which affects the local dynamics of actin [23, 24]. Several mechanisms that regulate cofilin activity in dendritic spines have been reported, such as Rho GTPase-dependent regulation of LIM-kinase (LIMK), which results in cofilin phosphorylation and suppression of its activity [32, 34, 35, 38, 43, 44].

The complex networks of actin-regulating players in spines make it difficult to effectively target a specific pathway and distinguish its effects. Therefore, an optogenetic approach is advantageous in allowing for immediate probing and acquisition of live changes in protein dynamics and subsequent spine morphology.

15.1.1 The Cofilin Switch Is Regulated by Two Opposing Pathways

The local concentration of cofilin in subcellular compartments determines its behavior and subsequent shifting of the G-actin/F-actin ratio [23, 24, 32, 33]. Several signaling pathways that control cofilin activity in dendritic spines have been reported, such as the influx of Ca^{2+} through N-methyl-D-aspartate receptors (NMDARs) that rapidly activates calcineurin [45] and leads to cofilin dephosphorylation by Slingshot phosphatase (SSH) [46]. This study proposes an optical modulation approach to counter this effect by activating an alternative signaling pathway using a photoactivatable Rac1 (PA-Rac1) probe. By activating a PA-Rac1-Pak1-LIMK pathway, the aim is to induce phosphorylation of cofilin and inhibit its F-actin-severing function. Figure 15.1 depicts the various pathways that regulate cofilin in dendritic spines, along with the study's proposed optogenetic modulation method.

Since the activity of cofilin depends on both its phosphorylation state and localization, photoactivatable probes are useful tools to instantaneously and precisely control the cofilin switch. In particular, PA-Rac1 allows for reversible switching of activity with fast kinetics and is conveniently tagged with a mCherry fluorophore for visualizing PA-Rac-positive neurons without premature photoactivation [47]. Current pharmacological methods are chemically invasive, non-precise, and difficult in targeting a single neuron. A strong correlation has been established between excessive cofilin activity and AD pathology; therefore, is it imperative to further investigate the transport mechanisms of cofilin in and out of spines. A live video tracking approach can help to monitor and quantify the spatiotemporal regulation of cofilin in the spines, to better aid the study of its transport and mechanistic role in synaptic loss.

15.2 Experimental Procedures

15.2.1 Materials and Methods

Primary hippocampal neurons were prepared from embryonic day E15–E16 mice and plated on glass-bottom dishes coated with poly-D-lysine and laminin as previously described [31] (Fig. 15.2a). After 12–14 days in vitro (DIV), hippocampal cultures were transfected with a pTriEx-mCherry-PA-Rac1 (a gift from Klaus Hahn, Addgene plasmid # 22027) using a calcium phosphate method to express photoactivatable Rac1. Previously, Wu and colleagues developed this light-switchable probe by steric inhibition of its binding site to Pak1 using the photo-reactive LOV (light-oxygen-voltage) domain [47] (Fig. 15.2c). PA-Rac contains mutations to

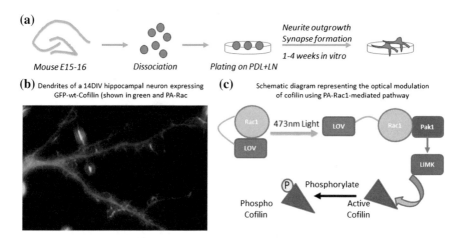

Fig. 15.2 Experimental schematics of cell culture, transfection, and photoactivation methods of the hippocampal neurons. **a** Primary hippocampal neurons were isolated from the hippocampus of embryonic day 15–16 (E15-16) mice, dissociated enzymatically, plated onto poly-D-lysine and laminin-plated glass coverslips, and allowed to mature in culture for 1–4 weeks. **b** The cultures were transfected at 14 days in vitro (DIV) using calcium phosphate method to overexpress PA-Rac and GFP-tagged-wt-Cofilin (shown in green). The live changes in cofilin localization were recorded by tracking the GFP signal, while simultaneously photoactivating PA-Rac with the same wavelength. **c** Exposure to ∼480 nm light results in conformational changes that expose Rac1 active binding site to Pak1. This results in Pak1 phosphorylation, which in turn triggers the activation of LIMK and subsequent phosphorylation (suppression) of cofilin

eliminate any dominant negative effects and reduce its interactions with alternative upstream effectors [47]. PA-Rac and a GFP-tagged-wt-Cofilin were co-expressed in cultured neurons (Fig. 15.2b), which were illuminated with approximately 473 nm light to optically trigger the signaling cascade shown below (Fig. 15.2c).

Rac1 activation was first confirmed in hippocampal neurons by illuminating PA-Rac-expressing neurons for 15 min and immunostaining for downstream effector, phospho-Pak1 (pPak1) [48]. As expected, higher levels of pPak1 were detected in PA-Rac-positive neurons in comparison to GFP-transfected controls [48]. Next, cofilin was optically modulated in live neurons by continuous whole-cell exposure, and time-lapse images were taken every 30 s. The live changes in the localization of GFP-wt-cofilin were recorded by tracking the GFP signal; this exposure to blue (480 nm) light allowed for simultaneous activation of PA-Rac and tracking of cofilin in spines. The live imaging was conducted with a Nikon Eclipse Ti inverted microscope made available by the University of California, Riverside (UCR) IGERT for Video Bioinformatics. Immunostaining results were analyzed using a Leica SP5 inverted confocal microscope made available by UCR's Microscopy and Imaging Core facility.

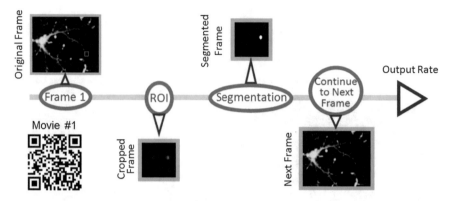

Fig. 15.3 Automated software was developed to track the dynamics of cofilin. The algorithm can quickly track and characterize regions of high intensity GFP-tagged cofilin in live cell videos. The resulting data can be used to determine parameters such as length, change in area, distance traveled, and speed of cofilin ROIs. Movie 1 quick response (QR) code shows a sample video of the segmented cofilin clusters

15.2.2 Video Tracking Algorithms

To quantify the movement of cofilin in and out of the dendritic spines, an automated cofilin tracking software was developed to segment regions of GFP-tagged

Table 15.1 Extracted Features after Segmentation of Videos

Features	Description
Area	Number of pixels in the segmentation
Perimeter	Number of pixels making the boundary of the segmentation
Centroid	Point representing the center of mass of the segmentation
Solidity	Ratio of pixels in segmentation to pixels in convex hull of segmentation
Orientation	Angle between the x-axis and major axis of the fitted ellipse with the same second-moments as the segmentation
Major axis	Major axis length of the fitted ellipse
Minor axis	Minor axis length of the fitted ellipse
Eccentricity	Ratio of distance between foci of the fitted ellipse and its major axis length. Takes a value between 0 and 1 with values closer to 0 being more circular
Max intensity	Pixel value of the most intensity pixel in the segmentation
Min. intensity	Pixel value of the least intensity pixel in the segmentation
Change in area	Difference in area between each successive frame
Change in perimeter	Difference in perimeter between each successive frame
Change in centroid	Pairwise difference in centroid between each successive frame
Velocity	Computed velocity of the segmented cofilin ROI

fluorescence (Fig. 15.3). To track the movement of cofilin, the video is thresholded to segment regions of high pixel intensity as shown in the Movie #1. Each video frame is then sorted into connected components. Extremely large and extremely small regions are removed due to their unlikelihood of being cofilin deposits. Next, the centroid of the silhouette is extracted and used to compute the pairwise distance of its coordinates between frames. The units of cofilin movement in pixels per frame can be translated to meters per second given the imaging parameters. The algorithm can extract an array of features (shown in Table 15.1), which can relate information about transport properties. Also, the diffusion coefficient of a freely moving cofilin region was extracted, as the mean square displacement of the ROI over a given time frame. The diffusion coefficient was calculated as 9.06E-3 μm^2/s, which is far slower than the typical 5–50 μm^2/s value of diffusion coefficient for proteins in the cytoplasm [49]. Therefore, diffusion was ruled out as the mechanism of cofilin transport in spines of hippocampal neurons.

The algorithm was used to segment regions of high cofilin intensity in time-lapse videos of PA-Rac-/wt-Cofilin-expressing neurons. Figure 15.4 illustrates an oscillatory type of cofilin transport in a subset of the dendritic spines. The software was used to approximate that every 5 min cofilin will accumulate in the spine heads, followed by a rapid export from dendritic spines, which can be visualized in attached Movie #2. Also, cofilin can cluster in the form of aggregated cofilin-actin rods, and can be distinguished as areas of cofilin saturation (high pixel intensity). To analyze the effect on the entire neuron as a whole, it was necessary to develop a statistical model of the different types of cofilin and changes occurring in the target video population over time.

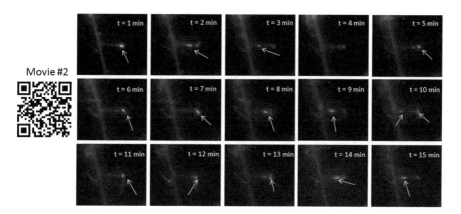

Fig. 15.4 Tracking cofilin in PA-Rac and wt-cofilin-expressing neurons using time-lapse imaging. Photoactivation of PA-Rac in neurons expressing GFP-wt-cofilin resulted in an oscillatory type of cofilin accumulation and export from the dendritic spines approximately every 5 min. Cofilin can cluster and move in spines and can be identified by areas of cofilin saturation and high pixel intensity. White arrows denotes cofilin clusters in the dendritic spine. Movie 2 QR code provides a link to the corresponding video

15.2.3   Video Bioinformatics Analysis

The different species of cofilin were classified over time in the entire video, using the cofilin tracking algorithm and additional bioinformatics tools (Fig. 15.5). Specifically, kymographs can provide a graphical representation of dynamics over time, where the spatial axis represents time [50]. Kymographs of dendritic regions were used to map the trajectory of the ROIs and visualize features such as length

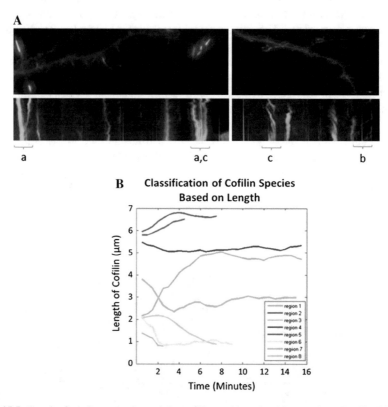

Fig. 15.5 Results from kymographs and the cofilin tracking algorithm used to classify different cofilin species. **A** Kymographs of cofilin-expressing neurons used to illustrate spatiotemporal trajectory of GFP-cofilin regions (*green*) in response to various stimuli. Time-lapse imaging reveals different cofilin classifications: (*a*) cofilin-actin aggregates, characterized by long regions of high cofilin intensity and minimal movement; (*b*) freely moving cofilin regions, which exhibit an oscillatory type of dynamics; and (*c*) hybrid structures, which alternate between rods and freely moving cofilin. The bottom row represents the kymographs as maximum intensity projections of the top images along the y-axis. As you move down vertically on the kymograph, time increases. **B** Classification results from the cofilin tracking algorithm, and their regions of high cofilin-GFP intensity were segmented and their length (major axis line) was extracted. Regions 3, 6, and 7 can be classified as freely moving cofilin regions, since their lengths are ≤3 μm. However, regions 2, 4, and 5 can be classified as rods due to longer lengths, and regions 1 and 7 as hybrids since they display a significant shift in length

and movement of cofilin. Kymographs were created by taking a maximum intensity projection of each frame and concatenating them on top of each other to create a time lapse image. The y-axis represents time in frames and the x-axis is the projection image of each frame. Kymographs of wt-Cofilin-expressing neurons were used to illustrate spatiotemporal translocation of GFP-cofilin regions (Fig. 15.5A, green) in response to the PA-Rac activation. Time-lapse imaging revealed different classes of cofilin: cofilin-actin aggregates, which can be characterized by long regions of high cofilin intensity and minimal movement (Fig. 15.5Aa); freely moving cofilin regions, which exhibit an oscillatory type of dynamics (Fig. 15.5Ab); and hybrid structures, which can alter between rods and freely moving cofilin (Fig. 15.5Ac). Subsequently, the cofilin tracking algorithm was used to segment regions of high cofilin-GFP intensity, followed by extraction of various morphological features listed in Table 15.1.

15.2.4 Cofilin Classification and Features

As previously mentioned, hyperactive cofilin can lead to the formation of cofilin-saturated actin filaments called rods [6]. Rods disrupt synaptic function and are a good indicator of synaptic loss prior to neuronal loss [6]. Therefore, it is desirable to be able to automatically classify the various types and relative ratios of the various cofilin forms, such as the accumulated rods in neurites. This can be done with a cofilin rod classification algorithm, which uses a combination of features for detection of various cofilin forms. This software is able to classify the segmented tracks as cofilin-actin rods, freely moving cofilin, or hybrids cofilin regions by using a decision tree. The length of the segmented region is a useful indicator and has been previously used to identify rods in hippocampal neurons [50]. The length of a cofilin region is found by extracting the segmented region's major axis length. Next, regions <3 μm length were classified as freely moving cofilin, and regions of ≥3 μm length were classified as either rod or hybrid cofilin forms. However, additional features were required to make further distinctions between rods and hybrids.

Another key feature is the change in centroid of the segmented ROI, which detects the distance traveled (pair-wise displacement) between each frame. If the region is not freely moving, then the change in centroid is checked to determine if it is a rod or hybrid cofilin form. Figure 15.6b demonstrates the average change in centroid values between each frame for the different cofilin classes. All three curves experience a drop in the first and last few frames, which may be due to fact that not all tracks are equivalent in length. The algorithm can be developed in the future to recognize a returning cyclic cofilin region that has previously disappeared. Therefore, the data may be skewed at both extremes of the curve. The freely moving curve also shows an elevated change in centroid values and a noisier curve, which may be a good indicator of its oscillatory behavior. Conversely, the rod and hybrid cofilin groups have lowered change in centroid values and smoother curves,

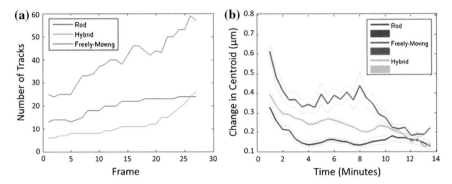

Fig. 15.6 A cofilin classifier algorithm was used to distinguish the different cofilin forms based on extracted characteristics. **a** The number of segmented cofilin regions were tracked in each frame, and classified as either rods, hybrids, or freely moving cofilin forms. Note that the number of freely moving cofilin (*red curve*) significantly increases over time; whereas, the rods (*blue curve*) and hybrids (*green curve*) predominately remain unchanged. **b** A change of centroid feature was used to further differentiate between the different types of cofilin. Note that the freely moving cofilin curve has several peaks, which may correlate with its oscillatory type of behavior. In contrast, the rod and hybrid curves are smooth, which is in line with the fact that they move less

which correlate with the observation that they move less. The hybrid curves have a steady decline in change of centroid over time. It is interesting that the number of freely moving cofilin regions increased over time (Fig. 15.6a). This may signify that the total pool of cofilin is being converted into more dynamic, freely moving types. However, the number of rods and hybrids are relatively stable, indicating that fewer new structures of these two cofilin types appear over time.

15.2.5 Monitoring the Remodeling of Spines

A spine tracking algorithm was developed to monitor and quantify the synaptic remodeling effects. Extracting the spines begins by focusing on a subregion in the video and segmenting the entire dendrite and its spines from the video's background. The software allows the user freedom to select a region of intensity and intensity threshold. Once extracted, the segmentation silhouette is automatically divided into two parts: the main dendritic branch and its individual spines. Next, spines are removed with an algorithm using morphological operators. The algorithm begins by removing boundary pixels from the silhouette until only a general skeleton of its shape remains. Afterwards spur pixels (pixels with exactly one, 8-connected neighbor) are removed from the skeleton. Spur pixels are connected diagonally to other pixels and connected in only one direction. This allows us to remove all diagonally connected spines. Also, spines that diverge more than a specified angle (45°) from the main dendrite were removed. These two operations enable removal of all spines from the skeleton leaving the main dendritic branch.

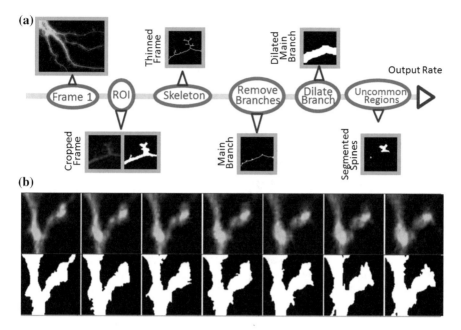

Fig. 15.7 a Automated spine tracking software was developed to study the effects of PA-Rac activation on density, size, and shape of spines in live videos. The algorithm tracks live video recordings of GFP-tagged cofilin in neurons to visualize the remodeling effects of spines. First, the entire dendrite and its spines are segmented by a user-specified threshold. Once extracted, the segmentation silhouette is divided into the main dendritic branch and individual spines. Next, the algorithm uses morphological operators to remove boundary pixels from the silhouette, leaving the main dendritic branch. The algorithm then dilates the branch by a function of its border pixels to create an estimation of the main dendritic shaft. This estimation is subtracted from the original silhouette to give the spine segmentation. By monitoring the volume and morphology of dendritic spines, their changes over the video timeframe can be calculated. **b** The *top row* displays the dendritic spine over time, while the *bottom row* displays the segmentation of the time points

The algorithm than dilates the branch by a function of its border pixels, giving us a general idea of the size of the main dendritic branch. This result is subtracted from the original silhouette to give only the segmentations of dendritic spines. By monitoring the volume and morphology of spines, changes over time, and other extracted features can be acquired for subsequent spine analysis (Fig. 15.7).

15.3 Discussion

Here, an optogenetic method was developed to regulate cofilin-mediated remodeling of dendritic spines in hippocampal neurons via a Rac1-Pak1-LIMK signaling pathway. A genetically encoded PA-Rac1 mutant was used to transiently photoactivate live hippocampal neurons expressing GFP-tagged-wt-cofilin. The phosphorylation state of cofilin was modulated and its translocation in spines was simultaneously recorded by exposure to approximately 473 nm light. An automated tracking algorithm was developed to study cofilin dynamics by classifying various forms of cofilin over time. PA-Rac activation resulted in cyclic export of GFP-wt-cofilin from a subset of spines. Furthermore, a classification algorithm was developed to show that the number of freely moving cofilin regions increased over time.

Dephosphorylation (activation) of cofilin can be stimulated by initiators of neuronal degeneration, such as oxidative stress, excitotoxic glutamate, and soluble forms of Aβ [6, 27, 51]. Also, LTD-like mechanisms can lead to a rapid translocation of active cofilin to dendritic spines [31]. The number of neurotransmitter receptors at the PSD is crucial for synaptic plasticity [12–16, 21]; therefore, cofilin may be involved in actin-mediated recycling of receptors. Furthermore, cofilin-actin rods block normal intracellular trafficking, which may further implicate cofilin in neurodegeneration. The software can be used to track cofilin and study its transport properties in response to various experimental conditions, such as PA-Rac activation or $Aβ_{(1-42)}$ treatments. Furthermore, the software can be expanded to encompass multi-channel recording for the purpose of co-localization studies and can be used to investigate the role of other aiding molecules in the cofilin transport machinery. Lastly, the spine tracking algorithm can analyze high-resolution, spatiotemporal modifications of the subsequent actin remodeling in various treatments. This can aid the study cofilin-mediated regulation of spine morphology, composition, and stability, which likely contribute to the long-lasting changes in synaptic efficacy, with significant implications for AD pathology and normal spine plasticity.

Controlling excessive activity of cofilin could serve as a therapeutic approach for rescuing the Aβ-induced synaptic loss associated with AD. While individuals who are predisposed to AD can do nothing to change various risk factors such as age or genetics, research aimed at the degenerative effects of the disease leave hope for alternative treatments. Although many studies have been directed at generating drug candidates that reduce levels of Aβ peptides and plaque, none have been successful in sustaining memory and cognitive functions in AD patients. Therefore, the goal is to target the disease in its earliest stages by modulating synaptic connections. The optogenetic method proposed in this chapter is useful for achieving simultaneous and harmless optical manipulation of cofilin in dendritic spines. Moreover, using live video segmentation and other bioinformatics tools, it is possible to study the transport of cofilin in dendritic spines. Other imaging modalities can be used to further study and measure spatiotemporal dynamics. By utilizing advanced bioinformatics techniques, the mechanisms of cofilin regulation in spines can be elucidated with a potential for clinical applications to prevent or slow the progression of Aβ-mediated synaptic and cognitive deficits in AD patients.

15.4 Conclusions

This chapter describes a method for the multichannel recording and activation of a photoactivatable Rac1 (PA-Rac1) probe in cultured hippocampal neurons. Since elevated cofilin activity has been linked to loss of synapses and dendritic spines, modulating cofilin via a Rac1-Pak1-LIMK1-mediated pathway can serve as a therapeutic approaches that target synaptic loss in disorders such as Alzheimer's disease. Live imaging experiments revealed that PA-Rac activation triggers the cyclic export of cofilin in a subset of dendritic spines. Next, a video bioinformatics approach was developed to track the GFP-labeled-cofilin clusters in live neuronal cultures. After segmenting the cofilin clusters, their spatiotemporal dynamics were evaluated using kymographs. Furthermore, the different cofilin types were classified based on various morphological features. Additionally, a spine segmentation algorithm was developed in order to track the size, number, and shape of dendritic spines.

Future work in this study will focus on automatic classification of spine shape and correlation to the cofilin dynamics, such as to its translocation trajectory relative to the dendrite. Also, the two algorithms can be consolidated to study the effect of cofilin translocation on spine density and morphology. In follow-up studies, tracking the movements of the constitutively active cofilin^{S3A} (Ser3 mutated to an alanine) and phospho-mimetic cofilin^{S3D} (Ser3 mutated to aspartate) mutants and the resulting effects on the various forms of cofilin can be explored. This will determine whether phosphorylation alone is responsible for directed transport and elucidate the mechanisms regulating cofilin transport in dendritic spines. By optogenetically probing and studying the cofilin system, mechanistic insights can be explored regarding the regulation of cofilin activity and translocation, which underlie actin remodeling in dendritic spines.

Acknowledgment This work was supported in part by the National Science Foundation Integrative Graduate Education and Research Traineeship (IGERT) in Video Bioinformatics (DGE-0903667). Atena Zahedi and Vincent On are IGERT Fellows.

References

1. Gervais F et al (2007) Targeting soluble Abeta peptide with Tramiprosate for the treatment of brain amyloidosis. Neurobiol Aging 28(40):537–547
2. Palop JJ, Mucke L (2010) Amyloid-beta-induced neuronal dysfunction in Alzheimer's disease: from synapses toward neural networks. Nat Neurosci 13(7):812–818
3. Scheff SW et al (2007) Synaptic alternations in CA1 in mild Alzheimer disease and mild cognitive impairment. Neurology 68(18):1501–1508
4. Halpain S, Spencer K, Graber S (2005) Dynamics and pathology of dendritic spines. Prog Brain Res 147:29–37
5. Ethell IM, Pasquale EB (2005) Molecular mechanisms of dendritic spine development and remodeling. Prog Neurobiol 75:161–205
6. Bamburg JR et al (2010) ADF/Cofilin-actin rods in neurodegenerative diseases. Curr Alzheimer Res 7(3):241–250

7. Rao A, Craig AM (2000) Signaling between the actin cytoskeleton and the postsynaptic density of dendritic spines. Hippocampus 10(5):527–541
8. Sorra KE, Harris KM (2000) Overview on the structure, composition, function, development, and plasticity of hippocampal dendritic spines. Hippocampus 10(5):501–511
9. Hering H, Sheng M (2001) Dendritic spines: structure, dynamics and regulation. Nat Rev Neurosci 2(12):880–888
10. Yuste R, Bonhoeffer T (2004) Genesis of dendritic spines: insights from ultra-structural and imaging studies. Nat Rev Neurosci 5(1):24–34
11. Matsuzaki M et al (2001) Dendritic spine geometry is critical for AMPA receptor expression in hippocampal CA1 pyramidal neurons. Nat Neurosci 4(11):1086–1092
12. Segal M (2005) Dendritic spines and long-term plasticity. Nat Rev Neurosci 6(4):277–284
13. Yasumatsu N et al (2008) Principles of long-term dynamics of dendritic spines. J Neurosci 28(50):13592–13608
14. Matsuzaki M et al (2004) Structural basis of long-term potentiation in single dendritic spines. Nature 429(6993):761–766
15. Okamoto K et al (2004) Rapid and persistent modulation of actin dynamics regulates postsynaptic reorganization underlying bidirectional plasticity. Nat Neurosci 7(10):1104–1112
16. Zhou Q, Homma KJ, Poo MM (2004) Shrinkage of dendritic spines associated with long-term depression of hippocampal synapses. Neuron 44(5):749–757
17. Zito K et al (2009) Rapid functional maturation of nascent dendritic spines. Neuron 61(2):247–258
18. Kasai H et al (2010) Structural dynamics of dendritic spines in memory and cognition. Trends Neurosci 33(3):121–129
19. Takumi Y et al (1999) Different modes of expression of AMPA and NMDA receptors in hippocampal synapses. Nat Neurosci 2(7):618–624
20. Racca C et al (2000) NMDA receptor content of synapses in stratum radiatum of the hippocampal CA1 area. J Neurosci 20(7):2512–2522
21. Minamide LS et al (2000) Neurodegenerative stimuli induce persistent ADF/cofilin-actin rods that disrupt distal neurite function. Nat Cell Biol 2(9):628–636
22. Fukazawa Y et al (2003) Hippocampal LTP is accompanied by enhanced F-actin content within the dendritic spines that is essential for late LTP maintenance in vivo. Neuron 38(3):447–460
23. Andrianantoandro E, Pollard TD (2006) Mechanisms of actin filament turnover by severing and nucleation at different concentrations of ADF/cofilin. Mol Cell 24(1):13–23
24. Chan C, Beltzner CC, Pollard TD (2009) Cofilin dissociates Arp2/3 complex and branches from actin filaments. Curr Biol 19(7):537–545. (REPEAT OF 35)
25. Gungabissoon RA, Bamburg JR (2003) Regulation of growth cone actin dynamics by ADF/cofilin. J Histochem Cytochem 51(4):411–420
26. Whiteman IT, Gervasio OL, Cullen KM et al (2009) Activated actin-depolymerizing factor/cofilin sequesters phosphorylated microtubule-associated protein during the assembly of Alzheimer-like neuritic cytoskeletal striations. J Neurosci 29(41):12994–13005
27. Maloney MT et al (2005) Beta-secretase-cleaved amyloid precursor protein accumulates at actin inclusions induced in neurons by stress or amyloid beta: a feedforward mechanism for Alzheimer's disease. J Neurosci 25(49):11313–11321
28. Shankar GM et al (2007) Natural oligomers of the Alzheimer amyloid-beta protein induce reversible synapse loss by modulating an NMDA-type glutamate receptor-dependent signaling pathway. J Neurosci 27(11):2866–2875
29. Hsieh H et al (2006) AMPAR removal underlies Abeta-induced synaptic depression and dendritic spine loss. Neuron 52(5):831–843
30. Li S et al (2009) Soluble oligomers of amyloid Beta protein facilitate hippocampal long-term depression by disrupting neuronal glutamate uptake. Neuron 62(60):788–801
31. Pontrello CG et al (2012) Cofilin under control of β-arrestin2 in NMDA-dependent dendritic spine plasticity, long-term depression (LTD), and learning. PNAS

32. Condeelis J (2001) How is actin polymerization nucleated in vivo? Trends Cell Biol 11 (7):288–293
33. Sarmiere PD, Bamburg JR (2004) Regulation of the neuronal actin cytoskeleton by ADF/cofilin. J Neurobiol 58(1):103–117
34. Yang N et al (1998) Cofilin phosphorylation by LIM-kinase 1 and its role in Rac-mediated actin reorganization. Nature 393(6687):809–812
35. Gungabissoon RA, Bamburg JR (2003) Regulation of growth cone actin dynamics by ADF/cofilin. J Histochem Cytochem 51(4):411–420
36. DesMarais V et al (2005) Cofilin takes the lead. J Cel Sci 118(Pt 1):19–26
37. Bamburg JR (1999) Proteins of the ADF/cofilin family: essential regulators of actin dynamics. Annu Rev Cell Dev Biol 15:185–230
38. Shi Y et al (2009) Focal adhesion kinase acts downstream of EphB receptors to maintain mature dendritic spines by regulating cofilin activity. J Neurosci 29(25):8129–8142
39. Bamburg JR (1999) Proteins of the ADF/cofilin family: essential regulators of actin dynamics. Annu Rev Cell Dev Biol 15:185–230
40. Condeelis J (2001) How is actin polymerization nucleated in vivo? Trends Cell Biol 11(7):288–293
41. Sarmier PD, Bamburg JR (2004) Regulation of the neuronal actin cytoskeleton by ADF/cofilin. J Neurobiol 58(1):103–117
42. Chan C, Beltzner CC, Pollard TD (2009) Cofilin dissociates Arp2/3 complex and branches from actin filaments. Curr Biol 19(7):537–545
43. Arber S et al (1998) Regulation of actin dynamics through phosphorylation of cofilin by LIM-kinase. Nature 393(6687):805–809
44. Carlisle HJ et al (2008) SynGAP regulates steady-state and activity-dependent phosphorylation of cofilin. J Neurosci 28(50):13673–13683
45. Quinlan EM, Halpain S (1996) Postsynaptic mechanisms for bidirectional control of MAP2 phosphorylation by glutamate receptors. Neuron 16(2):357–368
46. Wang Y, Shibasaki F, Mizuno K (2005) Calcium signal-induced cofilin dephosphorylation is mediated by Slingshot via calcineurin. J Biol Chem 280(13):12683–12689
47. Wu YI, Hahn KM et al (2009) A genetically-encoded photoactivatable Rac controls the motility of living cells. Nature 461(7260):104–108
48. Zahedi A et al (2013) Optogenetics to target actin-mediated synaptic loss in Alzheimer's. In: Proceedings of SPIE 8586, optogenetics: optical methods for cellular control, vol 85860S
49. Elowitz MB, Surette MG, Wolf PE, Stock JB, Leibler S (1999) Protein mobility in the cytoplasm of Escherichia coli. J Bacteriol 1:197203
50. Chicon J et al (2012) Cofilin aggregation blocks intracellular trafficking and induces synaptic loss in hippocampal neurons. J Biol Chem 287:3919–3929
51. Davis RC et al (2009) Mapping cofilin-actin rods in stressed hippocampal slices and the role of cdc42 in amyloid-beta-induced rods. J Alzheimers Dis 18(1):35–50

Part VI
Software, Systems and Databases

Chapter 16
Integrated 5-D Cell Tracking and Linked Analytics in the FARSIGHT Open Source Toolkit

Amine Merouane, Arunachalam Narayanaswamy
and Badrinath Roysam

Abstract Modern optical microscopy is now a multi-dimensional imaging tool that enables recording of multiple structures and dynamic processes in living specimens in their three-dimensional (3-D) spatial context and temporal order, yielding information-rich 5-D images (3-D space, time, spectra). Of interest are complex and dynamic tissue microenvironments that play critical roles in health and disease, e.g., tumors, and immune system tissues. The task of analyzing these images exceeds human ability due to the volume of the data, its structural complexity, and the dynamic behaviors of cells. There is a need for automated systems to assist the human analyst to map the tissue anatomy, quantify structural, and temporal associations; identify critical events, map event locations, and timing to the tissue anatomic context; produce meaningful summaries of multivariate measurement data; and compare perturbed and normal datasets for testing hypotheses, exploration, and systems modeling. Beyond automation, there is a need for "computational sensing" of sub-visual tissue patterns and cell behaviors. Recent progress has resulted in algorithms for tracking migrating cells from time-lapse multi-photon image sequences. This chapter describes algorithms and an open source toolkit (www.farsight-toolkit.org) for end-to-end analysis of 5-D microscopy data including image pre-processing, cell segmentation, automated tracking, linked visualization of image-derived measurements and analyses in the context of the image data, and

---

Portions of this chapter appeared as: A. Narayanaswamy, E. Ladi, Y. Al-Kofahi, Y. Chen, C. Carothers, E. Robey, and B. Roysam, "5-D imaging and parallel automated analysis of cellular events in living immune tissue microenvironments," 2010 IEEE International Symposium on Biomedical Imaging: From Nano to Macro, 2010, pp 1435–1438.

A. Merouane · B. Roysam (✉)
University of Houston, Houston, TX, USA
e-mail: broysam@central.uh.edu

A. Merouane
e-mail: amerouane@uh.edu

A. Narayanaswamy
Google Inc., Mountain View, CA, USA
e-mail: arunachalam@google.com

© Springer International Publishing Switzerland 2015
B. Bhanu and P. Talbot (eds.), *Video Bioinformatics*,
Computational Biology 22, DOI 10.1007/978-3-319-23724-4_16

edit-based validation of cell movement tracks, and multivariate pattern analysis tools for translating the resulting quantitative measurements into biological insight. This framework provides the biologist with the necessary tools to achieve reliable quantification, and help accelerate the experimental and discovery cycles.

16.1 Introduction

Many complex and dynamic biological tissue microenvironments play critical roles in health and disease—examples include the immune system, stem-cell niches, tissues surrounding implants, developing/remodeling/injured tissues, tumors, and cancer stem-cell niches [1, 2]. Achieving a deep understanding of these microenvironments is a broadly recognized need that cuts across disease areas and disciplines. However, there is slow progress in characterizing these microenvironments compared to the need, because current methods require the investigator to painstakingly piece together knowledge from a large numbers of experiments, each of which yields a small amount of data (e.g., 2D slices of 3D tissue, static snapshots of dynamic tissue, small number of molecular entities per image, and low-throughput investigation). High-throughput methods (e.g., gene and protein arrays, and flow cytometry) help by enabling a larger number of molecules and cells to be studied at once. However, by disrupting the cell and tissue architecture, they miss vital information on sub-cellular protein localization, location of cells in tissue, dynamic interactions, collective behaviors, and spatiotemporal relationships between cells and structures. Indeed, our greatest knowledge gaps relate to the in vivo dynamic behaviors of cells in these microenvironments, and the influence of the microenvironment on the cell behaviors.

Advances in multi-photon, multi-channel, time-lapse microscopy, coupled with improved live-cell chamber designs and sensitive photodetectors have enabled 5-dimensional (x, y, z, t, λ) imaging of living tissue microenvironments over extended durations, providing information-rich datasets revealing dynamic processes in their native tissue context [3]. As an example, panel A of Fig. 16.1 shows one frame from a 5-D movie. Intact thymic lobes from a GFP transgenic mouse were imaged ex vivo using a custom-built two-photon microscope. In panel B, the lymph node of an adult mouse was imaged in vivo using a two-photon microscope. The lymph node of the rat was surgically exposed, mounted tightly on an apparatus and imaged. The aim of this study was to understand the role of the LFA1 integrin in the interaction of T cells and dendritic cells. The full movie captures the migrations and interactions occurring in a crowded field of wild-type and genetically modified (P14) thymocytes with dendritic cells, in relation to the vasculature (red), allowing the cell interactions to be studied in their vascular context for the first time.

The motivation of 5-D imaging is straightforward. A single 5-D movie is more informative than hundreds, or even thousands of 2-D images of fixed, thin specimens labeled with a couple of stains. These movies can provide more insight from

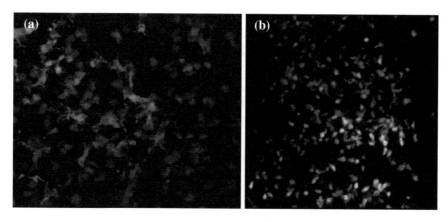

Fig. 16.1 Immune system microenvironments are dense and intricate in structure and the interactions between components of the niche are complex. **a** Ex vivo live image (2-D maximum intensity projection of a 3-D 2-photon image) of a developing thymus from a mouse revealing four channels consisting of wild-type thymocytes (*cyan*), F5 thymocytes (*green*), dendritic cells (*crimson*), and blood vessels (*red*) [3]. **b** In vivo live image of a lymph node in a mouse showing 3-channels—CD4$^+$ LFA-1^{KI} T cells (*green*), OTI cells (*cyan*), dendritic cells GP33+ (*red*) [30]

fewer experiments and animals by enabling direct observation of processes instead of having to infer them indirectly from multiple images (that may come from different animals). They can reveal subtle dynamic behaviors and rare events as they occur, which are nearly impossible to infer from fixed specimens. They can reveal the tissue context/influences behind observed processes (e.g., the presence of nearby vasculature or an interacting cell type), and expose more of the underlying variables simultaneously. This can lead to direct, comprehensive, and simpler systems modeling.

Achieving the potential of 5-D microscopy-driven research requires automated systems to quantify 5-D data, because the size and complexity of 5-D data precludes human analysis, and mere 5-D visualization is insufficient. Manual analysis is expensive, tedious, slow, and subjective. There is a need for automated tools to aid the researcher through all stages of experimentation starting with specimen preparation and microscopy, image segmentation and tracking, quality control, multivariate data analysis, and design of the next round of experiments based on insights from the previous round of experiments.

In order to achieve the aforementioned goals, one needs a synergistic combination of accurate, scalable cell segmentation, and tracking algorithm combined with powerful editing and analytics tools. In this chapter, we describe these tools that have been built as a part of the Farsight toolkit. Figure 16.2 depicts a flowchart of the toolkit's building blocks, workflow and the software components.[1]

---

[1]Parts of this figure appeared in Kevin W. Eliceiri, Michael R. Berthold, Ilya G. Goldberg, Luis Ibáñez, B.S. Manjunath, Maryann E. Martone, Robert F. Murphy, Hanchuan Peng, Anne L.

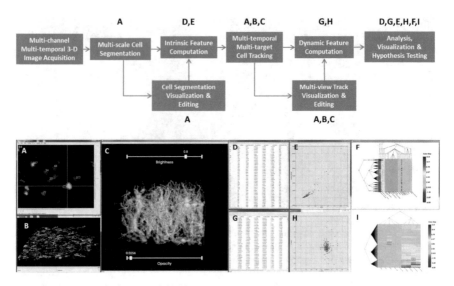

Fig. 16.2 Illustrating the Farsight-5D system. *Top* diagram illustrating a flowchart of the workflow and the components of the Farsight-5D system. *Bottom* screenshots of the system windows. Each module corresponds to one or more view (window) of the system. **a** Shows as an orthogonal (*x*, *y*, *z*, *t*) view. Cells are delineated and identified with *numbers* that correspond to rows of a table of cell measurements (**d**) and points in the scatter plot (**e**). **b, c** The cell tracking results are displayed in a *beads on strings* and *kymograph* view, showing the 3D movement paths of cells for detecting anomalies. The corresponding cell dynamic measurements are shown in table (**g**) and scatter (**h**). **f** Bi-cluster plots organize the cell data into groups based on the intrinsic measurements in (**d**). **i** Bi-cluster plot organizing the cell tracks into groups based on the dynamic measurements in table (**g**)

The organization of this chapter is as follows. Section 16.2 describes a segmentation algorithm that uses a combination of foreground extraction and seed detection algorithms to delineate cell boundaries. Section 16.3 describes a fully automated multi-temporal method for globally optimal cell tracking. The algorithm formulates the tracking problem as a second-order edge selection problem in directed hypergraph. 0–1 integer programing is used to solve the assignment problem. Section 16.4 discusses the multi-view editing and analytics framework and its implementation in the Farsight toolkit. Section 16.6 presents conclusions and future directions.

(Footnote 1 continued)

Plant, Badrinath Roysam, Nico Stuurman, Jason R. Swedlow, Pavel Tomancak, and Anne E. Carpenter, "Biological Imaging Software Tools," Nature Methods 9, 697–710 (2012).

16.2   Image Pre-processing and Cell Segmentation

Automatic segmentation and delineation of cells is typically the first step in analyzing time-lapse microscopy images. It is crucial for a cell segmentation algorithm
to be able to accurately detect the correct number of cells and delineate them
automatically. Designing a robust cell segmentation algorithm can be a difficult
task. One must address several challenges, some of which are related to the inherent
heterogeneity in cell morphology such as shapes and sizes; others are due to the
limitations present in the image acquisition process such as poor and varying image
contrast and image noise. Computational cost and algorithm complexity are also
important factors, which need to be taken into consideration.

Although designing a robust segmentation algorithm is essential, image
pre-processing is often required to facilitate the application of the following steps in
the analysis pipeline, i.e., segmentation and tracking. The choice of the
pre-processing algorithms usually depends on the type of artifacts present in the raw
image data. We typically apply a background subtraction step to correct for
non-uniform illumination in the image, followed by Gaussian or median filtering for
noise reduction. The type of the smoothing filter is usually based on the image
statistics. More sophisticated de-noising algorithms [4–6] can also be used although
sometimes at the expense of computational complexity.

The cell segmentation procedure based on the algorithm proposed by [7] provides important advantages for large-scale automated cell analysis. The most
important advantage is the absence of adjustable parameters for a default application of the algorithm to image data. The few parameters that remain are intuitive
and the algorithm is not overly sensitive to them. This algorithm starts by extracting
the image foreground using the Poisson distribution-based minimum error thresholding and the Graph Cuts algorithm [8]. The next step consists of identifying the
number of cells and cell centers. This is achieved by transforming the intensity
image to represent the cells by blobs. The local maxima points of these blobs can be
interpreted as cell centers, which will also be referred to as seed points in the rest of
this chapter. The transformation is based on a multi-scale Laplacian of Gaussian
(LoG) filtering of the intensity image [9]. Given a set of seed points and a transformed image, a fast clustering algorithm known as local maximum clustering [10]
is used to delineate each cell.

For the initial binarization of an image $I(\mathbf{x})$ where $\mathbf{x}$ denotes the 2-D/3-D pixel
coordinate; the normalized histogram denoted by $p(i)$ where i denotes the intensity
of a pixel in the range $[0, I_{max}]$ which is modeled as a mixture of two Poisson
distributions as follows:

$$p(i) = P_0 \times p(i\backslash 0) + P_1 \times p(i\backslash 1), \qquad (16.1)$$

P_0 and P_1 represent the a priori probabilities of the background and the fore-
ground regions. $p(i\backslash 0)$ and $p(i\backslash 1)$ are Poisson distributions with means μ_0 and μ_1,

respectively. The aim is to find an optimal threshold value $t*$ by minimizing the error function as given by [11]

$$t^* = \operatorname{argmin}_t (\mu - P_0(t)(\ln P_0(t) + \mu_0(t)\ln\mu_0(t)) - P_1(t)(\ln P_1(t) + \mu_1(t)\ln\mu_1(t))).$$
(16.2)

For a given threshold t, the parameters of the mixture model are determined by

$$P_0(t) = \sum_{i=0}^{t} p(i), \quad \mu_0(t) = \frac{1}{P_0(t)}\sum_{i=0}^{t} i \times p(i),$$

$$P_1(t) = \sum_{i=t+1}^{I_{max}} p(i), \quad \mu_1(t) = \frac{1}{P_1(t)}\sum_{i=t+1}^{I_{max}} i \times p(i),$$
(16.3)

The threshold t^* is found by iterating through all intensity values and then setting t^* to the intensity value that minimizes the objective function. This step results in a binary image with foreground pixels labeled 1 and background labeled 0. This initial binarization is only adequate in settings where images have uniform foreground and background regions with high signal-to-noise ratio. In cases where images suffer from low signal-to-noise ratio or have highly textured regions, the binary image will potentially contain a large number of mislabeled pixels. In order to obtain a clean binary image, the initial binarization is refined by incorporating spatial relationships between neighboring pixels. In other words, a pixel's label does not only depend on its intensity but also on the intensities of its neighboring pixels. Mathematically, an optimal pixel labeling $L(\mathbf{x})$ is obtained by minimizing the following energy function:

$$E(L(\mathbf{x})) = \sum_{(\mathbf{x})} D(L(\mathbf{x}); I(\mathbf{x})) + \sum_{(\mathbf{x})} \sum_{(\mathbf{x}')\in N(\mathbf{x})} V(L(\mathbf{x}), L(\mathbf{x}')).$$
(16.4)

The data term in $E(L(\mathbf{x}))$ represents the cost of assigning a label $j \in \{0, 1\}$ to a pixel. It can be written as follows:

$$D(L(\mathbf{x}); I(\mathbf{x})) = -\ln p(I(\mathbf{x})\backslash j).$$
(16.5)

The data term measures how likely a label is for a pixel given the observed image. The second term is called the smoothness term. It penalizes different labels for neighboring pixels and measures the degree to which the image is not piecewise smooth [8]. It can be written as follows [12]:

$$V(L(\mathbf{x}), L(\mathbf{x}')) = \rho(L(\mathbf{x}), L(\mathbf{x}')) \times \exp\left(-\frac{|I(\mathbf{x}) - I(\mathbf{x}')|}{2\sigma_L^2}\right),$$
(16.6)

where

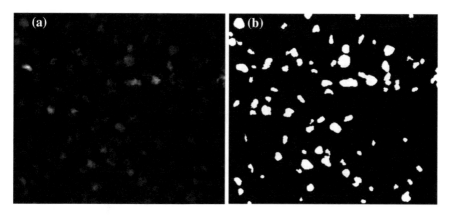

Fig. 16.3 a Projection of 3-D fluorescent image of Thymocyte channel at frame 1 of 5-D image shown in Fig. 16.1. **b** Projection of binarization results using the minimum error thresholding and the Graph Cuts algorithm

$$\rho(L(\mathbf{x}), L(\mathbf{x}')) = \begin{cases} 1, & \text{if } L(\mathbf{x}) \neq L(\mathbf{x}') \\ 0, & \text{if } L(\mathbf{x}) = L(\mathbf{x}') \end{cases} \tag{16.7}$$

where σ_L is a scale parameter. Optimization of the energy function in Eq. (16.7) is achieved via the max-flow/min-cut algorithm developed by [13]. Note that for this binary case, the Graph Cuts algorithm finds a global minimum of the energy function in low-order polynomial time. Figure 16.3 illustrates the binarization results using the described method.

To separate (split) such clusters, initial markers, i.e., seed points, are required. Detecting these points is analogous to finding cell centers. In some cases, the cells in an image appear as bright objects with increasing intensity values toward their centers and the seed points can be readily detected using a local maxima detection algorithm such as the *h-maxima* transform [14]. However, in instances where images contain textured cells or suffer from low signal-to-noise ratio, detecting local maxima points becomes a non-trivial task. In such scenarios, the intensity image needs to be transformed to a surface such that (1) the surface's peaks are representative of the cell centers (seed points), and (2) the surface is smooth enough so that maxima detection algorithms can be easily applied. The output of the convolution of a Laplacian of Gaussian filter $\text{LoG}(\mathbf{x}, \sigma)$ with an image containing blob-like objects with radius r when $r = \sqrt{2}\sigma$ is a smooth surface with a peak response at the center of each object. In order to detect objects of different sizes, the LoG filter is applied at different scales. Each LoG filter produces a response image with possibly a different peak location for each scale. To address this issue, the response at each pixel (voxel) location is set to be the maximum response over all scales.

Even so, this approach can only be used if the LoG filter is scale-normalized such that the response at different scales can be compared. The normalized LoG can be written as [9]

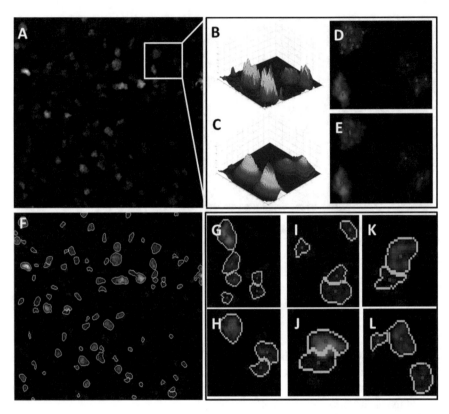

Fig. 16.4 Illustrating the seed detection and clustering steps of the segmentation algorithm. **a** Projection of 3-D fluorescent image of thymocyte channel at frame 1. **b** Surface plot of the highlighted region in *panel* (**a**). **c** Surface plot of the multi-scale LoG response of the highlighted region in *panel* (**a**). **d** Seed (in *red*) detection results using the local maxima clustering on the original image in (**b**). **e** Seed detection results using the local maxima clustering on the LoG response image in (**b**). **f** Segmentation results of the image shown in (**a**). **g–l** show enlarged regions from different parts of the result image in *panel* (**f**)

$$R(\mathbf{x}) = \mathrm{argmax}_\sigma(\mathrm{LoG}_{\mathrm{norm}}(\mathbf{x}; \sigma) \times I(\mathbf{x})). \qquad (16.8)$$

Finally, given a response image $R(\mathbf{x})$, the clustering algorithm presented in [10] is used to obtain the final segmentation. The foreground region is partitioned into clusters each of which represents a single cell. The algorithm starts by searching for local maxima points around the neighborhood of each point $\mathbf{x}$. The search neighborhood is defined by a resolution parameter vector $\mathbf{r} = [r_x, r_y, r_z]$. The result is a set of seed points representative of cluster centers, Fig. 16.4e. Given a set of N seed points labeled $\{1, \ldots, N\}$, the algorithm iterates through each foreground pixel (voxel) and assigns to it the same label assigned to the local maximum in its neighborhood. This process is repeated until all foreground pixels (voxels) have been assigned a value in the set $\{1, \ldots, N\}$. The output of this clustering algorithm is a label image where pixels belonging to the same cells are assigned the same value. Algorithm 1 provides a pseudocode of the described segmentation algorithm. Figure 16.4f–l illustrates the segmentation results.

Algorithm 1: Summary of the cell segmentation algorithm.

Input:

$I(\mathbf{x})$, smoothness parameter σ_L^2, LoG scale parameters $\sigma = [\sigma_{min}, …, \sigma_{max}]$, resolution parameter r.

Steps:

1. Compute $L(\mathbf{x})$ using equations (2) and (4)
2. Compute normalized LoG $R(\mathbf{x})$ using equation (8)
3. Compute seed point using $R(\mathbf{x})$ and the algorithm in [10]
4. Partition $L(\mathbf{x})$ using the algorithm in [10]

When applied to a sequence of images, the segmentation algorithm outputs a set of objects $\left\{ C_t^i \right\}_{t=1…T}^{i=1…N(t)}$ where T denotes the number of frames and $N(t)$ represents the number of cells in frame t. From these results, a collection of single-cell measurements, including location, intensity, area, shape factor, and molecular biomarker associations, are computed and concatenated in a single vector denoted as f_t^i. These features essentially capture cellular states (static) which can, for instance, be used in discovering heterogeneity in cell population. These features are also used in computing the temporal association costs of the tracking algorithm as will be described in the next section.

16.3 Cell Tracking

16.3.1 Prior Work

Tracking of motile objects in images has been a major focus in the field of computer vision for a long time. The vast interest shown in this topic has led a wide variety of algorithms ranging from overly highly tuned algorithms designed for a particular need to general algorithms that may not be suitable for a specific need. The primary goal behind a cell tracking algorithm is to find temporal correspondences between cells in a time-lapse video.

Most approaches try to approximate Multiple Hypothesis Tracking (MHT) to a few frames and find locally optimal solutions. Graph theoretic techniques solve the multi-target tracking as an assignment problem in some abstract space. In [15], the authors give an overview of assignment problems that arise in tracking. In [16], the authors used sequential MHT to construct cell lineages. Their modeling included mitosis events but not cells entering and leaving the field of view. The same approach was utilized by [17] to perform 5-D image analysis. In [18], the authors proposed a greedy method for multi-frame correspondence based on maximum path cover. The algorithm described in this chapter builds on this work and improves upon it in many ways. Multi-temporal correspondences are computed through second-order modeling and a globally optimal solution is provided. Jaqaman et al. [19] proposed a particle tracking algorithm that uses a two-part approach. In the first step, a frame-to-frame association is computed through linear programming as in [17]. Complete tracks are computed by combining track segments using gap closing, merging, and splitting.

Kalman and particle filtering techniques [20, 21] are often used for cell tracking. These sequential techniques have a high space requirement while handling joint distributions of states for multi-target tracking. Sophisticated re-sampling methods and measurement gating [20] are necessary to overcome the degeneracy problem. Another class of approaches is based on active contours [22]. These are unsuitable under high object density, and suffer from the well-known leakage problem. Topological constraints are necessary to avoid merging of contours [23]. These methods perform the association on frame-to-frame basis.

Many applications handle segmentation and tracking steps independently, enabling a modular approach. Improvement in segmentation techniques can easily be accommodated. The described method uses this approach to take advantage of the recent advances in cell segmentation.

16.3.2 The Algorithm

We formulate the cell tracking problem as an embedding in a hypergraph. A directed hypergraph is an ordered pair $HG = (V, E)$, where V is a set of N nodes and E is a multiset of hyperedges. A hyperedge $e = \{T(e), H(e)\}$ is an ordered pair of sets of vertices, where $T(e) \subset V$ is the *tail* of the edge e and $H(e) \subset V$ is the *head* of the edge e where $T(e) \cap H(e) = \phi$ (the null set). The nodes of the directed hypergraph are the cells and auxiliary vertices. Denote the ith vertex at time t by C_t^i. The feature vector for C_t^i is denoted f_t^i. The features include the centroid $(\mathbf{x}_c)$, time (t), volume (v), and bounding box. t varies between 0 and $T - 1$ when there are T frames in the image sequence. With the exception of the first and last time frames, every cell has a connected *appear* and a *disappear* vertex.

In our hypergraph, we represented the cells as vertices in the graph. We also augment this set of vertices with auxiliary vertices to model cellular events such as entry and exit of cells from the field of view of the microscope. We add hyperedges that model the possible cellular events. Then, we compute second-order hyperedges that characterize the second-order motion events and compute their utilities. Next, we formulate the tracking as a subset selection problem in the second-order hyperedge space. We represent the subset selection as an integer-programming problem and find a global optimum using a fast branch-and-cut method. Finally, we use a feature-based method to correct the segmentation errors found in the previous step and compute the corrected labels for the cells in the tracks.

The hypergraph contains three types of vertices: cells, appear, and disappear vertices. Appear and disappear vertices are auxiliary vertices that model the entry and exit of cells. Cell tracks only end on one of these auxiliary vertices. The hypergraph also contains hyperedges which are of five types, each signifying an event: *translation*, *appearance*, *disappearance*, *split*, and *merge*. Translation, appearance, and disappearance edges connect two vertices, whereas split and merge hyperedges connect three vertices. Translation indicates movement of a cell entirely within the field of view. Multi-frame edges are translation edges connecting cells that are more than one time point apart. Appearance edges connect an appear vertex to a cell and disappearance edges connect a cell to a disappear vertex. Merge hyperedges connect two cells from the same time point to another cell in the next time point. These hyperedges represent a segmentation error and not a mitosis event. Mitosis events are much slower events than the typical time scales we deal within our studies. Similarly, a split hyperedge represents a segmentation error leading to two cells in the next time point. Note that all the hyperedges discussed until now merely form the hypotheses set. The eventual selection problem identifies the correct hypotheses subset based on the utilities and imposed constraints.

Second-order hyperedges (SOH) are formed by combining *neighboring* first-order hyperedges (FOH). Mathematically, a second-order hyperedge S is an ordered pair $S = (E_1, E_2), E_{1Y} \cap E_{2X} \neq \phi$, where E_1, E_2 are two first-order edges. It is easy to see that $S \in \{E_{\text{in}}(V) \times E_{\text{out}}(V)\}$.

The second-order hyperedges are necessary to model second-order motion characteristics of a cell. In the following sections, we describe why the use of SOHs eliminates the need for pre-computing the local directions of cell movements to calculate the utility. Typically, the drawback of using SOH's is a combinatorial explosion of number of motion models. We overcome this by a principled approach to utility estimation.

The initial set of hyperedges represents a set of all possible cellular events, not all of which might have occurred in reality. The goal of our tracking algorithm is to find the correct subset of valid hypothesis. The utility for a hyperedge is the probability that the hyperedge corresponds to a real event. We propose two types of motion models (a) *parametric* motion model (b) *non-parametric* motion model. The difference between the two motion models is in the way the utility functions are generalizable. The non-parametric motion model has more degrees of freedom than the parametric counterpart and is thus more generalizable.

16.3.3   Motion Models

Parametric motion model The first-order utility $U(C_t^i, C_{t+1}^j)$ of a hyperedge connecting C_t^i and C_{t+1}^j is the a posteriori probability of the hyperedge corresponding to a real cellular event. Every hyperedge is associated with a set of features in an N-dimensional space. In the parametric motion model, the likelihood functions are represented by multi-dimensional Gaussian functions over the feature space of the hyperedges. The exact features used are described below. Formally, we the likelihood of a hyperedge given its feature vector (of dimensionality N) is

$$P\left(C_t^i, C_{t+1}^j\right) = \frac{1}{\sqrt{(2\pi)^N |\Sigma|}} e^{-\frac{1}{2}\left(\|f_t^i - f_{t+1}^j\| - \mu\right)\Sigma^{-1}\left(\|f_t^i - f_{t+1}^j\| - \mu\right)}, \qquad (16.9)$$

where Σ is the covariance matrix and μ is the mean vector. The mean and covariance matrix were estimated from a training set. Equation (16.9) defines the utility function of a translation edge. $U(C_t^i, C_{t+1}^j)$ is the product of the likelihood with prior ω. $\omega(T), \omega(MS)$ and $\omega(AD)$ are the prior probabilities of translation edges, merge/split edges, and appearance/disappearance edges, respectively.

$$U\left(C_t^i, C_{t+1}^j\right) = P\left(C_t^i, C_{t+1}^j\right) \times \omega(T). \qquad (16.10)$$

The likelihood of a split hyperedge is the sum of the likelihoods of the contributing hyperedges after accounting for the change in the volume of the split cells. The utility for a split hyperedge is dependent on the prior of a split hyperedge. Equation (16.11) formally defines the utility of a split hyperedge:

$$U\left(C_t^i, C_{t+1}^j, C_{t+1}^k\right) = \left(P\left(C_t^i, C_{t+1}^{\prime j}\right) + P\left(C_t^j, C_{t+1}^{\prime k}\right)\right) \times \omega(MS), \qquad (16.11)$$

where volume $(C_{t+1}^{\prime j})$ = volume $(C_{t+1}^{\prime k})$ = volume (C_{t+1}^j) + volume (C_{t+1}^k). In other words, for a split hypothesis, we consider the sum of the volume of the split cells to be equal to the volume of the cell prior to split to account for the change in volume. We compute it for the merge hypothesis in the same way.

We do not compute the first-order utility functions for appearance and disappearance edges because proximity to the boundary is not representative of the true likelihood of entry/exit of cells. The axial resolution is small in 2-photon microscopy (2 μm or more). We propose a technique based on forward and backward prediction of cell locations to compute second-order utilities for cells that appear/disappear.

SOHs have a second-order utility (U_s) associated with them. Figure 16.6 describes the calculation of U_s for the simplest case as a function of the first-order utility U. U_s is the product of first-order utilities and a correction factor for the alignment, which takes into account the deviation from a locally smooth linear motion model. Let A, B, C denote three cells at time points $t - 1, t, t + 1$, respectively. A vertex D is predicted at time t at the mean location of A, C. The distance between B and D is d_{align}. Alignment factor F_{align} is a normalized Gaussian function on d_{align}. Here, σ^2 is the distance variance in Σ. The following equation mathematically defines U_s for the simplest case:

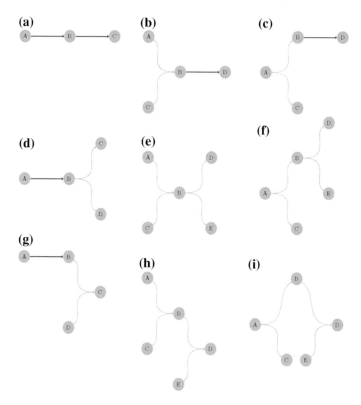

Fig. 16.5 An illustration of nine different motion models represented as a second-order hyperedge centered about a vertex B. The vertices need not be from consecutive time points. Each translation edge in the above figure could be a multi-frame edge in which case the cell was undetected over a short number of time frames

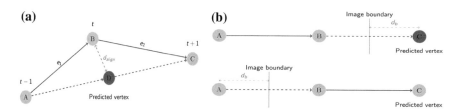

Fig. 16.6 a Illustration of the misalignment cost calculation. D is the predicted vertex location based on A and C. Misalignment cost is computed as in (3.5). **b** Forward and backward prediction for estimating the utility of cells entering and leaving the field of view

$$U_s(A, B, C) = U(A, B) \times U(B, C) \times F_{\text{align}}(e_1, e_2), \qquad (16.12)$$

$$F_{\text{align}} = e^{-\frac{d_{\text{align}}^2}{2\sigma^2}}. \qquad (16.13)$$

The calculation of U_s for the other cases illustrated in Fig. 16.5 is more involved. The basic idea behind computing the second-order utilities that contain first-order merge/split hyperedges is to enumerate all possible sub-paths. The rationale being that in the absence of segmentation errors, this would have been their true utilities. All second-order edges are centered about a vertex B. We consider all possible simple paths through this vertex. For example, in Fig. 16.5 Panel G, the path $A \rightsquigarrow B \rightsquigarrow C$ is the only simple path through B. Therefore, the second-order utility for the path (A, B, C, D) is defined as $U_s(A, B, C, D) = U_s(A, B', C)$, where B' is the volume-adjusted version of B as described earlier. Note that although vertex D is not directly involved in the computation, and the volume of B is adjusted to B' using the volume of D. In Panel E, not all paths through B can be chosen simultaneously. In this case, we consider the two possible sets of simple paths: (i) (A, B, E), (C, B, D) and (ii) (A, B, D), (C, B, E). Among the two possible set of simple paths, we choose the one that maximizes the utility, i.e.,

$$U_s = \max \left(U_s(A', B, D') + U_s(C', B, E'), U_s(A', B, E') + U_s(C', B, D') \right). \quad (16.14)$$

For all the other possible cases that have not been explicitly described here, the same principle is applied. For a disappearance edge, we need to predict the location of the cell in the future, using its known location for two consecutive time points. We call this *forward prediction*. For an appearance edge, we need to predict the location before the cell came into the field of view. We call this *backward prediction*. In Fig. 16.6 Panel A, we perform a forward prediction of the location of B, given the locations of A, B. We predict the location of C though a linear extrapolation from the locations of A, B. Let us define d_b as the shortest distance between the predicted cell C's location and the image boundary. This distance is measured along the closest of the six image boundaries. Equation (16.15) defines $U_s(A, B, C)$ for the forward prediction as a function of d_b. The correction factor is a Gaussian function, dependent on d_b:

$$U_s(A, B, C) = U(A, B) \times \left(1 - e^{-\frac{(d_b - \mu_b)^2}{2\sigma_b^2}} \right) \times \omega(AD). \qquad (16.15)$$

The second-order utility for the backward prediction is defined analogously. This symmetry, between forward and backward prediction, is a desirable property. Performing tracking on a movie or its time-reversed copy produces identical results by our formulation.

Non-parametric motion model Non-parametric estimation models the motion more accurately since it does not impose any constraint like locally smooth linear motion. Each motion type can have different numbers of features and number of participating cells. Each feature has a different distribution depending on the motion type. Therefore, we used a non-parametric kernel density estimator (NKDE) to compute the utility functions based on the method by Parzen and Rosenblatt [24, 25]. We used smooth Gaussian kernels with a bandwidth estimated using the method proposed by Silverman [26].

The goal of the NKDE is to estimate the probability density functions using a training dataset. Let $\mathbf{x}$ be any point in D-dimensional Euclidian space. Let $\mathbf{x^n} \in \mathcal{R}^{\mathbf{D}}$ refer to the $\mathbf{n}$th sample training point of a training dataset of N points. We used a product kernel independently in each dimension for simplicity. However, this does not limit the estimated probability density function to be feature independent [26]. We use a Gaussian kernel in each of the D dimensions with a bandwidth h_d. K is given by

$$K(\mathbf{x}, \mathbf{x}^n, h_1, h_2, \ldots, h_D) = \frac{1}{h_1 h_2 \ldots h_D} \prod_{d=1}^{D} K_d \left(\frac{x_d - x_d^n}{h_d} \right), \tag{16.16}$$

where $K_d(x)$ is a unit Gaussian kernel. The estimated density can be defined as the sum of the normalized kernels:

$$P_{\text{KDE}}(\mathbf{x}) = \frac{1}{N h_1 h_2 \ldots h_D} \sum_{n=1}^{N} \prod_{d=1}^{D} K_d \left(\frac{x_d - x_d^n}{h_d} \right). \tag{16.17}$$

The bandwidth parameter plays an important role in determining $P_{\text{KDE}}(\mathbf{x})$. It has been shown [26] that the optimal value of the bandwidth parameter h^* for a single-dimensional Gaussian density function is given by $h^* = 1.06 \sigma N^{-1/5}$.

We used the NKDE just described to compute one-dimensional and two-dimensional distributions of the marginal utility function distributions as shown. We call these marginal distributions as utility components. Each component is a uni- or multi-dimensional joint distribution. We normalized these marginal distributions to a maximum of 1. The utility function for a given type of motion is a product of utility components. This method avoids the curse of dimensionality because computing higher-order functions from the training data requires higher number of training samples [26].

The first-order utility values $U(.)$ for the translation merge and split events were calculated as follows:

$$U\left(C_t^i, C_{t+1}^j, \ldots\right) = \omega(type) \times \prod_{\forall k} P_{\text{KDE}}^k \left(C_t^i, C_{t+1}^j, \ldots\right), \tag{16.18}$$

where $C_t^i, C_{t+1}^j, \ldots$ form the vertices involved in the FOH and $\omega(.)$ denotes the prior for the first-order hyperdge. Table 16.1 summarizes the features used to calculate the second-order utility for the different motion types. In order to compute the complete second-order utility, i.e., utility value for a SOH, we used a combination of the first-order utilities and the alignment factor. In this new model, we calculated the first-order utility and the alignment factor using the non-parametric motion. For a simple translation–translation SOH, the utility is given by

Table 16.1 A table of utility function components and the features used to compute them

Utility type	Features used	Mathematical definition	Dimensions
Translation	Distance	$\|\mathbf{d}\|$	1
	Volume overlap ratio	$\dfrac{|b(v_1) \cap b(v_2)|}{|b(v_1)| + |b(v_2)|}$	1
	Time difference	$|t_2 - t_1| - 1$	1
Merge/Split	Distances	$\text{MIN}(\|\mathbf{d}_{13}\|, \|\mathbf{d}_{23}\|), \text{MAX}(\|\mathbf{d}_{13}\|, \|\mathbf{d}_{23}\|)$	2
	Avg. volume overlap ratio	$0.5\left(\dfrac{|b(v_1) \cap b(v_3)|}{|b(v_1)| + |b(v_3)|} + \dfrac{|b(v_3) \cap b(v_2)|}{|b(v_3)| + |b(v_2)|}\right)$	1
	Volume ratio	$\dfrac{V_3}{V_1 + V_2}$	1
	Time difference	$|t_3 - t_1| + 1$	1
Appear/ Disappearance	Boundary distances	$bd(A|D), bd(v_1)$	2
Alignment factor	Distance ratio, direction projection	$\dfrac{\text{MIN}(\|\mathbf{d}_{12}\|, \|\mathbf{d}_{23}\|)}{\text{MAX}(\|\mathbf{d}_{12}\|, \|\mathbf{d}_{23}\|)}, \dfrac{\mathbf{d}_{12} \cdot \mathbf{d}_{23}}{\|\mathbf{d}_{12}\|\|\mathbf{d}_{23}\|}$	2

Different components have different numbers of dimensions. The utility components are multiplied to compute the utility. The separation of independent components helps us avoid the curse of dimensionality. Glossary of terms: d distance, $b(v)$ bounding box volume of vertex v, t time point, V volume of the cell.

$$U_S(A, B, C) = U(A, B) \times U(B, C) \times F_{\text{align}}(A \to B, B \to C), \qquad (16.19)$$

One could add more features (like intensity, texture information) in the future to improve the accuracy of the utility functions. In our experiments, we trained the algorithm using the ground truth tracks from one of the synthetic datasets with a simulated track density of 180. We generated the test datasets independently at various densities. We now formulate the tracking problem as a subset selection problem over these second-order hyperedges.

16.3.4 Second-Order Hyperedge Subset Selection

Let $\mathcal{S}$ denote the set of all second-order edges in the tracking hypergraph G. A second-order matching M is a set of second-order hyperedges $M \subset \mathcal{S}$ such that for every vertex at most one first-order hyperedge from M is incident on it and at most one hyperedge from M emanates from it. Let $\mathcal{M}$ denote the set of all possible second-order matchings in G. A maximum second-order matching M_w is a second-order matching such that for every $M \in \mathcal{M}$, $\sum_{e \in M_w} U_s(e) \geq \sum_{e \in M} U_s(e)$. We also denote the second-order utility of a matching as $U_s(M) = \sum_{e \in M} U_s(e)$.

Based on the above definitions, the tracking problem is the problem of finding the maximum second-order matching in a directed hypergraph. Two things remain to be shown: (i) A second-order matching produces a valid set of correspondences for a tracking result and (ii) A maximum second-order matching maximizes tracking saliency. A second-order matching represents a set of track correspondences. Every SOH is composed of two FOHs. Therefore, a second-order matching corresponds to a set of first-order hyperedges too. Connected component labeling of the matching FOHs gives the individual tracks. For a set of tracking correspondences to be valid, a cell cannot belong to two different tracks. The constraint that every cell has an *in-* and *out-degree* of one guarantees this. Note that split hyperedge and a merge hyperedge contribute to an in-/out-degree of one. A maximum second-order matching maximizes $U_s(M)$. In turn, we are maximizing the total probability of true correspondences.

Every cell belongs to only one track, and they all start with an appearance hyperedge and terminate with a disappearance hyperedge. The only region where tracks split/merge are at the locations of split/merge hyperedges. We resolve these regions in the post-tracking step.

16.3.5 Integer Programming

We formulate the second-order hyperedge selection problem as the following integer program. The feasible region R_{IP} for the integer program is denoted $R_{IP} = \{x \in \mathcal{Z}_+^n | Ax \le b\}$. The optimal solution is given by $x^* \in R_{IP}$, where

$$\mathbf{c}^T \mathbf{x}^* \ge \mathbf{c}^T \mathbf{x}_0, \ \forall \mathbf{x}_0 \in R_{IP}. \tag{16.20}$$

This is the *global optimum* of the objective function and not a *local optimum*. Let $X(S)$ be a Boolean variable for every SOH $S \in \mathcal{S}$, the set of all second-order hyperedges:

$$X(S) = \begin{cases} 0, & S \in \mathcal{S} - M \\ 1, & S \in M \end{cases}. \tag{16.21}$$

The objective function $O(M)$ to maximize is given by

$$\text{Maximize} : O(M) = \sum_{\forall S \in \mathcal{S}} X(S) U_s(S), \tag{16.22}$$

It can be easily seen that $O(M) = U_s(M)$, because

$$\sum_{\forall S \in \mathcal{S}} X(S) U_s(S) = \sum_{\forall S \in M} X(S) U_s(S) = \sum_{\forall S \in M} U_s(S). \qquad (16.23)$$

We have two sets of constraints—vertex constraints and edge constraints. The vertex constraints avoid multiple chosen second-order edges being centered on a vertex. The second set of constraints avoids multiple overlapping SOHs on the out- and in-edges. The edge constraint also ensures that tracks can originate or terminate in one of the four cases (i) appearance edge, (ii) disappearance edge, (iii) beginning of the sequence, and (iv) end of the time sequence. The first constraint is given by

$$\sum_{\forall i} X(S_v^i) = 1, \quad \forall v \in V, \qquad (16.24)$$

where S_v^i is the ith SOH centered on v. Let us define two new sets of SOHs for every hyperedge. Let S_{E_1} and S_{E_2} denote the two FOHs that form the SOH $S = (S_{E_1}, S_{E_2})$. Let $S_{in}(e) = \{S \in \mathcal{S} | S_{E_2} = e)$ denote the set of SOHs that are incident on e. Let $S_{out}(e) = \{S \in \mathcal{S} | S_{E_1} = e)$ denote the set of SOHs that originate from e. The edge constraint is

$$\left(\sum_{\forall S \in S_{in}(e)} X(S) \right) \times |e_Y| = \left(\sum_{\forall S \in S_{out}(e)} X(S) \right) \times |e_X|, \forall \{e \in E | |S_{in}(e)| > 0 \text{ and } |S_{out}(e)| > 0\},$$

$$(16.25)$$

where $e = (e_X, e_Y)$. One can see that the edge constraints do not exist for appearance/disappearance or for the edges in the first and last frames. Note that the number of constraints is $O(|V| + |E|)$. If we assume that every vertex has an average in-degree/out-degree of d, then the total number of constraints is approximately $|V| + 2|V|d$. Therefore, for sparse graphs like ours, with a max degree of Δ, total number of constraints is $O(|V|\Delta)$. The number of variables is $|\mathcal{S}|$, which is $O(|V|\Delta^2)$.

We used the ILOG CPLEX optimization package (http://www-03.ibm.com/ibm/university/academic/pub/page/academic_initiative) to solve the integer program using a branch-and-cut method. Solving an integer program is computationally expensive, with the worst-case running time having an exponential growth rate. The branch-and-cut algorithm is a hybrid method that uses a combination of branch-and-bound, and cutting planes. Although theoretically it could take exponentially longer time, they seldom do because our integer program's solution is close to a linear program solution in most cases. This allows the branch-and-cut algorithm to use only a few hundred cutting planes to arrive at the solution. The typical running time for the optimization step is in the order of a few minutes for the largest of our datasets.

By ensuring that all the SOHs in the graph are matched simultaneously, we compute the global optimum of the objective function. The SOHs with $X(.) = 1$ gives the desired matching. An FOH is selected if any of the SOH containing it is present in the matching. After this step, we resolved all the merges/splits using a post-tracking step.

16.3.6   Post-tracking Identification and Correction of Segmentation Errors

The second-order matching contains merge and split hyperedges due to two reasons: (i) under-segmentation and (ii) over-segmentation. We identify these errors and correct the segmentations (and the corresponding tracking results). First, we identify all paths in the matching from a merge hyperedge to a split hyperedge through simple translations. We then identify if the split or the merged vertices are physically separated. If the segmentations do not share a boundary, it is an indication that the pair was correctly segmented. Over-segmented cells always share boundaries at some part of their segmentation. If the evidence is not found, one cannot conclude if the segmentations were correct or over-segmented. We continue to search beyond the split/merged vertices to see if there is an evidence of separation. If such evidence is found, we split all the cells in the path into two. We then identify the correct correspondences by maximizing the total second-order utility over all possible correspondences. We iterate this procedure over all possible paths formed between a merge and split hyperedge. Since the number of continuously under-segmented cells in a track is rather small, there is no concern about a combinatorial explosion. After this process, we merged all the remaining fragments that can be merged and repeat the process until it converges. Algorithm 2 summarizes the described cell tracking algorithm

---

Algorithm 2: Summary of the cell tracking algorithm

Input:

Feature vectors f_t^i for each cell C_t^i, and utility parameters $\Sigma^{-1}, \mu, \omega(T), \omega(MS), \sigma^2, \sigma_b^2, \mu_b$

Steps:

1. Construct the hypergraph: translation, appear/disappear, split, and merge edges.
2. Compute edge utilities using equations (10), (11),(14), and (15)
3. Solve the integer program in (22), (24), and (25)
4. Resolve merge/split

16.3.7 Validation

We have validated our tracking algorithm's accuracy both in terms of its segmentation correction ability and its spatiotemporal tracking accuracy. The segmentation errors were not manually corrected prior to tracking. For the real images obtained through two-photon microscopy, we performed manual human validation of the results of our tracking algorithm. We have compared our algorithm's performance against two previously published works. The first method is the 5-D image analysis framework published by [17]. The second method is the *u-track* algorithm used for particle tracking by [19]. For the synthetic datasets, we compared the non-parametric motion model with the fixed Gaussian model. For the real datasets, we used the fixed Gaussian-based model to make the comparison.

First, we computed the segmentation accuracy post-tracking to quantify the improvement achieved by the proposed tracking algorithm. The proposed tracking algorithm corrects the segmentation errors in three ways: (i) by merging over-segmented cells, (ii) by splitting under-segmented cells, and (iii) by correctly retaining the tracks across undetected time frames. We quote the results for Chen et al. from their work [17] for the same datasets. For [19] work, we have used the freely available MATLAB implementation of *u-track* provided by the authors.

We evaluated using four representative datasets with different cell densities. The image volumes were $256 \times 256 \times 21$ voxels with varied number of time points. The resolution in physical units was 0.64 μm $\times$ 0.64 μm $\times$ 2.0 μm for the three dimensions, respectively. The images were of low axial resolution, commonly seen in 2-photon microscopy.

Table 16.2 summarizes our findings in terms of cell segmentation accuracy. The initial error rate was low (2.83 %). However, in high-throughput studies, this error rate results in a large number of edit operations from a user's point of view. We found the accuracy of the segmentation post-tracking using our method to be 1.34 %. We improved the overall accuracy by about 53 % compared to a naïve segmentation computed independently at every time point. The spatiotemporal continuity exploited by our tracking algorithm accounts for this improvement. In high-throughput studies, this improvement in itself could be a game changer.

Table 16.2 A compilation of segmentation error rates comparing various algorithms

Dataset	Initial (%)	Proposed algorithm (%)	Jaqaman et al. 2008 (%)	Initial segmentation—Chen et al. 2009 (%)
1	1.64	0.00	0.00	6.56
2	2.10	0.00	0.00	16.60
3	2.89	2.28	2.59	13.39
4	3.13	1.51	2.12	14.21
Overall	2.83	1.34	1.76	13.84

The proposed algorithm improves segmentation accuracy better than the prior works by [17, 19]

It is clear that the proposed algorithm performs the best among the ones compared. U-track improves the accuracy of segmentation by 38 % from 2.83 to 1.76 %. Chen et al.'s reported average segmentation accuracy is 13.84 %. We also evaluated the tracking accuracy by computing the number of correctly identified *correspondences*. We define correspondence as an association between two cells at different time points. A correct correspondence is a mapping between the same cell captured at consecutive time points.

A single error in segmentation can lead to multiple errors in tracking. In our work, we do not manually edit the segmentation prior to tracking. The tracking error rate is the percentage of incorrect number of correspondences identified by the algorithm among all the correct correspondences.

In Table 16.3, we show the error rates for nine different datasets. The tracking algorithm's error rate is heavily dependent on the segmentation error rate. The proposed algorithm achieves an overall error rate of 2.23 %. The proposed algorithm performed significantly better than Chen et al.'s algorithm. The proposed algorithm had 28.8 % lesser error rate than Jaqaman et al.'s algorithm.

We generated synthetic datasets by producing an ensemble of 3-D cell images that vary in orientation, texture, brightness, shape, radius, elongation, average speed, instantaneous speed, start and end locations, etc. We picked these cells randomly to move in a 3-D volume in a curvilinear path. We chose the starting point for the tracks from a region 50 % larger than the image volume itself to simulate cells entering the field of view. To study the sensitivity, we varied the simulated density of the track (number of tracks simulated in a given volume) and also added additive Gaussian noise to the data. For every data point, we generated five datasets and found an ensemble average of the errors. Figure 16.7 shows the performance of the two proposed motion models under different values of track

Table 16.3 Summary of tracking error rates

Dataset	Total #	Proposed tracking algorithm (%)	u-track (%)	Chen et al. Segmentation, Tracking (%)
m12bp14pos	235	0.00	0.00	6.56 %, 7.23 %
m12cp14pos	472	0.00	0.00	16.60 %, 6.57 %
m13cp14pos	643	2.80	3.27	13.39 %, 5.29 %
m5bwt	1932	2.02	3.16	14.21 %, 4.92 %
120307m1s7*	692	3.76	3.61	–
120307m1s5*	589	3.40	3.40	–
120307m1s9*	959	2.71	3.55	–
120307m2s6*	748	2.54	4.41	–
120507m1s5*	943	1.38	3.39	–
	Combined	2.23	3.13	13.84 %, 5.39 %[†]

[†]We did not correct the segmentations prior to tracking in the first two methods. A single error in segmentation can potentially cause multiple errors in track correspondences. For the datasets indicated by a *, we validated only a sub-volume (25–50 %)

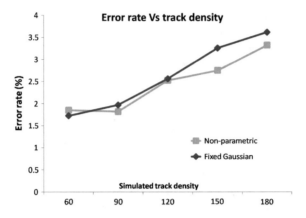

Fig. 16.7 A plot of error rate as a function of track density based on the synthetic dataset experiments (lower is better). Our method is robust even at higher track densities. The error rate scales gracefully. The non-parametric motion model outperforms the fixed Gaussian motion model Fixed Gaussian motion model

density. As expected, the non-parametric motion model is better because of superior modeling by 17 %. Our experiments with additive Gaussian imaging noise showed similar reduction in error rates.

16.4   Multi-view Editing and Analytics Framework

Automated segmentation and tracking algorithms are needed to analyze large image datasets with high accuracy and speed. However, even the best-available algorithms are prone to errors. Low signal-to-noise ratio and poor and varying contrast commonly introduce segmentation errors, which subsequently leads to tracking errors. Therefore, some amount of human intervention is necessary to bring the accuracy of the results to a sufficient level to permit biological inferences to be made confidently.

Human visual inspection is still the gold standard, and one of the most widely used techniques in verifying the correctness of the segmentation/tracking results. In the case of 2-D sequences, it is typical for a user to inspect the results by either scrolling through movie frames using a frame slicer or by visualizing a static 3-D (2-D + *t*) volume rendering, also called a *Kymograph*. For 3-D sequences, visualization of the complete movie becomes more difficult. There are two favored approaches for inspecting the automated analysis results in the context of the original image data: (1) the user is equipped with two slicers. The first slicer is used to scroll through the axial axis of the 3-D image, while the second is used to scroll through movie frames (time). Visual inspection with this method is tedious and can be time consuming especially in high-throughput studies. (2) The user is provided with a 3-D volume rendering of the image stack and a time slicer. Cell tracks are static and are overlaid on the volume rendering. Here, the observer scrolls through time frames while visualizing cells moving along their tracks. This way, the user has a complete perspective over his/her data and the segmentation and tracking results.

In addition to visual proofreading, correction of segmentation and tracking errors is sometimes required in order to improve the accuracy of the results. Although the tracking algorithm described in the previous section copes well with segmentation errors, correction of these errors is usually performed before the tracking step. There are four types of segmentation errors, namely, under-segmentation, over-segmentation, false detection, and missed detection. Under-segmentation means that two or more cells were detected as one cell. A split operation is required to correct these errors. Over-segmentation occurs when a cell is detected as two or more cells. In this case, a group of cells is merged together to form one single object. False/missed detection requires a (an) deletion/addition operation.

We categorize tracking errors into two types based on the edit operations required for correction. Two edit operations (arguably) can be performed, addition and deletion of tracks. Addition of sub-tracks is usually required when a track association is missed between frames. Deletion is necessary when a wrong track correspondence is present. Sometimes, both operations have to be applied to fix a tracking error of a single cell track. For instance, assume that a correspondence error for cell i occurred between frames t and $t + 1$, and that cell i has been mislabeled as cell j starting from frame $t + 1$. The track for cell i between frames t and $t + 1$ has to be deleted first, and then a track between cell i at frame t and cell j at frame $t + 1$ has to be added.

Edit operations can be difficult and time consuming, especially in high-throughput studies where thousands of objects and object tracks have to be validated. A better alternative would be to perform edits on a group of objects instead of one object at a time. Furthermore, analytics guided editing methods can facilitate the task of detecting hundreds of errors at a time. For example, a combination of features such as shared boundary and volumes of cells can be used to detect over-segmented objects. Incorrect track association might lead to larger values of instantaneous speed of the cell compared to the average speed of the cell in its entire duration. Based on the idea that similar errors would share similar features, editing can be performed on a large number of objects with just few clicks.

The FARSIGHT toolkit implements a multi-view segmentation and track editing framework with six views: (1) Image view with stack and frame slicers, (2) 3-D Kymograph view, (3) 3-D *String on Beads* view, (4) Bi-cluster view, (5) Table view, and (6) Scatter plot view. Figure 16.8 depicts a diagram illustrating the components and the organization of the software system. The data component consists of the original image data, the labeled images, and the cellular and dynamic measurements of each cell. In this implementation, the components of the system are actively linked together, meaning, any operation performed on any of the views will automatically reflect on all the other views. For instance, a delete operation in the Kymograph view sends a signal to the data and selection module. The module

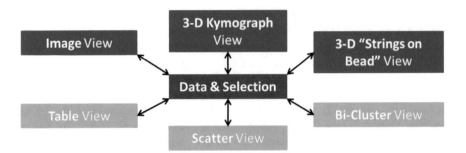

Fig. 16.8 Diagram illustrating the software organization of the multi-view editing framework. The modules colored in *blue* represent the raw data and result visualization modules. The modules in *green* constitute the analytics views of the system. The Data and Selection module represents the master component and is actively linked to all the other components

updates the edited data, i.e., features are recomputed and images are re-labeled. In turn, the data and selection module sends updated signals to all other modules. The other modules then fetch the new data and update their views. This design provides an efficient way of handling data while enabling low-memory implementation. The toolkit is aimed to provide the user with the ability to perform a comprehensive inspection and editing of the results. Each view has specific advantages that might bring new insights to the validation process. As an example, by selecting to display the *volume* of the cells in the *x*-axis and the length of their tracks in the *y*-axis of the scatter plot, it is possible to identify all cells that have small volume and exist only for a short duration in the lower left corner of the plot, which could potentially indicate an a segmentation error.

Figure 16.9 shows a screenshot of the different views. The image view in Fig. 16.9a displays the multi-channel image data and the cell boundaries on the 2-D slices of the 3-D image stacks at each frame. This view is equipped with a *z*-slicer and time slicer. This view is useful for identification and editing of segmentation errors. It also allows for interactive editing of all the types of segmentation errors described previously in this section.

The 3-D Kymograph view in Fig. 16.9b represents a volume rendering of the full movie in a single snapshot. In the case of 2-D sequences, the 2-D images are stacked together to form a 3-D volume. The axes of the Kymograph are defined by $(\mathbf{x}, t)$. For 3-D time series, each 3-D image is converted to a 2-D image by taking its maximum intensity projection. The projected images are then stacked into a 3-D volume to allow for kymograph rendering. The detected objects (cell centroids) are visualized as 3-D points and cell tracks are displayed as 3-D colored lines. Cell tracks in the raw image appear as tracking errors become easier to identify. This view also allows for correction of tracking errors. Tracks can be deleted or added. Once a track is selected, it is highlighted in white. The user can then delete the track by pressing the key (d) from the keyboard. After the edit is performed, the cell tracks are automatically assigned a new label. To add a track, the user selects two object centroids and then presses the key (a) from the keyboard. Again, the edited

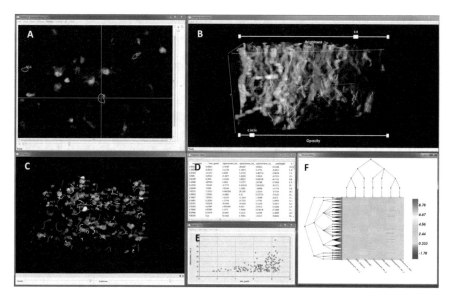

Fig. 16.9 Screenshot of the FARSIGHT-5D multi-view editor. **a** Image view of the original 5-D multi-channel data with cell segmentation results. The window has a time slider that allows the user to scroll through frames and a z-slider to scroll the 3-D stack. Cells are delineated and identified with a unique number. **b** 3-D Kymograph view shows cell tracks as *colored lines* overlaid on a spatiotemporal (x, y, t) projection of the original image. **d** Table of dynamic measurements. **e** Scatter plot showing the total displacement in the x-direction versus average speed. **f** Bi-clustering map of cell track measurements

track is automatically given a new label and the other views are subsequently updated. In addition to single-edit commands, group selections of tracks are allowed in order to perform batch operations.

In addition to the image views described above, the toolkit implements three analytics views. These views give the user the ability to explore the image data from a quantitative perspective. The table shown in Fig. 16.9d lists all the computed intrinsic and dynamic measurements for each cell. It is useful for visualizing the exact measurements of a cell. The scatter plot view in Fig. 16.9e displays an x–y plot of two cell measurements. The user can select which measurements to plot on the ordinate and the abscissa using a menu item. This view is particularly convenient for exploring the statistical correlation between object measurements. For instance, if object volumes are plotted against object intensities, one can immediately recognize the correlation between these two features. Batch selections and edits are also allowed in this view. The user can select a group of points by drawing a polygon around the points then presses the key (s) from the keyboard to delete the highlighted objects.

The heatmap view shown in Fig. 16.9f represents co-clusters of the entire cell population dynamics over the full movie period, i.e., co-clusters of complete cell tracks. Each row in the heatmap corresponds to a single cell and each column represents a measurement summarizing the dynamic behavior a cell. The dynamic features include, but are not limited to, total displacement magnitude, path length, maximum speed, minimum speed, and average speed. The partitioning of cell population and dynamic features illustrated in the heatmap is achieved via an extension of the unsupervised clustering method developed in [27, 28]. Given a data matrix $X^{n \times m}$, with n rows corresponding to n cells and m columns corresponding to m dynamic measurements, this algorithm alternates between hierarchical clustering of the rows and the columns of the matrix. This process is iteratively applied until convergence. The result is a simultaneous hierarchical partitioning of the data. The reader can refer to [27] for a detailed description of the algorithm. This view is particularly useful for visualizing groups of cells based on their dynamic phenotype. It equips the user with the ability to unveil and organize the hidden patterns present in the data, i.e., it is a powerful data-driven visualization and discovery tool.

16.5 Applications

The results of the described 5-D image analysis framework features was put to test in the 4-color dataset collected from Thymus [29]. The study was to understand the differences in motility between two different thymocytes, F5 thymocyte and the wild-type thymocytes. The goal of the study was also to understand the motility's relationship to other elements in the neighborhood like dendritic cells and blood vessels.

In individual cell features, the features that rejected the null hypothesis (that they are from the same population) were (in the order of most significant to least significant) surface area, volume, bounding box volume, shape, instantaneous speed, change in distance to blood vessel, etc. In total, 11 features rejected the null hypothesis, but many among them are strongly correlated-like volume-bounding box volume, volume–surface area, etc. Overall, track features, average speed, confinement ratios, path lengths, average contact to DC, and average change in distance to blood vessel rejected the null hypothesis. Note that previously published result has verified that normalized center to edge distance to DC is statistically different for the wild type and F5. We identified the thymocyte-DC contact by a threshold on this measure. The biological interpretation of the remaining features that rejected the null hypotheses remains to be investigated. We are also analyzing the datasets produced in vivo imaging in the lymph node (Fig. 16.10).

Our algorithm was tested against two datasets: (i) Thymus of a GFP transgenic mouse was imaged ex vivo using a custom-built two-photon microscope and (ii) Lymph node of a mouse was imaged in vivo. We conducted several experiments in the study, which involved imaging two to four channels. The aim of the study was to understand the interaction of genetically modified thymocytes with dendritic cells and blood vessels.

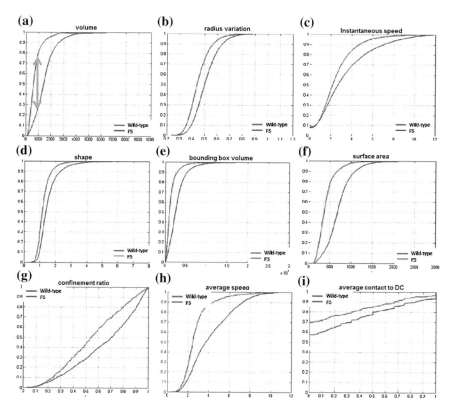

Fig. 16.10 A gallery of plots showing the cumulative distribution functions $F(x)$ for the distribution of the features among the wild-type and F5 thymocytes. An *arrow* in (**a**) marks the point of maximum disparity. **a–f** correspond to individual cell features and **g, h** correspond to whole track features

Two types of features were computed: intrinsic and associative. Intrinsic features are calculated purely based on a single channel of data post-tracking. Associative features involved calculating quantitative measures involving the interaction of multiple channels. We computed intrinsic and associative features for every cell and for every track. In order to compute features for a track, we averaged the cell-based features over the entire duration of the track. The time-based cell features included (but were not limited to): instantaneous speed, displacement, direction of movement relative to the nearest blood vessel, and change in distance to the blood vessel. For whole track features, we computed average values of distance to blood vessel, angle relative to blood vessel, contact to a dendritic cell, speed, confinement ratio and total track displacement, and total track path length. These features were computed both for the wild-type cells and the genetically modified cells. We then used the Kolmogorov–Smirnov test to identify the features that discriminate the two cell populations.

16.6   Conclusion and Future Directions

The FARSIGHT open source toolkit represents a synergistic and comprehensive combination of multiple image processing algorithms, ranging from image pre-processing, automated cell segmentation, cell tracking, feature extraction, and feature analysis tools. The image analysis algorithms were specifically designed and chosen for usability in the hands of a biologist. Specifically, we have prioritized algorithms with the fewest adjustable parameters. The few parameters that remain are minimally sensitive and intuitive. An operationally important aspect of the toolkit is the incorporation of practically effective tools for efficient inspection and corrective editing of the automated results. Another innovative aspect of the toolkit relates to the underlying software infrastructure. Specifically, all of the image-derived measurements, the visualization tools, and analytical tools are actively linked, allowing a user to work from the most favorable viewer. Importantly, the edits can be defined by the analytic tools (e.g., split cells based on size outliers), and also the fact that edits made in any one view are immediately reflected in all others. This framework aims to provide the biologist with the necessary tools to achieve rapid experimental and discovery cycles.

References

 1. Ladi E, Schwickert TA, Chtanova T, Chen Y, Herzmark P, Yin X, Aaron H, Chan SW, Lipp M, Roysam B, Robey EA (2008) Thymocyte-dendritic cell interactions near sources of CCR7 ligands in the thymic cortex. J Immunol 181(10):7014–7023
 2. Shen Q, Wang Y, Kokovay E, Lin G, Chuang SM, Goderie SK, Roysam B, Temple S (2008) Adult SVZ stem cells lie in a vascular niche: a quantitative analysis of niche cell-cell interactions. Cell Stem Cell 3(3):289–300
 3. Ladi E, Yin Z, Chtanova T, Robey EA (2006) Thymic microenvironments for T cell differentiation and selection. Nat Immunol 338–343
 4. Elad M, Aharon M (2006) Image denoising via sparse and redundant representations over learned dictionaries. IEEE Trans Image Process 15(12):3736–3745
 5. Chatterjee P, Milanfar P (2012) Patch-based near-optimal image denoising. IEEE Trans Image Process 21(4):1635–1649
 6. Li S, Yin H, Fang L (2012) Group-sparse representation with dictionary learning for medical image denoising and fusion. IEEE Trans Biomed Eng 59(12):3450–3459
 7. Al-Kofahi Y, Lassoued W, Lee W, Roysam B (2010) Improved automatic detection and segmentation of cell nuclei in histopathology images. IEEE Trans Biomed Eng 57(4):841–852
 8. Boykov Y, Veksler O, Zabih R (2001) Fast approximate energy minimization via graph cuts. IEEE Trans Pattern Anal Mach Intell 23(11):1222–1239
 9. Lindeberg T (1998) Feature detection with automatic scale selection. Int J Comput Vis
10. Wu X, Chen Y, Brooks BR, Su YA (2004) The local maximum clustering method and its application in microarray gene expression data analysis. EURASIP J Adv Signal Process 1:53–63
11. Fan J (1998) Notes on Poisson distribution-based minimum error thresholding. Pattern Recogn Lett 19(5–6):425–431

12. Boykov GF-LY (2006) Graph cuts and efficient N-D image segmentation. Int J Comput Vis 70 (2):109–131
13. Boykov Y, Kolmogorov V (2004) An experimental comparison of min-cut/max-flow algorithms for energy minimization in vision. IEEE Trans Pattern Anal Mach Intell 26 (9):1124–1137
14. Soille P (1999) Morphological image analysis: principles and applications. Springer, Berlin
15. Poore AB, Gadaleta S (2006) Some assignment problems arising from multiple target tracking. Math Comput Model 43(9–10):1074–1091
16. Al-Kofahi O, Radke RJ, Goderie SK, Shen Q, Temple S, Roysam B (2006) Automated cell lineage construction: a rapid method to analyze clonal development established with murine neural progenitor cells. Cell Cycle 5(3):327–335
17. Chen Y, Ladi E, Herzmark P, Robey E, Roysam B (2009) Automated 5-D analysis of cell migration and interaction in the thymic cortex from time-lapse sequences of 3-D multi-channel multi-photon images. J Immunol Methods 340(1):65–80
18. Shafique K, Shah M, Florida C A non-iterative greedy algorithm for multi-frame point correspondence graph theoretical formulation
19. Jaqaman K, Loerke D, Mettlen M, Kuwata H, Grinstein S, Schmid SL, Danuser G (2008) Robust single-particle tracking in live-cell time-lapse sequences 5(8)
20. Smal I, Draegestein K, Galjart N, Niessen W, Meijering E (2008) Particle filtering for multiple object tracking in dynamic fluorescence microscopy images: application to microtubule growth analysis. IEEE Trans Med Imaging 27(6):789–804
21. Cardinale J, Rauch A, Barral Y, Sz G (2009) Bayesian image analysis with on-line confidence estimates and its application to microtubule tracking. Institute of Theoretical Computer Science and Swiss Institute of Bioinformatics, ETH Zurich, Switzerland Institute of Biochemistry, ETH Zurich, Switzer. In: IEEE international symposium on biomedical imaging: from nano to macro, 2009. ISBI '09, pp 1091–1094
22. Paragios N, Deriche R (2000) Geodesic active contours and level sets for the detection and tracking of moving objects. Pattern Anal Mach 22(3):266–280
23. Perera A, Tsai C-L, Flatland RY, Stewart CV (2000) Maintaining valid topology with active contours: theory and application. In: IEEE Computer Society Conference on Computer Vision and Pattern Recognition, vol 1, pp 496–502
24. Parzen E (1962) On estimation of a probability density function and mode. Ann. Math. Stat. 33(3):1065–1076
25. Rosenblatt M (1956) Remarks on some nonparametric estimates of a density function. Ann Math Stat 27(3):832–837
26. Silverman BW (1986) Density estimation for statistics and data analysis. CRC Press
27. Coifman RR, Gavish M (2011) Harmonic analysis of digital data bases. Wavelets Multiscale Anal 1–37
28. Gavish M, Coifman RR, Haven N (2010) Multiscale wavelets on trees, graphs and high dimensional data: theory and applications to semi supervised learning. ICML 367–374
29. Narayanaswamy A, Ladi E, Al-Kofahi Y, Chen Y, Carothers C, Robey E, Roysam B (2010) 5-D imaging and parallel automated analysis of cellular events in living immune tissue microenvironments. In: 2010 IEEE international symposium on biomedical imaging from nano to macro, pp 1435–1438
30. Park EJ, Peixoto A, Imai Y, Goodarzi A, Cheng G, Carman CV, von Andrian UH, Shimaoka M (2010) Distinct roles for LFA-1 affinity regulation during T-cell adhesion, diapedesis, and interstitial migration in lymph nodes. Blood 115(8):1572–1581

Chapter 17
Video Bioinformatics Databases and Software

Ninad S. Thakoor, Alberto C. Cruz and Bir Bhanu

Abstract Video bioinformatics focuses on understanding biological events on various timescales by analyzing spatiotemporal images. Fundamental to the success of this venture is the ability to automatically analyze these images to extract relevant information. For evaluation of an automated image analysis technique, one needs to have the image data, the ground truth (i.e., the expected outcome), and a statistically meaningful testing process. The field of computer vision, which deals with the automated analysis of all sorts of images, has been making steady progress for years. It has benefited immensely by the availability of public datasets and shared software. This chapter surveys these databases and software.

Multiple public datasets are available for key computer vision problems such as segmentation [1], tracking [2], and object recognition [3] and there are many software options for automating analysis [4, 5]. Sharing of the data and software has a variety of benefits such as

- it avoids duplication of effort thereby speeding up the research;
- it allows fair analysis of performance of alternate techniques;
- shared image data can be re-purposed for other applications; and
- shared software allows experiments to be reproduced.

This chapter surveys publicly available video bioinformatics (VBI) databases and software.

N.S. Thakoor (✉) · A.C. Cruz · B. Bhanu
Center for Research in Intelligent Systems, University of California,
Winston Chung Hall Suite 216, 900 University Ave, Riverside, CA 92507, USA
e-mail: nthakoor@ee.ucr.edu

A.C. Cruz
e-mail: albert.cureg.cruz@gmail.com

B. Bhanu
e-mail: bhanu@ee.ucr.edu

© Springer International Publishing Switzerland 2015
B. Bhanu and P. Talbot (eds.), *Video Bioinformatics*,
Computational Biology 22, DOI 10.1007/978-3-319-23724-4_17

17.1 VBI Databases

17.1.1 Cell Tracking Challenge Dataset

The Cell Tracking Challenge was organized by members of three institutions: the Center for Biomedical Image Analysis, Masaryk University, Brno, Czech Republic; the Center for Applied Medical Research, University of Navarra, Pamplona, Spain; and Erasmus University Medical Center, Rotterdam, The Netherlands. The dataset comprises 24 time-lapse sequences each for training and testing. Half of these are real microscopy sequences, while the rest are computer generated. These sequences have varying cell density and noise levels. The real data includes GFP transfected rat mesenchymal stems cells (2D), GFP transfected H157 human squamous lung carcinoma cells (3D), MDA231 human breast carcinoma cells infected with a pure murine stem cell virus vector including GFP (3D), GFP transfected GOWT1 mouse embryonic stem cells (2D), Histone 2B-GFP expressing HeLa cells (2D), and Chinese hamster ovarian cells overexpressing proliferating cell nuclear antigen tagged with GFP (3D). Ground truth was generated by manual labeling by experts. The dataset and challenge are documented in detail in [6]. The dataset is available at: http://www.codesolorzano.com/celltrackingchallenge.

17.1.2 Cardiac Motion Tracking Challenge Dataset

These data were acquired at the Division of Imaging Sciences and Biomedical Engineering, King's College London, United Kingdom, and the Department of

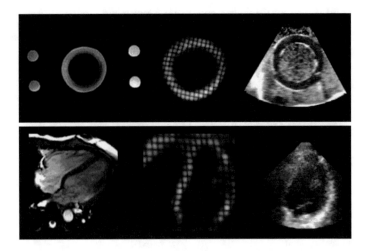

Fig. 17.1 Sample images from cardiac motion tracking challenge dataset

Internal Medicine II—Cardiology, University of Ulm, Germany. The database includes magnetic resonance (MR) and 3D ultrasound images from a dynamic phantom and 15 healthy volunteers. This dataset and the challenge are documented in [7]. http://www.cardiacatlas.org/web/guest/stacom-2011-motion-tracking-challenge-data (Fig. 17.1).

17.1.3   4D Left-Ventricular Segmentation Challenge Dataset

This dataset was made available as a part of the Medical Image Computing and Computer Assisted Intervention (MCCAI) 2011 workshop entitled: "Statistical Atlases and Computational Models in the Heart (STACOM): Imaging and Modeling Challenges." The data consist of a full set of 4D cine-MRI from short-axes and long-axes views at nearly 20 frames through the cardiac cycle in 200 patients with myocardial infarction. The dataset is available online at: http://www.cardiacatlas.org/web/guest/stacom2011.

17.1.4   Particle Tracking Challenge Dataset

This dataset was provided for a Particle Tracking Challenge Workshop as a part of International Symposium on Biomedical Imaging. Because generating ground truth for real particle tracking data can be challenging, the workshop chose to use simulated data. The set includes simulated data for viruses, vesicles, receptors, and microtubules with varying dynamics, particle density, and signal quality. The data can be downloaded at: http://www.bioimageanalysis.org/track/. Detailed discussion of the dataset and evaluation is provided in [8].

17.1.5   Broad Bioimage Benchmark Collection

The Broad Bioimage Benchmark Collection (BBBC) is a collection of microscopy image sets [9]. Apart from the images, each set includes a description of the biological application and some type of ground truth. At the time of writing of this chapter, there were 22 different sets of images available in the collection. The collection is available at: http://www.broadinstitute.org/bbbc/.

17.1.6 BITE: Brain Images of Tumors for Evaluation Database

This database is created by McConnell Brain Imaging Centre (BIC) of the Montreal Neurological Institute, McGill University to facilitate development of new techniques and evaluation [10]. The database provides pre- and postoperative MR and intraoperative ultrasound images acquired from 14 brain tumor patients. The database can be accessed at: http://www.bic.mni.mcgill.ca/Services/ServicesBITE.

17.1.7 The Kahn Dynamic Proteomics Database

The Kahn Dynamic Proteomics project is being conducted by Alon lab, Weizmann Institute of Science. The project aims to monitor the position and amounts of endogenous proteins in living human cells. This site contains a database of the proteins tagged by yellow fluorescent protein. Each protein entry includes detailed sequence and functional annotation, images of protein localization, movies, and protein dynamics. The database site is: http://www.weizmann.ac.il/mcb/UriAlon/DynamProt/.

17.1.8 Worm Developmental Dynamics Database

This database is being developed by the Laboratory for Developmental Dynamics, RIKEN Quantitative Biology Center, Kobe. The database holds a collection of quantitative information about cell division dynamics in early *Caenorhabditis elegans* embryos. The images were captured by four-dimensional differential contrast interference (DIC) microscopy. The database can be accessed at: http://so.qbic.riken.jp/wddd/cdd/.

17.1.9 The Plant Organelles Database 3 (PODB3)

PODB3 provides static images and movie data of plant organelles, protocols for plant organelle research, and is provided by National Institute for Basic Biology [11]. Specifically, it is made up of 'The Electron Micrograph Database', 'The Perceptive Organelles Database', 'The Organelles Movie Database', 'The Organellome Database', 'The Functional Analysis Database', and 'External Links to other databases and Web pages'. The database is available at: http://podb.nibb.ac.jp/Organellome.

17.1.10 Mouse Behavior Data

This dataset provided by the Center for Biological and Computational Learning has over 10.6 h of continuously labeled video (8 day videos and 4 night videos) for the eight mouse behaviors of interest: drinking, eating, grooming, hanging, micro-movement, rearing, resting, and walking. The dataset is described in [12] and is available at: http://cbcl.mit.edu/software-datasets/mouse/.

17.1.11 BIOCHANGE Challenge

National Institute of Standards and Technology (NIST) organized benchmarking study of lung CT change measurement algorithms and computer-assisted diagnosis (CAD) Tools in 2008 and 2011 called BIOCHANGE. Goal of the study was to develop methodology and performance measures to validate change analysis using publicly available lung CT data. The data provides CT images pairs of 96 lesions. The challenge webpage is: http://www.nist.gov/itl/iad/dmg/biochangechallenge.cfm.

17.1.12 Sample Datasets

Various software packages provide sample imagery which is primarily to experiment with the functionality of the package. The imagery can to be of diverse types, but it is generally too limited to perform experiments and thorough evaluation.

17.1.12.1 Cell Profiler Example Images

Cell profiler is an open source cell image analysis software designed for biologists. The sample images from various pipelines are provided at: http://www.cellprofiler.org/examples.shtml.

17.1.12.2 bioView3D Sample Datasets

bioView3D volume renderer is an open source and cross-platform software for biologists to visualize fluorescence microscopy images in 3D. The software along with sample data is available at: http://biodev.ece.ucsb.edu/webpages/software/bioView3D/.

Fig. 17.2 Interface of re3data

17.1.13   Database Registries

While this chapter lists various VBI databases available at the time of publication, the list of the datasets is ever growing. Due to this, we also provide information about database repositories which should provide up-to-date information about datasets. However, note that these repositories are not limited to VBI.

17.1.13.1   re3data

In 2013, the registry of research data repositories (re3data) was established with the goal of creating a global registry of research data repositories from various disciplines (Fig. 17.2). The registry currently holds nearly 650 repositories, available at: http://www.re3data.org/.

17.1.13.2   BioSharing

BioSharing combines registries of policies and standards along with databases. The databases are described in accordance with BioDBcore guidelines (The guidelines can be found at: http://biodbcore.org/). There are 631 databases in the BioSharing repository. Available at: http://www.biosharing.org/.

17.2 Software

There are three options when automating a VBI task. If you can program or if you collaborate with programmers, you can create your own software using image processing and computer vision libraries. If you cannot program, there are free programs designed to be used by researchers with limited programming and image processing experience. These programs allow the creation of plugins so that programming savvy researchers can share their algorithms with others to be used in a VBI assay. However, these plugins are sometime unstable. For researchers able to purchase expensive software, there are commercially available software packages that are more stable than their free counterparts. A list of VBI software is given in Table 17.1.

17.2.1 Software for Programmers

For programmers in the field of image processing and computer vision, a VBI task is addressed by creating a new program specific to that task. The program is usually created in Matrix Laboratory (MATLAB) or in C++ with Open Source Computer Vision (OpenCV).

17.2.1.1 Matrix Laboratory (MATLAB)

MATLAB was originally designed to assist students with linear algebra problems but has since expanded its scope to address engineering, programming, and statistics problems. Its current version includes image processing and computer vision toolboxes, and its use is ubiquitous in the field. It is a command-line-based system, but the user can enter a list of commands into a script to create a program. MATLAB comes with a feature called GUIDE to create stand-alone executables. It is a high-level programming language and comes with many features commonly used in the field of image processing and computer vision.

17.2.1.2 Open Source Computer Visions (OpenCV)

OpenCV is a library to be used in C++, and it is not a stand-alone program. It does not offer as many features as MATLAB, but it is faster than MATLAB for the following reasons:

- MATLAB is a script language, but C ++ is compiled;
- MATLAB was originally designed to address linear algebra problems, and for the best performance, must be written in terms of matrices without iterative

Table 17.1 List of VBI software

Name	Type	Programming expertise required	Features	Platforms	License	Created
Open Source Computer Vision (OpenCV)	C++ Library	Expert programmer	Standard library used by image processing and computer vision scientists	Windows, Linux, Mac OS, iOS and Android	BSD license, open source	1999
Matrix Laboratory (MATLAB)	Command-line interface and scripting language	Knowledgable programmer	Users script their own programs	Windows, Mac OS, Linux	Proprietary	1984
ImageJ and ImageJ2	Broad-scope image processing program	Not necessary	Plugins, image processing operations	Windows, Mac OS, Linux, Sharp Zaurus PDA	Public domain, open source	1997
FIJI is Just ImageJ (FIJI) [5]	Broad-scope image processing program	Not necessary	Community curated plugins, streamlined version of ImageJ, allows many languages	Windows, Mac OS, Ubuntu	Base program: GPL, open source. Plugins: Varies	2008
Icy [4]	Broad-scope image processing program	Not necessary	Community curated plugins, work flows	Windows, Mac OS, Linux	GPL, open source	2011
Amira	Medical image visualization program	Not necessary	3D and 4D visualization, imaging segmentation, registration	Windows, Mac OS, Linux	Proprietary	1999
CL Quant	Bioimage processing program	Not necessary	N-D image processing operations	Windows 64-bit	Proprietary	–
Bitplane	Bioimage processing program	Not necessary	Up to 4D images, image processing suite, image processing operations, special filament program	Windows 64-bit	Proprietary	1993

control flow such as for and while loops to run in a comparable time to a C++ program;

- memory in MATLAB is dynamically allocated and can unnecessarily increase memory costs with no measures for the programmer to prevent it.

A program may be prototyped in MATLAB because of its ease of use, but if speed or memory becomes a bottleneck the final version should be programmed in C++ with OpenCV.

17.2.2 Free Software for Non-programmers

Writing a new program for each task is time-consuming, and not feasible for researchers who are not programmers or do not collaborate with programmers. However, there is an existing collection of software to automate VBI assays without a need to create a new program to address every VBI task.

17.2.2.1 ImageJ

The most widely used program is ImageJ, which is a stand-alone program that was created by the National Institutes of Health in 1997 (Fig. 17.3). It works much like an image manipulation program, such as Adobe Photoshop. The user opens the program, and then opens an image within ImageJ, performs operations on the image, and gets visual feedback for operations performed on the image. However, instead of photo-manipulation, the focus of ImageJ is biomedical imaging applications in 2D and 3D. Since its inception, there are numerous plugins that can be used with ImageJ from 3D cell imaging to radiological image processing. ImageJ also includes image processing commands such as filtering, Fourier analysis, convolution, and binary image operations, much like the features offered by MATLAB and OpenCV. However, no knowledge of programming is required to perform these operations because of its GUI interface. There are so many plugins for ImageJ that attempting to list them all would be beyond the scope of this chapter. One source estimated 500 plugins in 2012 [4]. While there are many plugins available, bioimaging tasks are sometimes very specific, requiring unique software to be created for each task. For this reason, ImageJ may require some programming knowledge to create a new plugin if an existing plugin cannot be applied to a task. ImageJ is in the public domain. ImageJ is still being improved, and the current version of ImageJ is ImageJ2, which is a complete overhaul of ImageJ but maintains backward compatibility. ImageJ2 focuses on multidimensional data. It offers a suite of community curated plugins, much like FIJI, described below [4].

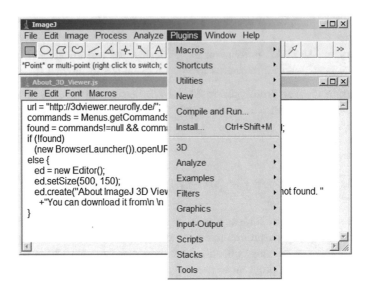

Fig. 17.3 The ImageJ interface. Note the lack of built-in features; the user is responsible for finding an appropriate plugin from the 500 plugins offered by developers, or programming the plugins themselves

17.2.2.2 FIJI Is Just ImageJ (FIJI)

FIJI started as a plugin for ImageJ but expanded to the point where the developers decided to *fork* ImageJ (Fig. 17.4). A fork is when developers take an existing source code of a program and decide to create their own version of the software from that source code. In this case, the developers created FIJI from the ImageJ source code. It is currently supported by 12 institutions including the National Institutes of Health, UW-Madison, MIT, Harvard, and more. It differs from ImageJ in the following ways [5]:

- There is a suite of plugins that come with the FIJI distribution that are approved by the community. In the original implementation of ImageJ, anyone can create a plugin for ImageJ and host it on the Internet. Because there is no quality control for an ImageJ plugin released in this way, there is no guarantee that the plugin a user downloads will work. FIJI addresses this issue by bundling plugins with the FIJI distribution that have been approved by the community. A developer can submit a plugin to be reviewed by the FIJI community, and, if it is approved, it is bundled with the program.
- The FIJI community has an active group of computer scientists who actively work to improve the efficiency of the program.
- It is designed to be a streamlined version of ImageJ. It is designed to be used by users without much programming experience.
- It comes with the Java runtime, unlike ImageJ.

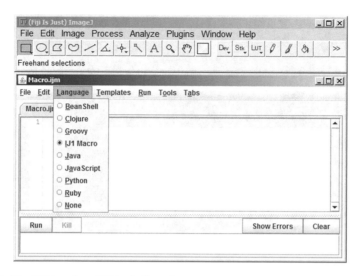

Fig. 17.4 The FIJI interface. FIJI is similar to ImageJ. However, it comes with community vetted plugins, and allows you to program the plugins in many languages

- It automatically updates.
- It allows programmers to use Beanshell, Javascript, JRuby, Jython and Clojure, whereas ImageJ programming of plugins is limited to Java.

FIJI is slightly easier to use than ImageJ, because if a community curated plugin is applicable for a user's problem, it will be more reliable than an equivalent ImageJ plugin.

17.2.2.3   Icy

Icy is a different program not based on ImageJ, but is similar in operation (Fig. 17.5). It has a smaller community than ImageJ or FIJI because it is newer, but it focuses heavily on user experience with a professional GUI and is the easiest to use. In the current version, Icy has a built-in chat room. Icy was created to promote extended reproducible research. The concept of reproducible research was outlined in [13] and was expanded by the developers of Icy:

1. Researchers should provide all the information necessary for others to repeat their experiments;
2. plugins should be presented in a modular way so that researchers can visualize algorithms in terms of a work flow, improving reproducibiliy;
3. and plugins should be standardized and interoperable.

Item 1 is defined in [13]. Items 2 and 3 are defined by the Icy developers in [4]. The developers of Icy wanted to address item 1, which they felt was not specifically

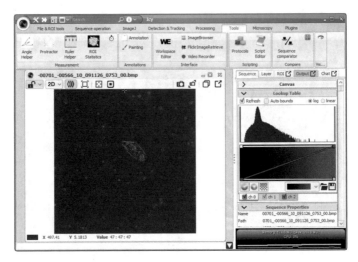

Fig. 17.5 The Icy interface. Unlike ImageJ and FIJI, Icy has a more aesthetically pleasing graphical user interface, and comes with many plugins

addressed by the sharing of plugins in ImageJ and FIJI. While a plugin could be created in ImageJ and FIJI, it was up to the researcher to disclose use of the plugin so that others could reproduce their work. It is possible that a researcher, intending to distribute their research freely as a plugin, does not properly instruct others on its use or programs poorly, thus preventing reproduction of their work when others attempt to use the plugin. Icy addresses this by enforcing points 2 and 3. While the user can script in JavaScript, and perform image operations in the same way as FIJI and ImageJ, Icy offers a protocol editor that describes complex algorithms in terms of work flows which are much easier to use by programmers and non-programmers alike. Both FIJI and Icy have plugins that automatically update.

17.2.3 Software Licensing

Software license controls ability of a person to use, study, modify, and distribute a software. We review two popular licenses for open source software here.

17.2.3.1 Berkeley Software Distribution (BSD)

A BSD license is a license that guarantees a person's right to use the software freely, to modify the software, and to redistribute it as they wish. OpenCV comes with a *Berkeley Software Distribution* (BSD) license. A derivation of an OpenCV program can be sold by a third party not related to the authors. Certain versions of the BSD license have an *attribution* requirement. This requires that the user acknowledge the authors. The OpenCV version of BSD does not have an attribution requirement. Specifically, it is a *BSD 3-clause*. With a *public domain* license, the user is allowed the same freedom as a BSD license. However, the key differences are that:

- with a BSD, the authors retain the copyright and are giving it freely, whereas with a public domain license, the authors are giving up all rights to the software;
- certain versions of BSD may require attribution (*BSD 4-clause*);
- and public domain is not supported in some countries, so a BSD may be required to ensure freedom of distribution.

For this reason we recommend that developers use a BSD license instead of a public domain license.

17.2.3.2 General Public License (GPL)

A GPL is nearly the same as a BSD license except that it has a *copyleft*. Copyleft requires that derivations of the work also carry a GPL, requiring free distribution. This means that a work derived from the FIJI main program cannot be sold. It must be free because it must carry a GPL. This is in contrast to a BSD 3-clause, which is *permissive*, where there are no restrictions on how derivations of the software can be redistributed or modified. The main FIJI program carries a *General Public License* (GPL), though plugins can carry whatever license the author issues with the plugin. Icy is licensed with a GPL and ImageJ is built into the program.

17.2.4 Commercial Software

Unlike the previous section, the software in this section is available commercially. They are not freely distributed and are closed-source. Typically, this software has a very high monetary cost. Free software is buggy, which is an issue highlighted by the developers of Icy and FIJI. While commercial software may be very expensive, there is an expectation that the software will be stable.

17.2.4.1 Amira

Amira is a program for visualizing and processing 3D and 4D data. It is supported by the Zuse Institute Berlin. It has a focus on medical imaging modalities such as CT, PET and ultrasound, but could be applied to other imaging modalities as well. It comes with a suite of commonly used image processing techniques, but focuses on segmentation, registration, and professional visualizations of results.

17.2.4.2 CL-Quant

Nikon offers a bioimage image processing program for operators of the Biostation CT called CL-Quant. This software is specifically designed for bioimaging analysis of images from the Biostation CT. It allows users to create recipes to segment the regions-of-interest to collect results. The program emphasizes creating a recipe, a specific set of steps, to apply to images, promoting reproducibility. CL-Quant's ease of use is comparable to Icy. Some algorithms are *supervised* so results from this software are less reproducible than others.

17.2.4.3 Bitplane

Bitplane offers a suite of programs that address many VBI tasks: Imaris, MeasurementPro, ImarisTrack, Imaris Cell, FilamentTracer, ImarisColoc, ImarisXT, Imaris Vantage, and Imaris Batch. These programs can analyze microscopy images in 2D, 3D, and 4D. FilamentTracers is a specific program to track filaments in cells. ImarisXT allows bitplane to interface between other programs. Bitplane focuses on analysis of bioimaging software and can produce stunning visual results.

17.3 Conclusions

A vast amount of image data is generated by biologists during their experiment. However, only a very small fraction of this raw image data is shared publically. One major obstacle in sharing the image data is it sheer size, which can be many gigabytes to even terabytes. Although many funding agencies and institutes require that data used in experiments be made public, only higher level data is made public as it is much more compact and directly usable by other biologists. Biological experiments are conducted in a controlled environment. While it is quite possible that the conditions of these experiments might reduce reusability of the image data for top level biological inference, for design of automated image analysis techniques this raw data can be immensely useful. Need for annotation adds additional burden for the biologists especially when they are needed to label hundreds of instances of the same objects/events so that the computer vision algorithms can learn from these annotations. This additional effort needed can prove to be a deterrent to image data sharing. Better computer vision algorithms which can learn from a very small number of objects/instances and in an interactive way would reduce the burden of annotation significantly.

Of chief concern to any researcher is reproducibility of work. When creating your own software solution to a VBI problem, it should later be shared with others to promote reproducibility. Currently, this is not the case with most research. This is sometimes not the fault of the researcher: it takes time to polish software for others to use and to create manuals and instructions so others can properly use the software. Rather than creating a software package from scratch, it is advisable to use one of the plugin friendly softwares to create a plugin which solves your problem. This allows one to reuse many components such as user interface, preprocessing and input/output making it easier for biologists to create and subsequently share the software.

Acknowledgment This work was supported in part by the National Science Foundation Integrative Graduate Education and Research Traineeship (IGERT) in Video Bioinformatics (DGE-0903667). Alberto Cruz is an IGERT Fellow.

References

1. Berkeley segmentation data set and benchmarks 500 [Online]. http://www.eecs.berkeley.edu/Research/Projects/CS/vision/grouping/resources.html
2. PETS 2009 benchmark data [Online]. http://www.cvg.reading.ac.uk/PETS2009/a.html
3. The PASCAL visual object classes homepage [Online]. http://host.robots.ox.ac.uk/pascal/VOC/
4. de Chaumont F, Dallongeville S, Chenouard N, Hervé N, Pop S, Provoost T, Meas-Yedid V, Pankajakshan P, Lecomte T, Le Montagner Y et al (2012) Icy: an open bioimage informatics platform for extended reproducible research. Nat Methods 9(7):690–696
5. Schindelin J, Arganda-Carreras I, Frise E, Kaynig V, Longair M, Pietzsch T, Preibisch S, Rueden C, Saalfeld S, Schmid B et al (2012) Fiji: an open-source platform for biological-image analysis. Nat Methods 9(7):676–682
6. Maka M, Ulman V, Svoboda D, Matula P, Matula P, Ederra C, Urbiola A, Espaa T, Venkatesan S, Balak DM, Karas P, Bolckov T, Treitov M, Carthel C, Coraluppi S, Harder N, Rohr K, Magnusson KEG, Jaldn J, Blau HM, Dzyubachyk O, Kek P, Hagen GM, Pastor-Escuredo D, Jimenez-Carretero D, Ledesma-Carbayo MJ, Muoz-Barrutia A, Meijering E, Kozubek M, Ortiz-de Solorzano C (2014) A benchmark for comparison of cell tracking algorithms. Bioinformatics [Online]. http://bioinformatics.oxfordjournals.org/content/early/2014/03/04/bioinformatics.btu080.abstract
7. Tobon-Gomez C, Craene MD, McLeod K, Tautz L, Shi W, Hennemuth A, Prakosa A, Wang H, Carr-White G, Kapetanakis S, Lutz A, Rasche V, Schaeffter T, Butakoff C, Friman O, Mansi T, Sermesant M, Zhuang X, Ourselin S, Peitgen H-O, Pennec X, Razavi R, Rueckert D, Frangi A, Rhode K (2013) Benchmarking framework for myocardial tracking and deformation algorithms: an open access database. Med Image Anal 17(6):632–648 [Online]. http://www.sciencedirect.com/science/article/pii/S1361841513000388
8. Chenouard N, Smal I, de Chaumont F, Maska M, Sbalzarini IF, Gong Y, Cardinale J, Carthel C, Coraluppi S, Winter M, Cohen AR, Godinez WJ, Rohr K, Kalaidzidis Y, Liang L, Duncan J, Shen H, Xu Y, Magnusson KEG, Jalden J, Blau HM, Paul-Gilloteaux P, Roudot P, Kervrann C, Waharte F, Tinevez J-Y, Shorte SL, Willemse J, Celler K, van Wezel GP, Dan H-W, Tsai Y-S, de Solorzano CO, Olivo-Marin J-C, Meijering E (2014) Objective comparison of particle tracking methods. Nat Methods 11(3):281–289 [Online]. http://dx.doi.org/10.1038/nmeth.2808
9. Ljosa V, Sokolnicki KL, Carpenter AE (2012) Annotated high-throughput microscopy image sets for validation. Nat Methods 9(7):637–637 [Online]. http://dx.doi.org/10.1038/nmeth.2083
10. Mercier L, Del Maestro RF, Petrecca K, Araujo D, Haegelen C, Collins DL (2012) Online database of clinical mr and ultrasound images of brain tumors. Med Phys 39(6):3253–3261 [Online]. http://scitation.aip.org/content/aapm/journal/medphys/39/6/10.1118/1.4709600

11. Mano S, Nakamura T, Kondo M, Miwa T, Nishikawa S, Mimura T, Nagatani A, Nishimura M (2014) The plant organelles database 3 (podb3) update 2014: integrating electron micrographs and new options for plant organelle research. Plant Cell Physiol 55(1):e1 [Online]. http://pcp. oxfordjournals.org/content/55/1/e1.abstract
12. Jhuang H, Garrote E, Yu X, Khilnani V, Poggio T, Steele AD, Serre T (2010) Automated home-cage behavioural phenotyping of mice. Nat Commun 1:68 [Online]. http://dx.doi.org/ 10.1038/ncomms1064
13. Fomel S, Claerbout JF (2009) Reproducible research. Comput Sci Eng 11(1):5–7

Chapter 18
Understanding of the Biological Process of Nonverbal Communication: Facial Emotion and Expression Recognition

Alberto C. Cruz, B. Bhanu and N.S. Thakoor

Abstract Facial emotion and expression recognition is the study of facial expressions to infer the emotional state of a person. A camera captures video or images of a person's face and algorithms automatically, without the help of a human operator, detect his/her expressions to infer his/her underlying emotional state. There has been an increased interest in this field in the past decade, and a system that accomplishes these tasks in unconstrained settings is a realizable goal. In this chapter, we will discuss the process by which a human expresses an emotion; how it is perceived by the human visual system at a low level; how prediction of emotion is made by a human; and publicly available datasets currently used by researchers in the field.

Portions of this chapter are © IEEE 2013, 2014 and appeared in "Background suppressing Gabor energy filtering," "Score-based facial emotion recognition," and "Vision and attention theory-based sampling for continuous facial emotion recognition".

A.C. Cruz (✉) · B. Bhanu · N.S. Thakoor
Center for Research, Intelligent Systems, University of California, Riverside,
900 University Ave, Riverside, CA 92521, USA
e-mail: acruz37@csub.edu

B. Bhanu
e-mail: bhanu@ee.ucr.edu

N.S. Thakoor
e-mail: nthakoor@ee.ucr.edu

© Springer International Publishing Switzerland 2015
B. Bhanu and P. Talbot (eds.), *Video Bioinformatics*,
Computational Biology 22, DOI 10.1007/978-3-319-23724-4_18

18.1   Projecting Emotions Is a Biological Process

Among the first researchers in the field of facial emotion and expression recognition was Charles Darwin who, in 1872, published, "Expression of the Emotions in Man and Animals" [1]. In that text, Darwin connected human expressions and bodily movements to underlying emotional states. Communication between two humans is a complex process that involves more than just speech. Humans communicate nonverbally with gestures, pose, and expressions. *Gestures* are a general term for motion by the body. For example, a person giving thumbs up is considered a gesture. *Pose* refers to the position and orientation of the body. For example, a person could turn their face away from another person while communicating and this would be considered posing his/her face. *Expressions* are facial muscle movements. For example, a person moving their facial muscles to open their mouth is considered an expression. Expressions and emotions are not the same. *Emotions* are the underlying feelings of a person, which may be revealed, or concealed by his/her expressions, pose, and gestures during communication.

The understanding of human expressions and emotions is a biological process—particularly when framed in the context of their origins in mammals as Charles Darwin studied them. When two humans communicate, they use their hands to gesture, they use their facial muscles to form expressions, they focus their gaze, and they pose their face toward the other human. Facial expressions are critically important in this nonverbal communication between humans. Understanding of the emotions of others can be a difficult task for humans. This is highlighted by the existence of lie detectors: it is difficult for a human to determine if another human is being deceitful. *Video-based facial emotion recognition* is an important field of study where face video of a human is captured and computer algorithms must detect their facial expressions to infer their underlying emotional state.

Expression/emotion recognition has applications in medicine (Asperger's Syndrome [2], autism spectral conditions [3]), video games (Xbox Kinect [4]), human–computer interaction (embodied conversational agents [5]), deception detection [6], and *affective computing*. Affective computing is a field where computer interfaces can both understand and project human facial expressions to facilitate nonverbal communication with a human. An example is given in Fig. 18.1. In video-based facial emotion recognition, a system must automatically detect apparent facial expressions of a person and infer their underlying emotional state. There has been an increased interest in facial emotion recognition and the field has seen great advances. In Sect. 2.1, we discuss the process by which a human projects their emotions via facial expressions, and how it is interpreted by another human. In Sect. 2.2, we discuss how a current computer algorithm detects emotion.

Fig. 18.1 Aff-ective
computing

18.1.1   Projecting Emotions via Facial Expressions

If emotions and expressions are to be detected, they must first be quantified. However, the categorization and classification systems of emotion are an ongoing field of research in psychology and neuroscience. We will highlight three ways to quantify emotions and expressions: action units [10], emotions based on the Ekman big six [10] and the Fontaine emotional model [12], which is an extension of affective dimensions.

Expressions are facial muscle movements. Ekman and Friesen [10] defined the minimal set of facial muscle movements, or action units (AUs), that are used in expressions. This is the Facial Action Coding System. For example, a smile consists of AU 6 and AU 12. AU 6 indicates that a person's cheeks are raised. AU 12 indicates that the corners of a person's lips are being pulled outward. This often occurs when smiling. Emotion differs from expressions in that they are the underlying mental states that may illicit expressions. A common system for discrete emotional states is the Ekman big six: happiness, sadness, fear, surprise, anger, and disgust. Ekman posits that these six emotions are basic emotions that span across different cultures [10].

A different system for emotion labels is the Fontaine emotional model [12] with four affect dimensions: valence, arousal, power, and expectancy. An emotion occupies a point in this four-dimensional Euclidean space. Valence, also known as evaluation-pleasantness, describes positivity or negativity of the person's feelings or feelings of situation, e.g., happiness versus sadness. Arousal, also known as activation-arousal, describes a person's interest in the situation, e.g., eagerness versus anxiety. Power, also known as potency-control, describes a person's feeling of control or weakness within the situation, e.g., power versus submission. Expectancy, also known as unpredictability, describes the person's certainty of the situation, e.g., familiarity versus apprehension.

An Ekman big six emotion occupies a region in each of these four dimensions. For example, happiness would be positive-valued valence because the person

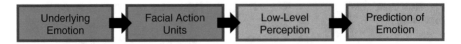

Fig. 18.2 Overview of how emotions are projected and perceived by other humans (*orange*). Prediction of another humans' emotion is a two-step process of perceiving a person's face a low level and predicting what emotion is being expressed

would feel positive about the situation. It would have positive arousal, because the person would enjoy the situation. It would have positive power, because a person would likely need to feel in control of himself to feel happy. It may be any value for expectancy, because a person may or may not be both surprised and happy. For a more detailed explanation, the reader is referred to [12]. With this system, multiple emotions can be expressed at the same time.

An overview of how humans communicate their emotions nonverbally is given in Fig. 18.2. First, a person has an underlying emotion. He or she may move their facial muscles based on their emotion. The facial muscle movements can be explicitly described by AUs. Groups of AUs form a *gesture*, such as a smile. These muscle movements are projected and perceived by another human. It is processed by the human visual system at a low level, and a judgment is made by another human as to what emotion the person is expressing. It is possible that the expressions projected by the human are not the underlying emotion, such as when a person is acting or when a person is being deceptive. In certain cultures, outward displays of emotion are frowned upon. This is why detecting the emotions of another human can be a difficult task.

18.1.2   Prediction of Emotions with Computer Algorithms

A computer algorithm may not predict emotion in the same way as the human visual system in Fig. 18.2 (Orange). Typically, the system will process the video of a frontal face on a frame-by-frame basis. The face region-of-interest (ROI) is extracted to determine where the face is in the video frame. The most commonly used method is the Viola and Jones cascade of Haar-like features [10]. This detector is ubiquitously used in both facial expression and emotion recognition, and face recognition. Because a person can rotate their face (called *out-of-plane* head movement) and because the distance between the person in the frame and the camera can vary (causing *scale* issues) the ROI of the faces must be aligned. This procedure is called *registration*. Popular registration algorithms are active appearance models (AAM) [14], a similarity transform (RST) using control points, such as eyes, that are detected with Viola and Jones [15], and SIFT-based registration [11]. Then, *features* are extracted. A feature method describes the face in a meaningful way such that you can describe the differences between two faces. Popular features are control points, a set of vectors describing the position of facial features in a

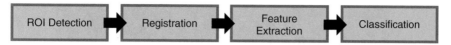

Fig. 18.3 Overview of a system for facial emotion recognition

Euclidean space; textures, and edges, noting where there are edges on the face that may correspond to facial features; and motion. The feature is represented as a vector. After feature extraction, a machine learning algorithm predicts the emotion of a person (*classification*) by receiving a feature vector (*testing sample*). It refers to a data set of other feature vectors where the emotion is known (*training data*), and it attempts to find the best match for the received feature vector. Performance is gauged by predicting the emotion of many testing samples (*testing data*) by the percent accuracy (*classification accuracy*). An overview of this process is given in Fig. 18.3.

While detecting the emotions of other humans can be a difficult task computer systems struggle to detect authentic emotions in humans, i.e., computers struggle with emotions that are not posed or acted. When actors and animators project emotion they intend for their emotions to be understood by an audience, thus, it is easier to detect acted or posed emotions. It makes sense then, that a computer system should emulate the process by which the human visual system understands the emotions projected by another human. In the following chapters, we will discuss: (1) feature extraction at a low level that resembles the way the human visual system processes images and (2) predicting human emotions in the same way the human visual system understands the face.

18.2   Low-Level Perception by the Human Visual System

In this chapter, we discuss the way the human visual system perceives images at a low level. Specifically, we will discuss the Gabor energy filters and the nonclassical receptive field. The understanding of the behavior of the human brain is a frontier of research. However, facial processing by the human visual system is one of the most well understood pathways of the brain. Visual information in the human visual system is processed by the eyes. Its first stop in the human visual system is the visual cortex, and its first stop within the visual cortex is the V1 area. In the V1 area, there are neurons that detect the presence of specific presence of an edge at a specific orientation, and at a specific width. Examples are given in Fig. 18.4.

This somewhat resembles how a computer algorithm represents an image: it is broken down into small meaningful parts. In the human visual system, an image is broken down into edges of specific orientations and widths. There is a facial feature descriptor that breaks down an image into meaningful edges in the same way as the V1 cortex, and it is called the *Gabor filter*.

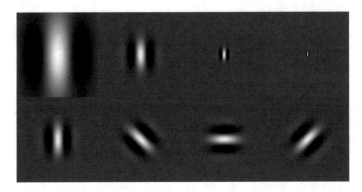

Fig. 18.4 Examples of a gratings. As you perceive these gratings, a different, specific part of your V1 area is responding to each grating

18.2.1 The Classical Receptive Field

A Gabor filter is a band-pass filter that can be applied to detect edges of a specific orientation and scale. An image is filtered by many Gabor filters under different parameters, called a filter *bank*. Each filter in the bank is attuned to a different orientation and scale. The V1 cortex's response when presented with edges is called the Classical Receptive Field [13]. Let $I(x,y)$ be an input image. Contours of the image are measured with a Gabor filter:

$$g(x, y; \gamma, \theta, \lambda, \sigma, \phi) = e^{\frac{\tilde{x}^2 + \gamma^2 \tilde{y}^2}{2\sigma^2}} \cos\left(2\pi \frac{\tilde{x}}{\lambda} + \phi\right) \tag{18.1}$$

where x and y are the pixel location; γ is the spatial aspect ratio; θ is the angle parameter where that attunes the filter to specific orientations; λ is the wavelength; σ^2 is the variance; ϕ is the phase offset taken to be 0 and π. To attune the Gabor filter as a local appearance filter with a small neighborhood [13]: $\sigma/\lambda = .56$, $\gamma = 0.5$ and vary θ. Henceforth in the paper, $g(x, y; \theta, \phi)$ is shorthand for the above values, i.e., $g(x, y; .5, \theta, 7.14, 3, \phi)$. $\tilde{x}$ and $\tilde{y}$ are defined as

$$\tilde{x} = x\cos\theta + y\sin\theta \tag{18.2}$$

$$\tilde{y} = -x\sin\theta + y\cos\theta \tag{18.3}$$

Typically, values of θ and ϕ are selected such that they detect edges of different orientations and magnitudes and do not overlap in their detection. $f(x,y)$ is filtered, first by $g(x, y; \theta, 0)$, then by $g(x, y; \theta, \pi)$ and its magnitude is taken to be the response. This is called the Gabor energy:

$$E(x, y; \theta) = \sqrt{((f * g)(x, y; \theta, 0))^2 + ((f * g)(x, y; \theta, \pi))^2} \qquad (18.4)$$

where $(f * g)(x, y; \theta, \phi)$ is the convolution of $f(x,y)$ and $g(x, y; \theta, \phi)$.

In summary, many Gabor filters are created by varying θ and ϕ. Each Gabor filter would resemble a grating given in Fig. 18.4. The collection of these filters is called a bank. If f is the image that needs description, it would be filtered by each filter in the bank. The resulting images indicate the parts of f that resembled the gratings in the bank. However, the human visual system takes one more step after computing the locations of gratings. It removes noise due to background texture.

18.2.2 The Nonclassical Receptive Field

One drawback of the Gabor filter is that it captures all edges in an image, including edges that are due to background texture. The human visual system automatically removes these edges. An example is given in Fig. 18.5, where, when perceiving the image, edges that are not consistent with the background texture will pop out. This is called the *pop-out effect*. It can be incorporated with the Gabor filter to improve robustness when images have background texture.

Equation 18.4 successfully captures the edge information of a specific orientation. It responds to edges in the same way a simple cell in the human visual system responds to a grating (an edge or a set of edges). However, the human visual system is able to detect edges in the presence of background texture. The background texture is referred to as the Nonclassical Receptive Field [13]. It is estimated as

$$t(x, y; \theta) = (E * w)(x, y) \qquad (18.5)$$

where the weight function w is

$$w(x, y) = \frac{1}{\|g(\text{DoG}(x, y))\|_1} g(\text{DoG}(x, y)) \qquad (18.6)$$

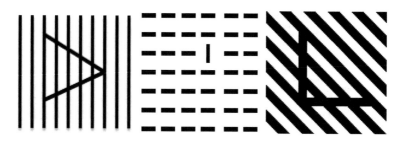

Fig. 18.5 Three examples illustrating the pop-out effect

where $g(z) = H(z) * z$, where $H(z)$ is the Heaviside step function. DoG(x,y) is a Difference of Gaussians:

$$DoG(x, y; K, \sigma) = \frac{1}{2\pi K^2 \sigma^2} e^{-\frac{x^2+y^2}{2K^2\sigma^2}} - \frac{1}{2\pi\sigma^2} e^{-\frac{x^2+y^2}{2\sigma^2}} \tag{18.7}$$

where K is a weight. σ^2 is the variance, the same as Eq. 18.1. This ensures that the filter is bounded within the original Gabor filter. $w(x,y)$ resembles the ridges of a Mexican Hat filter. When applied as the weight, Eq. 18.5 captures the edge information surrounding the current pixel. This allows background texture to be estimated on a per-pixel basis. It is removed via

$$\tilde{b}(x, y; \theta) = g(E(x, y; \theta) - \alpha t(x, y)) \tag{18.8}$$

where α is a parameter that affects how much of the background texture is removed. When $\alpha = 0$, there is no background texture suppression, and the filter is a Gabor energy filter. An example of background texture suppression is given in Fig. 18.6. Note that for all examples the dominant contours from the eyes and mouth are extracted, but the classical receptive field detects edges in places where there is no significant, perceivable edge.

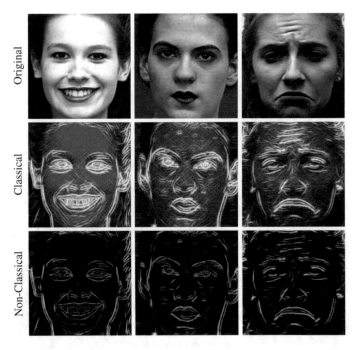

Fig. 18.6 Examples of faces processed with a the classical receptive field (Gabor) and the nonclassical receptive field

Equation 18.8 has retrieved the contours of $f(x,y)$ less background texture. It is computed for N different orientations. Conventionally, the responses from the N orientations would be concatenated and taken to be the feature vector. A method is needed to reduce the feature dimensionality. A representation of $\tilde{b}(x,y;\theta)$ is created that retains the maximal response for each pixel:

$$b(x,y) = max\big\{\tilde{b}(x,y;\theta)|\theta = \theta_1,\ldots,\theta_N\big\} \qquad (18.9)$$

Separately, an orientation map $\Theta(x,y)$ is constructed that contains the dominant orientation for each pixel:

$$\Theta(x,y) = argmax_\theta\big\{\tilde{b}(x,y;\theta)|\theta = \theta_1,\ldots,\theta_N\big\} \qquad (18.10)$$

Equations 18.9 and 18.10 retain the maximal edge. $b(x,y)$ retains the value of the maximum edge intensity, across all orientations, and $\Theta(x,y)$ stores the specific orientation of the maximal edge. The image, $f(x,y)$, is broken into M^2, equally sized, nonoverlapping regions. This process is referred to as *gridding*, and is used in facial emotion recognition [14]. It is often used in local binary patterns [15] and local phase quantization [12]. Local binary patterns and local phase quantization use a histogram to count the response of specific microtextures. We use a soft histogram, where votes are weighted by their maximal representation:

$$h(\theta_i) = \sum\nolimits_{\forall(x,y)|\Theta(x,y)=\theta_i} b(x,y) \qquad (18.11)$$

where $h(\theta)$ is an N bin histogram. A histogram is computed in each grid. The $M \times M$ grids are concatenated to form the feature vector for $f(x,y)$. Note that the Gabor filter detected many more edges that were not meaningful, and that removing background texture leaves only the significant edges.

18.2.3   Results for Prediction of Emotion

To test the effectiveness of the Gabor filter with background texture removal as a feature to represent facial information, we conducted experiments using the Audio/Visual Emotion Challenge 2012 development set frame-level sub-challenge (to be described in 5.4. The development set was split using threefold random cross-validation. Face ROI was extracted with Viola and Jones [13], faces were registered with SIFT-based registration [16] and a linear SVM was used as a classifier. We tested many features, and the proposed feature, Gabor filter with background texture removal is the best on average. Results are given in Table 18.1, in terms of the Pearson Product-moment Correlation Coefficient, where higher is better. The best performer is the nonclassical receptive field, followed by

Table 18.1 Results on AVEC 2012 development set frame-level sub-challenge

Feature	Pearson product-moment correlation coefficient				
	Arousal	Expectancy	Valence	Power	Average
DCT [17]	0.034	0.078	0.076	0.063	0.063
FPLBP [30]	**0.425**	0.108	0.291	0.093	0.229
LBP [6]	0.434	0.072	0.257	0.088	0.213
LBP-TOP [11]	0.389	0.092	0.177	0.084	0.186
LPQ [27]	0.032	0.085	0.072	0.076	0.066
SIFT [19]	0.037	0.038	0.073	0.048	0.049
TPLBP [30]	0.024	0.047	0.086	0.039	0.049
Classical receptive field (Gabor)	0.059	0.019	0.063	0.012	0.036
Nonclassical receptive field	0.417	**0.143**	**0.347**	**0.124**	**0.258**

For correlation, higher is better. *Bold* best performer. *Underline* second best performer

FPLBP. LBP [6], FPLBP [30], LBP-TOP [11], and TPLBP [30] are in the LBP family of feature descriptors and are currently the most commonly used feature for facial emotion recognition.

18.3 Face Perception by the Human Visual System

In this section, we discuss the pathways that the human visual system uses to combine visual information to understand the face. In their work on human face recognition, O'Toole, Roark and Abdi [28] group facial feature information processed by the brain as either static information or dynamic information. Static information refers to invariant facial features that could be generated from a single frame, such as eyebrows, iris color, geometric relationships, etc. Dynamic information refers to spatiotemporal facial motion information that could be generated from video, as well as gesture and pose, e.g., the facial muscle dynamics associated with smile. A crossover exists where, when perceiving a face, dynamic information and static information are combined under nonoptimal conditions and its contribution increases proportionally with the amount of experience the viewer has with that face. This phenomenon is called the supplemental information hypothesis (SIH) and it posits that humans represent characteristic facial motions or idiosyncratic gestures, titled, dynamic facial signatures, to a specific person. A previous model (Haxby et al.'s distributed neural system for face perception [29]) described this difference between static and dynamic information, but it was O'Toole, Roark and Abdi [28] who posited that there is a crossover from the motion-computing emotion-oriented middle temporal visual area to the appearance-oriented fusiform face area (see Fig. 18.7).

Fusion is a method in computer vision that resembles this same procedure. It is used in multimodal biometrics, but applicable to all computer vision applications.

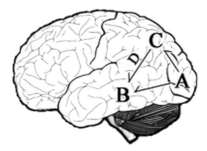

Fig. 18.7 Static face information in the visual cortex (**a**) is carried in the ventral stream destined for the facial fusiform area (**b**), while dynamic facial information is carried in the dorsal stream destined for the middle temporal area (**c**). The supplemental information hypothesis posits that there is a forward feedback (**d**) where dynamic facial signatures are combined with static information

In multimodal biometrics, there are typical multiple sensors, called modes. An example is a facial emotion recognition system employing infrared-depth and visible light sensors. Fusion is the method by which these two modes are combined. The benefit of combining multiple sensors addresses the issue if one of the sensors fails. As you increase the number of modes of different types the more accurate your results will be. However, as the number of modes increases, so too does the complexity of the system. A method must be carefully balance these two aspects. There are many ways to combine the information from different sensors, and this book focuses on the following methods: feature-level, match-score level, decision level, and microsimilarity fusion. These methods can be distinguished by the stage that fusion occurs in the pipeline. Feature-level fusion occurs in pre-classification, and combines the feature vectors from the different modes. Match-score and decision-level fusion occur post-classification. In that type of fusion, the posterior probabilities, decision values, or estimated labels are estimated from each mode. The initial results from these classifiers are fed into a second classifier or optimization process that produces the final result. Microsimilarity fusion [17, 20] differs from the other fusion methods by comparing the similarities between samples. Applied to the model of the human visual system, static and dynamic information are separate modes that are fused when perceiving a face.

In facial emotion and expression recognition, fusion is applied after features are extracted. The modes can be different features. For example, a method can fuse the results from many different features. In one example [19], Gabor features and optical flow features are combined with a HMM. The modes can also be entirely different recognition pipelines, with different registration methods and classifiers. When using too many modes, the complexity of a system results in high computational time and memory cost. For some data sets, such as the AVEC, that has roughly one and a half million frames, just a few modes are enough to cause an undesirable memory cost. A comparison of fusion methods is given in Table 18.2. k is the number of modes.

Table 18.2 Comparison of fusion methods

Type	Stage	Advantages	Disadvantages
Feature	Early (pre-classifier)	Easy to implement. Single classifier.	Increases number of features.
Match-score or decision	Late (post-classifier)	Aggregating process can account for failure of one of the modes.	Runs of classifier increased by k.
Microsimilarity	Early (replaces classifier)	Overcomes situations where training data does not properly represent testing data.	Problem must be framed in terms of comparisons to other samples.

18.3.1 Feature-Level Fusion

Among all the methods, feature-level fusion is the most commonly used. It is used so often that many times it is not referred to by name. If there are k modes, and X_i are the feature vectors from some mode i:

$$Y_F = [X_1 \quad \ldots \quad X_k] \tag{18.12}$$

where Y_F is the feature-level fusion of all of the modes. The feature vector size of Y is the sum of the size of all the modes. The objective of the method is that there is some combination of the features that may yield better classification result. To find this manifold, the concatenation in Eq. 18.12 is coupled with a feature reprojection and selection process, such as principle component analysis (PCA), independent component analysis (ICA), nonnegative matrix factorization (NMF) [20], or singular value decomposition (SVD) [21]. The purpose of these algorithms is to determine which features are useful from the feature vector. Some of the algorithms create combinations of features that are more useful than the features themselves.

There are two ways to perform coupling of feature selection and feature concatenation. In the first way, the feature vectors are reprojected before concatenation in Eq. 18.12. In the second way, Y is reprojected. The second method ensures that the most variant features from all modes are used. However, because PCA is applied after concatenation, there is no guarantee that all of the modes will be utilized in the reprojection. It could be that the eigenvectors favor a specific mode in training, and that this performance does not continue in testing. Overall, the benefit of feature-level fusion is that it is easy to implement. The negatives are that it increases feature vector length, and that, even if a reprojection and reduction method is applied, the number of feature vectors and their length may be so high that even computing the reprojection of the features is undesireable.

18.3.1.1   Q-Statistics for Feature Vector Disparity

The objective of feature-level fusion is to combine modes that are good in disparate conditions. That is, if there are two modes, A and B, mode A works well for samples where B fails, and vice versa. There is metric called the Q-statistic that predicts whether or not the features from two different modes will work well in a fusion scheme. It is

$$Q_{ij} = \frac{(n_{00}n_{11} - n_{01}n_{10})}{(n_{00}n_{11} + n_{01}n_{10})} \tag{18.13}$$

where Q_{ij} is the Q-statistic metric that measures how disparate the performance is between mode i and mode j; n_{00}, the number of samples where i and j misclassified the same sample; n_{11}, the number of samples where i and j correctly classified the same sample; n_{01}, the number where i misclassified and j correctly classified; n_{10}, the number where j misclassified and i correctly classified. Note that $Q_{ij} \in [-1, 1]$. The larger the value of Q_{ij}, the more disparate the feature vectors are, and the better it is to pair i and j.

18.3.2   Match-Score and Decision-Level Fusion

Match-score and decision-level fusion differ from feature-level fusion in that the fusion occurs post-classification. In match-score fusion, a classifier generates posterior probabilities, or decision values, both referred to as match-scores, for each mode separately. After estimation, a second classifier utilizes the match-scores as features. Decision-level fusion is similar but it utilizes the estimated labels as opposed to the match-scores. In one example [22], a single mode is used but neighboring time points are taken to be different modes. A classifier generates an initial hypothesis for each frame in terms of posterior probabilities, and an aggregator combines the posterior probabilities. $p(c|X_i)$ is the posteriori probability of membership to class c, given X_i, the features vectors for mode i. Concentionally, the posterior probability of a sample belonging a class, given a single mode X_0:

$$\tilde{C} = \text{argmax}_c \{\tilde{p}(C, X_0) | C = c_1, \ldots, c_m\} \tag{18.14}$$

where $\tilde{C}$ is the estimated label, m is the number of classes, and conventionally $\tilde{p}(C, X_0) = p(C|X_0)$. However, in fusion, we have more than one mode, and must modify $\tilde{p}(C, X_0)$:

$$\tilde{p}(C, X_0) = K(\{p(C|X_i)|X_i = X_1, \ldots, X_k\}) \tag{18.15}$$

where K is the aggregation rule. When K is the average, it takes the average of all the match-scores. K can be any rule, such as the harmonic mean, minimum, maximum, and mode. While the averaging rule could be proved optimal as the minimum variance unbiased solution, it is not always the best in practice. In one case, it was found that the maximum rule was the best performer [22]. The rule should be selected empirically.

18.3.3 Microsimilarity Fusion

In the training phase, a manifold is trained from the training data that best describes the samples encountered in training. With respect to facial emotion and expression recognition, in testing, a person may be encountered that is not in training. Because each person expresses an emotion in a different way, there is no guarantee that a person will express their emotions in the same way as others, and there is no guarantee that the training model will properly model the new, unforeseen emotions and expressions. Indeed, in some of the data sets to be detailed in Sect. 18.5, persons in onefold are not the same persons in another fold. In the presence of these technical challenges, current methods do not perform well. This situation is similar to the problems presented in unconstrained face recognition. Wolf et al. [17] proposed "learning with side information" to overcome these generalization problems. This concept can be exploited in facial emotion and expression recognition, and is realized by neutral and temporal microsimilarities [18]. Features are created that describe the relation of a sample to other samples, without relying on a manifold created by features.

One method of microsimilarity fusion is neutral similarity. It measures the intensity of an emotion. A feature is computed as the difference between the current face and neutral face, some estimate of a face that is emotion and expression neutral. This can be estimated with Avatar Image Registration [11]. The metric differentiates when facial expressions are close to neutral and when facial expressions are intense. This is computed by comparing a given frame, I to a reference or a neutral face. In another method, temporal similarity, the change between frames is measured. This microsimilarity provides additional information of how expressions are changing temporally. It is computed by comparing $I(\mathbf{x},t)$ with $I(\mathbf{x} - \delta, t)$, where δ is some offset. The microsimilarities can be computed post-feature extraction by an L1-norm vector difference between the two samples [17]. In one work, the energy of a modified SIFT-Flow algorithm is taken to be the feature [18]. The microsimilarities are taken to be the features for classification, and can be combined the features to yield better classification results.

18.3.4 Results for Prediction of Emotion

We conducted experiments to test the efficacy of fusing static and dynamic information with microsimilarities. We conducted experiments using MMI-DB and CK. We used the entire data set where there was video of a frontal face. We used the

Table 18.3 Interdatabase testing for score-based facial emotion recognition on CK +, MMI-DB and JAFFE databases

Method	CK+	MMI	C2 M	C2 J	MC2 J	M2C
Feature fusion	89.8	62.5	43.4	44.1	45.5	64.9
N/T score only	89.0	78.0	56.8	51.0*	52.0*	72.0
OSE score only	90.9	86.4	57.6	56.1	57.1	85.8
Microsimilarity fusion	**92.4**	**90.5**	**61.9**	58.1	**60.1**	**88.5**

Acronym indicates which dataset was used for training and which was used for testing. N/T: Neutral and temporal score. *C* CK+. *M* MMI. *J* JAFFE. For example, C2 M indicates that CK+ was used for training and MMI was used for testing
*Temporal score not used because dataset is images

most intensely expressed frame. Face ROI was extracted with Viola and Jones [13], faces were registered with SIFT-based registration [16] and a linear SVM was used as a classifier. The results are given in Table 18.3. Feature fusion only indicates that a fusion of LBP [6], FPLBP [30], and TPLBP [30] was used. Fusion of all microsimilarities is the best performer.

18.4   Data Sets

The field of facial emotion and expression recognition has advanced with the help of publicly available data sets. Among the first was the first Japanese female expression data set [23]. Since then, there have been many data sets available: Cohn-Kanade+ (CK+), MMI facial expression database (MMI-DB), the first facial expression recognition and analysis grand-challenge (FERA), the audio/visual emotion challenge (AVEC), ordered by date. The field has moved toward more spontaneous, naturally collected data. A comparison of publicly available data sets is given in Table 18.4. Examples of images from each data set are given in Table 18.4.

Expressions and emotions are not the same. Expressions are facial muscle movements and emotional states are the underlying mental states that may illicit expressions. Different data sets will use different types of labels. Ekman and Friesen defined the minimal set of facial muscle movements, or Action Units (AUs), used in expressions. This is the Facial Action Coding System [25]. The most commonly known system for discrete emotional states is the Ekman big six: happiness, sadness, fear, surprise, anger, and disgust. These categories have since been expanded beyond the original six [26].

18.4.1   Cohn-Kanade

It consists of 593 videos of 123 different individuals. A person faces a video camera and acts out one or more expressions. The majority of the videos are black and

Table 18.4 Comparison of publicly available data sets

Name	Date	Labels	Pose change	Acted
Japanese female facial expression database (JAFFE) [23]	1998	Emotions	No	Yes
Cohn-Kanade+ (CK+) [24]	2010	Emotions and AU	No	Yes
MMI facial expression database (MMI-DB) [7]	2010	Emotions	No	Yes
Facial emotion recognition and analysis grand-challenge 2011 (FERA) [8]	2011	Emotions and AU	Limited	Professionally
Audio/visual emotion challenge 2012 (AVEC) [9]	2012	Emotions	Yes	No

white, though there are some color sequences. The sequences follow a neutral-apex pattern, where the sequences tart with a neutral representation of the persons face, and ends when the persons expression is most intense. There is no pose change. Expressions are quantized in terms of facial action units (AU) [25]. There are many more negative samples than positive samples. Such a disproportionate a priori rate causes a high overall classification rate, so more attention should be given to the true positive and false negative rates.

18.4.2   MMI Facial Expression Database

The MMI facial expression database (MMI-DB) is frontal face video data. We use Part I and II of the database. It contains 736 videos. MMI-DB is acted and posed. Unlike CK+, the classes are 26 emotional states [7]. The videos follow an offset-onset-offset pattern. That is, emotion is neutral at frame one. It then peaks at the emotional apex near the middle of the video. Finally, the emotion returns to neutral at the end of the video.

18.4.3   Facial Emotion Recognition and Analysis Grand-Challenge 2011

The facial emotion recognition and analysis grand-challenge consists of 155 training videos and 134 test videos. The labels for the test videos are withheld from the public, and results must be submitted to another party for evaluation. There are seven subjects in the training data and six subjects in the test set, three of which are not in the training set. This data set differs from previous data sets in that the data was acted by professionals. It is less constrained than previous data sets, and the persons in the data may exhibit some pose change. The results are provided in terms of person independent, person specific, and overall.

18.4.4 Audio/Visual Emotion Challenge

The audio/visual emotion challenge [27] is also a grand-challenge data set. The challenge was hosted in 2011 and 2012, and continues to be offered. A person engages in conversation with a console displaying an embodied agent. There are no constraints. A person may turn away from the camera, causing an extreme pose change. An example is available online at [28]. The videos are recorded at $\sim$ 49fps and for so long that one and a half million frames are recorded. There are three sets: training, development, and testing. In training, a training model is trained on the training set and the development set is classified. The official results are given when classifying the testing set instead of the training set. Because AVEC has such a high number of frames, it is computationally undesirable to load all the frames into memory. AVEC is unique in that it quantizes emotional states in terms of the Fontaine emotional model [29], and that it reports results in terms of weighted accuracy, unweighted accuracy and correlation with ground truth.

Emotion is described in terms of valence, arousal, power, and expectancy. An emotion, such as happiness or sadness, occupies a point in this four-dimensional Euclidean space. Each dimension represents one of the four emotions. Valence, also known as evaluation-pleasantness, describes positivity or negativity of the subject's feelings or feelings of situation, e.g., happiness versus sadness. Power, also known as potency-control, describes a user's feeling of control or weakness within the situation, e.g., power versus submission. Arousal, also known as activation-arousal, describes a user's interest in the situation, e.g., eagerness versus anxiety. Expectancy, also known as unpredictability, describes the user's certainty of the situation, e.g., familiarity versus apprehension. For a more detailed explanation, please refer to Fontaine et al. [29].

18.5 Conclusion

In this chapter, we discussed the process by which a human expresses an emotion. There is an underlying emotional state which may elicit facial expressions. We discussed how it is perceived by the human visual system at a low level with the classical and nonclassical receptive field. We discussed how prediction of emotion is made by a human with information crossovers from different parts of the human visual system. It was found that computer algorithms modeled after the human visual system performed better than their heuristic counterparts. This makes sense because the projection of facial expressions understandable by other humans. Computer algorithms should emulate the human process for understanding facial expressions.

Acknowledgment This work was supported in part by the National Science Foundation Integrative Graduate Education and Research Traineeship (IGERT) in Video Bioinformatics (DGE-0903667). Alberto Cruz is an IGERT Fellow.

References

1. Darwin C (1872) The expression of the emotions in man and animals. John Murray
2. el Kaliouby R, Robinson P (2005) The emotional hearing aid: an assistive tool for children with asperger syndrome. Univ Access Inf Soc 4(2):121–134
3. Schuller B, Marchi E, Baron-Cohen S, O'Reilley H, Robinson P, Davies I, Golan O, Friedenson S, Friedenson S, Tal S, Newman S, Meir N, Shillo R, Camurri A, Piana S (2013) ASC-Inclusion: interactive emotion games for social inclusion of children with autism spectrum conditions. In: Intelligent digital games for empowerment and inclusion
4. Shotton J, Fitzgibbon A, Cook M, Finocchio M, Moore R, Kipman A, Blake A (2011) Real-Time human pose recognition in parts from single depth images. In: Proceedings of the IEEE computer vision and pattern recognition, Colorado Springs
5. McKeown G, Valstar M, Cowie R, Pantic M, Schröder M (2012) The SEMAINE database: annotated multimodal records of emotionally colored conversations between a person and a limited agent. IEEE Trans Affect Comput 3(1):5–17
6. Elkins AC, Sun Y, Zafeiriou S, Pantic M (2013) The face of an imposter: computer vision for deception detection. In: Proceedings of the Hawaii international conference on system sciences, Grand Wailea
7. Valstar MF, Pantic M (2010) Induced disgust, happiness and surprise: an addition to the MMI facial expression database. In: Proceedings of the international language resources and evaluation conference, Malta
8. Valstar MF, Mehu M, Jiang B, Pantic M, Scherer K (2012) Meta-analysis of the first facial expression recognition challenge. IEEE Trans Syst Man Cybern Part B 42(4):966–979
9. Shuller B, Valstar M, Eyben F, Cowie R, Pantic M (2012) AVEC 2012—the continuous audio/visual emotion challenge. In: Proceedings of the ACM international conference on multimodal interaction, Santa Monica
10. Viola P, Jones M (2001) Rapid object detection using a boosted cascade of simple features. In: IEEE CVPR
11. Yang S, Bhanu B (2012) Understanding discrete facial expressions in video using an emotion avatar image. IEEE Trans Syst, Man, Cybern, Part B 42(4):980–992
12. Heikkila J, Ojansivu V (2008) Blur insensitive texture classification using local phase quantization. In: Image and signal processing. Springer, New York, pp 236–243
13. Grigorescue C, Petkov N, Westenberg MA (2003) Contour detection based on nonclassical receptive field inhibition. IEEE Trans Image Process 12(7):729–739
14. Jiang B, Valstar MF, Pantic M (2012) Facial action detection using block-based pyramid appearance descriptors. In: Proceedings of the IEEE international conference on social computing, Amsterdam
15. Pietikainen M, Zhao G (2007) Dynamic texture recognition using local binary patterns with an application to facial expressions. IEEE Trans Pattern Anal Mach Intell 29(6):915–928
16. Pietikainen T, Ahonen A, Hadid M (2006) Face description with local binary patterns: application to face recognition. IEEE Trans Pattern Recogn Anal 28(12):2037–2041
17. Wolf L, Hassner T, Taigman Y (2011) Effective unconstrained face recognition by combining multiple descriptors and learned background statistic. IEEE Trans Pattern Recogn and Anal 33 (10):1978–1990
18. Cruz AC, Bhanu B, Thakoor NS (2013) Facial emotion recognition with expression energy. In: Proceedings of the ACM international conference on multimodal interaction, Santa Monica
19. Glodek M, Tschechne S, Layher G, Schels M, Brosch T, Scherer S, Kächele M, Schmidt M, Neumann H, Palm G, Schwenker F (2011) Multiple classifier systems for the classification of audio-visual emotional states. In: Proceedings of the affective computing and intelligent interaction, Memphis
20. Gupta MD, Jing X (2011) Non-negative matrix factorization as a feature selection tool for maximum margin classifiers. In: IEEE CVPR

21. Brunzell H, Eriksson J (2000) Feature reduction for classification of multidimensional data. Pattern Recogn 33(10):1741–1748
22. Cruz AC, Bhanu B, Yang S (2011) A psychologically inspired match-score fusion model for video-based facial expression recognition. In: Proceedings of the affective computing and intelligent interaction, Memphis
23. Lyons M, Akamatsu S (1998) Coding facial expressions with Gabor wavelets. In: Proceedings of the IEEE conference on automatic face and gesture recognition, Nara
24. Lucey P, Cohn JF, Kanade T, Saragih J, Ambadar Z (2010) The extended Cohn-Kanade dataset (CK+): A complete dataset for action unit. In: IEEE CVPR
25. Ekman P, Friesen W (1978) Facial action coding system: a technique for the measurement of facial movement. Consulting Psychologists Press, Palo Alto
26. Ekman P (1999) Basic emotions. In: The handbook of cognition and emotion. Wiley, New York, pp 45–60
27. Schuller B, Valstar M, Eyben F, Cowie R, Pantic M (2012) AVEC 2012—the continuous audio/visual emotion challenge. In: Proceedings of the ACM international conference on multimodal interaction, Santa Monica
28. McKeown G (2013) Youtube, 24 February 2011. http://www.youtube.com/watch?v=6KZc6e_EuCg. Accessed 21 June 2013 (Online)
29. Fontaine J, Scherer K, Roesch E, Ellsworth P (2007) The world of emotions is not two-dimensional. Psychol Sci 18(12):2050–1057
30. Soladie C, Salam H, Pelachaud C, Nicolas Stoiber RS (2012) A multimodal fuzzy inference system using a continuous facial expression representation for emotion detection. In: Proceedings of the ACM international conference on multimodal interaction, Santa Monica

Chapter 19
Identification and Retrieval of Moth Images Based on Wing Patterns

Linan Feng, Bir Bhanu and John Heraty

Abstract Moths are important life forms on the planet with approximately 160,000 species discovered. Entomologists in the past need to manually collect moth samples, take digital photos, identify the species, and archive into different categories. This process is time-consuming and requires a lot of human labors. As modern technologies in computer vision and machine learning advance, new algorithms have been developed in recognizing objects in digital images based on their visual attributes. The methods can also be applied to the entomology domain for recognizing biological identities. The *Lepidoptera* (moths and butterflies) in general can be identified and classified by their body morphological features; wing visual patterns that can be obtained using various image processing approaches in automated diagnostic systems. In this chapter, we describe a system for automated moth species identification and retrieval. The core of the system is a probabilistic model that infers semantically related visual (SRV) attributes from low-level visual features of the moth images in the training set, where moth wings are segmented into information-rich patches from which the local features are extracted, and the SRV attributes are provided by human experts as ground truth. For the testing images in the database, an automated identification process is evoked to translate the detected salient regions of low-level visual features on the moth wings into meaningful semantic SRV attributes. We further propose a novel network analysis-based approach to explore and utilize the co-occurrence patterns of SRV attributes as contextual cues to improve individual attribute detection accuracy. The effectiveness of the proposed approach is evaluated in automated moth identification and attribute-based image retrieval. In addition, a novel image descriptor called

L. Feng (✉) · B. Bhanu
Center for Research in Intelligent Systems, Bourns College of Engineering,
University of California at Riverside, Riverside, CA 92521, USA
e-mail: fengl@cs.ucr.edu

B. Bhanu
e-mail: bhanu@cris.ucr.edu

J. Heraty
Entomology Department, University of California at Riverside, Riverside, CA 92521, USA
e-mail: john.heraty@ucr.edu

© Springer International Publishing Switzerland 2015
B. Bhanu and P. Talbot (eds.), *Video Bioinformatics*,
Computational Biology 22, DOI 10.1007/978-3-319-23724-4_19

349

SRV attribute signature is introduced to record the visual and semantic properties of an image and is used to compare image similarity. Experiments are performed on an existing entomology database to illustrate the capabilities of our proposed system.

19.1 Introduction

Moths are important life forms on the planet with approximately 160,000 species discovered [1], compared to 17,500 species of butterflies [1], which share the same insect Order with Lepidoptera. Although most commonly seen moth species have dull wings (e.g., the Tomato Hornworm moth, see Fig. 19.1a), there are a great number of species that are known for their spectacular color and texture patterns on the wings (e.g., the Giant Silkworm moth and the Sunset moth, see Fig. 19.1b, c, respectively). As a consequence, much research on identifying the moth species from the entomologist side has focused on manually analyzing the taxonomic attributes on the wings such as color patterns, texture sizes, spot shapes, etc., in contrast with the counterpart biological research that classifies species based on DNA differences.

As image acquisition technology advances and the cost of storage devices decreases, the number of specimen images in entomology is grown at an extremely rapid rate both in private database collections and over the web [2–4]. Species identification, relying on manually processing images by entomologists and highly trained experts, is time-consuming and error-prone. The demand for more automated and efficient methods, to meet the requirements of real-world species identification such as agriculture and border control, is increasing. Given the lack of manually annotated text descriptors to the images, and the lack of consensus on the annotations caused by the subjectivity errors of the human experts, engines for archiving, searching, and retrieving insect images in the databases based on keywords and textual meta-data face great challenges in feasibility.

The progress in computer vision and pattern recognition algorithms provides an effective alternative for identifying the insect species and many computer-assisted systems that incorporate these algorithms have been invented in the past two decades [5, 6]. In the image retrieval domain, one of the common approaches

(a)   **(b)**   **(c)**

Fig. 19.1 Moth wings have color and texture patterns at different levels of complexity based on their species: **a** tomato hornworm, **b** giant silkworm and **c** sunset. Photo courtesy of Google Image search engine

introduced to complement the difficulties in text-based retrieval relies on the use of Content-Based Image Retrieval (CBIR) systems [7–9], where sample images are used as queries and compared with the database images based on visual content similarities [10, 11] (color, texture, object shape, etc.). In both the identification and retrieval scenarios, visual features that are extracted to represent morphological and taxonomic information play an important role in the final performance. Context information is often used to help improve individual detection performance of the visual features [12].

These intelligent systems provide a number of attractive functions to entomologists; however, the drawback is: most systems only extract visual features at image pixel level which do not contain human understandable information. However, recent research [13] shows that humans are more expecting to access images at *semantic* level. For example, users of a system are more likely to search by *finding all the moths containing eye spots on the dorsal hind wings* rather than *finding all the moth containing a region with dark blue pixels near the bottom of the image*. An intermediate layer of image semantic descriptor that can bridge the gap between user information need, and low-level visual feature is absent in most existing systems.

In this chapter, we present a new system for automated moth identification and retrieval based on the detection of visual attributes on the wings. The objective of our method is to mimic human behavior on differentiating species by looking at specific *visual contexts* on the wings. More specifically, the notion of "context" refers to discovering certain attribute relationships by taking into account their co-occurrence frequencies. The main motivation of our system relies on the conjecture that the attribute co-occurrence patterns encoded on different species can provide more information for refining the image descriptors. The approaches used are summarized as follows: We build image descriptors based on so-called *Semantically Related Visual (SRV) attributes*, which are the striking and stable physical traits on moth wings. Some examples of SRV attributes are shown in Fig. 19.2. The probabilistic existence of these attributes can be discovered from images by trained detectors using computer vision and pattern recognition

(a)   **(b)**

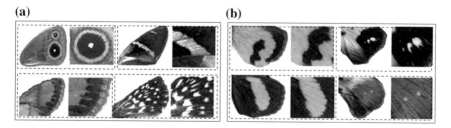

Fig. 19.2 Sample moth wings illustrate the Semantically Related Visual (SRV) attributes. **a** Four sets of SRV attributes on the dorsal fore wings: eye spot (*top left*), central white band (*top right*), marginal cuticle (*bottom left*) and snowflake mosaic (*bottom right*). In each set, the *right image* is the enlarged version of the *left image*. **b** Four sets of SRV attributes on the ventral hind wings. Note it is harder to described the images in a semantic way with simple texts compared to the images in group (**a**)

techniques. Our system detects and learns SRV attributes in a supervised way. The SRV-attributes are manually labeled by human experts to a small subset of the image database that is used for training the attribute detectors. The core of the detector is a probabilistic model that can infer SRV-attribute occurring scores from the unlabeled testing images. We characterize individual images by presenting SRV attributes into a so-called *SRV-attribute signature*. The species identification is performed by comparing the SRV-attribute signature similarity.

19.2 Related Work

Insect species identification recently has received great attention due to the urgent need for systems that can help in biodiversity monitoring [14], agriculture, and border control [15, 16], as well as conservation and other related research [17]. Likewise, identifying species is also the prerequisite to conducting more advanced biological research such as species evolution and developmental studies. However, the vast number of insect species and specimen images is a challenge for manual insect identification. The request for automated computer systems is only likely to grow in the future.

Several attempts have been made in the last two decades to design species identification systems from any type of available data. There have been sophisticated applications to solve problems in classifying orchard insects [5], recognizing the species-specific patterns on insect wings [6]. Qing et al. [18] developed an automatic identification system for rice pests based on the pictorial features of the images. It has been recognized that these computer-aided systems can overcome the manual processing time and errors caused by human subjectiveness.

One common property of these systems is that they all rely on images taken from carefully positioned target under consistent lighting conditions, which reduces the difficulty of the task to some extent. One interesting aspect of automated species identification is that the data are not limited to images. For example, the paper proposed by Ganchev et al. [19] describes the acoustic monitoring of singing insects that applies sound recognition technologies into the insect identification task. Meulemeester et al. [20] report on the recognition of bumble bee species based on statistic analysis of the chemical scent extracted from the cephalic secretions. A challenging competition on multimedia life species identification [21] was recently held on identifying plant, bird, and fish species using image, audio and video data separately.

With the increase of insect images, there is a growing tendency in the field of entomology using image retrieval systems to help archive, organize, and find images in an efficient manner. Great efforts have been made using content-based image retrieval technique to find the relevant images to a query based on the visual similarity; the prototype systems for retrieving Lepidoptera images include "butterfly family retrieval" [22] and a part-based system [10].

19.3 Technical Approach

19.3.1 Moth Image Dataset

The dataset used in this study is collected from an online library of moth, butterfly, and caterpillar specimen images created by Dr. Dan Janzen [23] over a long-term and ongoing project started in 1977 in northwestern Costa Rica. The goal of the inventory is to have records for all the 12,500+ species in the area. As of the end of 2009, the project had collected images of 4,500 species of moths, butterflies, and caterpillars. We use a subset of the adult moth images under the permission of Dr. Dan Janzen. The dataset is publicly available at http://janzen.sas.upenn.edu.

The images are available for both the dorsal and ventral aspects of the moths. Each image was resized into 600 × 400 pixels in resolution and is in RGB colors. Our complete dataset contains 37,310 specimen images covering 1,580 species of moth, but a majority of the species has less than twenty samples. Because our feature and attribute analysis are based on regions on the wings, and some specimens show typical damage ranging from age-dependent loss of wing scales (color distortion), missing parts of wings (incomplete image), or uninformative orientation differences in the wings or antennae, this makes the number of qualified samples even less, and we carefully selected fifty species across three family groups and six subfamily groups: *Hesperiidae* (*Hesperinae, Pyrginae*), *Notodontidae* (*Dioptinae, Nystaleinae*), and *Noctuidae* (*Catolacinae, Heterocampinae* [=*Rifargiriinae*]) from the original dataset. This new subcollection has a total of 4,530 specimens of good quality (see Table 19.1 for the distribution of the species used in our work).

We show sample images of 20 representative species out of the 50 species used in our work in Fig. 19.3. The moth specimens have been photographed against an approximate uniform (usually white or gray) background, but often with shadow artifacts. The specimens are curated in a uniformed way with the wings horizontal and generally with the hind margin of the forewing roughly perpendicular to the longitudinal axis, which facilities the subsequent image processing and feature extraction steps.

19.3.2 System Architecture

The flowchart of the proposed moth identification and retrieval system is shown in Fig. 19.4. The system architecture contains five major parts: (1) information extraction of moth images, (2) SRV attribute detection on moth wings, (3) co-occurrence network construction and co-occurrence pattern detection for the SRV attributes, (4) image signature building and refinement based on SRV attributes and their co-occurrence patterns, and finally (5) applications in moth species identification and retrieval. We give the details about each part in the following sections.

Table 19.1 Families, species and the number of samples in each species used in our work

Sub-families	Species	Images	Sub-families	Species	Images
Catolacinae	Ceroctenaamynta	101	Nystaleinae	Bardaximaperses	74
Catolacinae	Eudocimamaterna	85	Nystaleinae	Dasylophiabasitincta	78
Catolacinae	Eulepidotisfolium	76	Nystaleinae	Dasylophiamaxtla	98
Catolacinae	Eulepidotisrectimargo	57	Nystaleinae	Nystaleacollaris	85
Catolacinae	Hemicephalisagenoria	121	Nystaleinae	Tachudadiscreta	112
Catolacinae	Thysaniazenobia	79	Pyrginae	Atarnessallei	101
Dioptinae	Chrysoglossanorburyi	75	Pyrginae	Dyscophellusphraxanor	86
Dioptinae	Erbessaalbilinea	98	Pyrginae	Tithraustesnoctiluces	96
Dioptinae	Erbessasalvini	117	Pyrginae	Entheusmatho	99
Dioptinae	Nebulosaerymas	69	Pyrginae	Hyalothyrusneleus	82
Dioptinae	Tithrausteslambertae	87	Pyrginae	NascusBurns	94
Dioptinae	Polypoetesharuspex	92	Pyrginae	Phocidesnigrescens	104
Dioptinae	Dioptislongipennis	92	Pyrginae	Quadruscontubernalis	69
Hesperiinae	Methionopsisina	122	Pyrginae	Urbanusbelli	88
Hesperiinae	Neoxeniadesluda	107	Pyrginae	MelanopygeBurns	76
Hesperiinae	SalianaBurns	70	Pyrginae	Myscelusbelti	103
Hesperiinae	Salianafusta	97	Pyrginae	Mysoriaambigua	93
Hesperiinae	TalidesBurns	70	Rifargiriinae	Dicentriarustica	78
Hesperiinae	Vettiusconka	96	Rifargiriinae	Farigiasagana	84
Hesperiinae	Aromaaroma	135	Rifargiriinae	Hapigiodessifredomarini	93
Hesperiinae	Carystoidesescalantei	88	Rifargiriinae	Malocampamatralis	100
Nystaleinae	Lirimirisguatemalensis	95	Rifargiriinae	MeragisaJanzen	65
Nystaleinae	Isostylazetila	99	Rifargiriinae	Naprepahoula	74
Nystaleinae	Oriciadomina	101	Rifargiriinae	Pseudodryaspistacina	83
Nystaleinae	Scoturaleucophleps	117	Rifargiriinae	Rifargiadissepta	69

Fig. 19.3 Sample images for 20 moth species selected from all the species used in this work. We do not show all the species due to the space limit

19.3.3   Feature Extraction

19.3.3.1   Background Removal

It is important to partition the images into "background" and "foreground" because the background usually contains disturbing visual information (such as shadows created by the lighting device, bubbles and dirts on the specimen holder, etc.) that can affect the performance of the detector. We adopted the image symmetry-based approach [24] for background and shadow removal. The moth image dataset used in this chapter have the properties of moth wings with high reflection symmetry (Fig. 19.5a). Because the shadows have the most salient influence on the following processing steps, and they are not symmetric in the images, we use symmetry as the key constraint to remove the shadow.

The SIFT points of the image are detected (Fig. 19.5b) and symmetric pairs of the points are used to vote for a dominant symmetry axis (Fig. 19.5c). Based on the axis, a symmetry-integrated region-growing segmentation scheme is applied to remove the white background from the moth body and shadows (Fig. 19.5d), and the same segmentation process is run with smaller thresholds to partition the image into shadows and small local parts of the moth body (Fig. 19.5e). Finally, symmetry is used again to separate the shadows from the moth body by computing a symmetry affinity matrix. Since the shadows are always asymmetric with the axis of

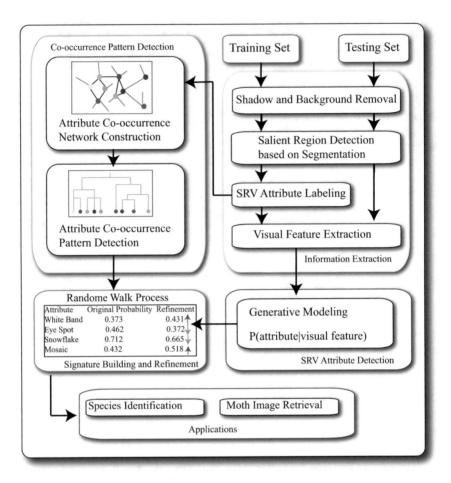

Fig. 19.4 The flowchart of the proposed moth species identification and retrieval system. It consists of: (1) information extraction, (2) SRV attribute detection, (3) attribute co-occurrence pattern detection, (4) attribute signature building and refinement, and (5) moth identification and retrieval applications

reflection, their symmetry affinity will have higher values than the parts of moth body, which is used as the criterion to remove the shadows (Fig. 19.5f).

19.3.3.2 SRV Attribute Labeling

A subregion of the moth wing is considered an SRV attribute if: (1) it repeatedly appears on moth wings across many images, (2) it has salient and unique visual properties, and (3) it can be described by a set of textual words that are descriptive for the subregion.

Fig. 19.5 Steps for background and shadow removal. **a** Original image (with shadow), **b** detected SIFT points, **c** detected symmetry axis, **d** background removed image, **e** segmentation for small parts, and **f** image after shadow removal

We scan the moth images and manually pick a group of SRV attributes. Similar ways have been utilized for designing "concepts" or "semantic attributes" in image classification and object recognition tasks. For example, building nameable and discriminative attributes with human-in-the-loop [25, 26]. However, compared to their semantic attributes, our SRV attributes cannot be described with concise semantic terms (e.g., "A region with scattered white dots on the margin of the hind wing on the dorsal side"). Therefore, we propose to index the SRV attributes by numbers, e.g., "attribute_1," "attribute_2," and so forth. We also explicitly incorporate the positions of the SRV attributes into the attribute index. Each moth has two types of wings: the forewing and the hindwing, and each type of wing has two views: the ventral view and the dorsal view; the SRV attribute index is finally defined in an unified format "attribute_No./wing_type/view," e.g., "attribute_1/forewing/dorsal," "attribute_5/hindwing/ventral," etc. Furthermore, as the moths are symmetrical to the center axis, we only label one side of the moth with the index of SRV attributes.

In order to acquire reliable attribute detectors, SRV attributes are labeled by human experts to the regions in the training images. The regions are represented by the minimum bounding rectangles (MBRs) which are produced using the online open source image labeling tool "LabelMe" [27].

19.3.3.3 Salient Region Detection by Segmentation

For the test images, we use the *salient region detector* to extract small regions or patches of various shapes that could potentially contain the interested SRV

Fig. 19.6 Results from salient region detection. **a** Symmetry-based segmentation, **b** segmentation without using symmetry. Two more results are shown in (**c**) and (**d**) using symmetry-based segmentation

attributes. A good region detector should produce patches that capture salient discriminative visual patterns in images. In this work, we apply a hierarchical segmentation approach based on reflection symmetry introduced in [24] to jointly segment the images and detect salient regions.

We apply symmetry axis detection on the moth images to compute a symmetry affinity matrix, which represents the correlation between the original image and the symmetrically reflected image. Each pixel has a continuous symmetry affinity value between 0 (perfectly symmetric) and 1 (totally asymmetric), which is computed by the Curvature of Gradient Vector Flow (CGVF) [28]. The symmetry affinity matrix of each image is further used as the symmetry cue to improve the region-growing segmentation. The original region-growing approach considers aggregating pixels into regions by pixel homogeneity. In this chapter, we modified the aggregation criterion to integrate the symmetry cue. More details about the approach are explained in [24].

Comparison between Fig. 19.6a, b indicates that using symmetry, more complete and coherent regions are partitioned. The result in Fig. 19.6b is obtained using the same region growing, but without symmetry, so it has many noisy and incomplete regions. The improvements are obtained using the symmetry cue only. Two more results on salient region detection using symmetry-based segmentation are shown in Fig. 19.6c, d.

19.3.3.4 Low-Level Feature Extraction

We represent the above detected salient regions by the minimum bounding rectangles (MBRs). The local features of each bounding rectangular are extracted and pooled into numeric vector descriptors. We have three different types of features used to describe each region: (a) color-based feature, (b) texture-based feature, and (c) SIFT keypoint-based feature.

(1) HSV color feature. The color feature is insensitive to changes of size and direction of regions. However, it suffers from the influence of illumination variations. For the color feature extraction, the original RGB (Red-Green-Blue) color image is first transformed into HSV (Hue-Saturation-Value) space, and only the hue and saturation components are used to reduce the impact from lighting conditions. We then divide the interval of each component into 36 bins, the image pixels inside the salient region are counted for each bin, and the histogram of the 72 bins is concatenated and normalized into the final color feature vector.

(2) Gray Level Co-occurrence Matrix (GLCM-based texture feature. Texture feature is useful to capture the regular patterns of the spatial arrangement of pixels and the intrinsic visual property of regions. We adopt the gray level co-occurrence matrix (GLCM) proposed by Haralick in [29] to extract the texture features. The GLCM is a pixel-based image processing method.

(3) SIFT (Scale Invariant Feature Transform)-based keypoint feature. SIFT [30] proposed by Lowe is a very popular feature used in computer vision and pattern analysis. SIFT feature has the advantage that it is invariant to changes in scale, rotation, and intensity. The major issues related to extracting this feature include selecting the keypoints and calculating the gradient histogram of pixels in a neighboring rectangular region. In this work, we apply the Difference-of-Gaussians (DoG) operator to extract the keypoints. For each keypoint, the 16×16 pixels in the neighboring region are used. We divide a region into 16 4×4 subregions. For each pixel in a subregion, we calculate the direction and magnitude of its gradient. We quantize the directions into 8 bins and build a histogram of gradient directions for each subregion. The magnitude of the gradient is used to weight the contribution of a pixel. Finally, the 8-dimensional feature vectors from the eight-bin direction histogram of each subregion are combined and weighted into a 128-dimensional vector to record local information around the keypoint.

19.3.4 SRV Attribute Detector Learning Module

In this module, the SRV attribute detector is trained using a generative approach based on probability theory. To illustrate the basic idea, consider a scenario in which an image region depicted by an N-dimensional low-level feature vector $\overrightarrow{X^N}$ is to be assigned into one of the K SRV attributes $k = 1, \ldots, K$ in a higher level of semantics. From probability theory, we know that the best solution is to achieve the *posterior probabilities* $p(k|X)$ for a given X and each attribute category k, and assign the attribute with the largest probability score to the region. In the generative model, we model the joint probability distribution $p(k, X)$ of image region features and attributes, and Bayes' theorem provides an alternative to derive $p(k|X)$ from $p(k, X)$:

$$p(k|X) = \frac{p(k, X)}{p(X)} = \frac{p(X|k)p(k)}{\sum_{i=1}^{K} p(X|i)p(i)} \tag{19.1}$$

As the sum in the denominator takes the same value for all the attribute categories, it can be viewed as a normalization factor over all the attributes. Equation (19.6) can be rewritten as:

$$p(k|X) \propto p(k, X) = p(X|k)p(k) \tag{19.2}$$

which means we only need to estimate the attribute prior probabilities $p(k)$ and the likelihood $p(X|k)$ separately. The generative model has the advantage that it can augment the large amount of unlabeled data in a dataset from a small portion of the labeled data.

As defined earlier K denotes the pool of SRV attributes. Let k_i be the ith attribute in K. According to the previous section, k_i is assigned to a set of image regions $R_{k_i} = \{r_1^i, r_2^i, \ldots, r_{n_{k_i}}^i\}$ along with the corresponding feature vectors $X_{k_i} = \{x_1^i, x_2^i, \ldots, x_{n_{k_i}}^i\}$, where n is the number of regions in an image. We assume the feature vector is sampled from some underlying multivariate density function $p_X(\cdot|k_i)$. We use a nonparametric kernel-based density estimate [31] for the distribution p_X. Assuming region r_t to be in the test image with feature vector x_t, we estimate $p_X(x_t|k_i)$ using a Gaussian kernel over the feature vectors X_{k_i}:

$$p_X(x_t|k_i) = \frac{1}{n} \sum_{j=1}^{n} \frac{\exp\{-(x_t - x_j)^T \Sigma^{-1}(x_t - x_j)\}}{\sqrt{2^n \pi^n |\Sigma|}} \tag{19.3}$$

Σ is the covariance matrix of the feature vectors in X_{k_i}.

$p(k_i)$ is estimated using Bayes estimators with a prior beta distribution, the probability distribution of $p(k_i)$ is given by:

$$p(k_i) = \frac{\mu \delta_{k_i, r} + N_{k_i}}{\mu + N_r}, \tag{19.4}$$

where μ is the smoothing parameter estimated from the training set, $\delta_{k_i, r} = 1$ if attribute k_i occurs in the training region r and 0 otherwise. N_{k_i} is the number of training regions that contain attribute k_i and N_r is the total number of training regions.

Finally, for each test region with feature vector x_t, the *posterior probability* of observing attribute k_i in K given x_t, $p(k_i|x_t)$ is given by multiplying the estimates of the two distributions:

$$p(k_i|x_t) = (\frac{1}{n}\sum_{j=1}^{n}\frac{exp\{-(x_t - x_j)^T\Sigma^{-1}(x_t - x_j)\}}{\sqrt{2^n\pi^n|\Sigma|}}) \times (\frac{\mu\delta_{k_i,r} + N_{k_i}}{\mu + N_r}) \qquad (19.5)$$

For each salient region extracted from a test image I, the occurrence probability of each attribute in that region is inferred by Eq. (19.5). The probabilities for all attributes are combined into a single vector which is called *region SRV attribute signature*. For a test image with several salient regions, we combine the region SRV attribute signature into a final vector by choosing the max score for each attribute. We name this vector as the *image SRV attribute signature*, and it is used as the semantic descriptor for images.

19.3.5   SRV Attribute Co-occurrence Pattern Detection Module

Attribute labels given by human experts as ground-truth semantic descriptions across the entire training image set are used to learn the contextual information based on the attribute label co-occurrences. In this section, we devise a novel approach to discover the co-occurrence patterns of the individual attributes based on network analysis theories. More specifically, we construct an attribute co-occurrence network to record all the pairwise co-occurrence between attributes. The patterns are detected as the communities in a network structure. A similar concept is used in social network to describe a group of people that have tightly established interpersonal relationships.

19.3.5.1   SRV Attribute Co-occurrence Pattern Detection

We first introduce the notion of community structure from the network perspective. One way to understand and analyze the correlations among individual items is to represent them in a graphical network. The nodes in the network corresponds to the individual items (attributes in our case), the edges describe the relationships (attribute co-occurrence in our case), and the edge weights denote the relevant importance of the relationship (co-occurrence frequency in our case).

A very common property of a complex network is known as the community structure, i.e., groups of nodes may have tight internal connections in terms of a large number of internal edges, while they may have less edges connecting each other. These groups of nodes constitute the communities in the network. The existence of community structure reflects underlying dependencies among elements in the target domain. If a group of individual attributes always occur together in the training image set, then an underlying co-occurrence pattern can be defined by these attributes, and this pattern can be used as a priori knowledge in the attribute detection for the test images.

19.3.6   Identification Module

The attribute detector learned from the training data is used in the identification module for the testing images. The inputs to the detector are the detected salient regions from the test images as well as the extracted low-level visual features. The output of the detector is the so-called "image SRV attribute signature". The species identification of testing images is performed by comparing testing image signatures with the training image signatures. Therefore, we also build the attribute signatures for the training images. For a training image I, the attribute signature is $S^{|A|}$ with each element $s(a_i) \in \{0, 1\}$ and $s(a_i) = 1$ when image I has regions labeled with attribute a_i and $= 0$ otherwise. We further divide the training images into groups based on their scientific species designation. The element values are averaged across the signatures within each species group for each individual attribute, and the obtained signature is called the *species prototype signature*.

The testing image of a species is identified by comparing its image attribute signature with the species prototype signatures of the fifty species. The distance between the two signatures is calculated by the Euclidean distance. The testing image is finally identified as the species with the smallest distance value. If several species have very similar distance values to the testing image, we assign all the species labels to that image, and let the image retrieval system give the final decision on the species based on the feedback from the users who are determined as experts by the retrieval system.

19.3.7   Retrieval and Relevance Feedback Module

We implement a query by example (QBE) paradigm for our retrieval system. QBE is widely used in conventional content-based image retrieval (CBIR) systems when the image meta-data, such as captions, surrounding texts, etc. are not available for keyword-based retrieval.

19.3.7.1   Image Retrieval Using Query by Example

In the QBE mode, the user is required to submit query in terms of an example specimen image to the system. Finding an appropriate query example, however, is still a challenging problem in the research area of CBIR. In our system, we provide an image browsing function in the user interface, and the user is allowed to browse all the images in the database and submit a query. Images are compared by their content similarity. Each image in the database is represented by a low-level visual feature vector F and a high-level SRV attribute signature S, for a query image Q and a database image Y. The distance between them is calculated by fusing the

Euclidean distance over the visual feature vectors and the Earth Mover's distance [32] over the SRV attribute signatures:

$$\text{Dist}(Q, Y) = \eta D_{\text{Euc}}(F_Q, F_Y) + (1 - \eta) D_{\text{EMD}}(S_Q, S_Y), \tag{19.6}$$

where η is the adjusting parameter between the two distance measures and is determined by the long-term cross-session retrieval history working on the subset of training images [33]. If the precision for a particular query is increased when more importance is put on the feature distance, then η is adjusted to a larger value, otherwise it becomes smaller.

19.3.7.2 Relevance Feedback

The Relevance feedback (RF) scheme has been verified as a performance booster for our retrieval system. The reason is that RF can capture more information about user's search intention, which can be used to refine the original image descriptors from feature extraction and attribute detection [34].

Our RF approach follows the Query Point Movement (QPM) paradigm as opposed to the Query Expansion (QEX) paradigm. We move the query point in both the feature space and the attribute space toward the center of the user's preference using both the relevant and irrelevant samples marked by the user at each retrieval iteration. However, before the users' decisions are used to refine the descriptors, their expertise in identifying moth species are evaluated by sample tests when they first enter the system. If an user has 90 % accuracy in identifying the species, their relevance feedback will take effect.

19.4 Experimental Results

We implemented the system on Microsoft Windows platform using C#net with the Windows Presentation Foundation application development framework. The image database with relevant features and attributes are deployed on MySQL server. The database is set up by importing.txt files with numeric values of the attributes and features, and textual information describing the image properties of the moth images.

19.4.1 Species Identification Results

We randomly sampled the images into 10 subsets, one subset was held out for testing and the rest of the subsets was used for training the model. This process was repeated ten times using each subset of images as the testing set. The average of these results on 10 subsets is reported for each combination of parameters $\{Q, \alpha, \eta\}$ on the testing set. We chose the parameter set that maximized the overall performance averaged over the ten testing subsets. The value of the selected parameters are: $Q = 0.3$, $\alpha = 0.6$, $\eta = 0.5$. The performance of the automated species identification is evaluated by the *accuracy* measure. A test image is assigned to the species category for which prototype signature has the smallest distance to the image's SRV attribute signature. The accuracy measure is defined for each species as the number of correctly identified individuals divided by the total number of specimens of that species in the testing set. A testing image is considered as a correct identification if the species label generated by the program matches with the ground-truth label.

To demonstrate the effectiveness of our proposed framework for the moth species identification application, we compare with the following approaches as baselines: **Baseline-I**: The most basic model that only uses the visual features extracted from Sect. 19.3.3.4. No SRV attributes and the signature representation have been used. The images are identified purely based on the visual feature vector similarity calculated using the Euclidean distance. **Baseline-II**: Our generative model for individual attribute detection unified with the attribute signature representation serves as the Baseline-II model. However, this model does not include attribute co-occurrence pattern detection and random walk refinement on the SRV attribute signatures. **VW-MSI**: We reimplemented a visual words-based model based on the available code (http://people.csail.mit.edu/fergus/iccv2005/bagwords.html) online for image classification [35] and name it as "Visual Words based Moth Species Identification" (VW-MSI). We only implemented the appearance model in the approach and ignored the complex spatial structures. **SRV-MSI**: Our proposed approach integrated with co-occurrence pattern detection and SRV attribute signature refinement. We name it as "SRV attribute based Moth Species Identification" (SRV-MSI).

We compared the species identification results of the proposed approach with other three approaches in Table 19.2. The mean and standard deviation of the accuracy of the experiments conducted for ten times are computed and shown for twenty species. As we can observe from Table 19.2, our system performs the best for almost all the species. This demonstrates the effectiveness of SRV attributes and the co-occurrence patterns used for signature refinement.

The total number of SRV attributes manually given to the images by the human experts is 450. As a result, the maximum length of the SRV attribute signature for the images is 450. In order to compare the impact from the vocabulary size of the attributes and the visual words for VW-MSI and SRV-MSI, we set the maximum size of the visual words vocabulary to 450 as well. The SRV attributes and the visual words are ranked in the relative vocabulary based on the number of appearance in the image collection.

Table 19.2 Identification accuracy for the 50 species (20 are shown for space reason)

Species	Baseline I		Baseline II		VW-MSI		SRV-MSI	
	Mean	Std	Mean	Std	Mean	Std	Mean	Std
Ceroctena amynta	0.2965	0.0321	0.4176	0.0169	0.4318	0.0196	0.4582	0.0174
Eudocima materna	0.4968	0.0257	0.5483	0.0275	0.5764	0.0319	0.5944	0.0209
Eulepidotis folium	0.3910	0.0279	0.4141	0.0264	0.4219	0.0267	0.4482	0.0371
Eulepidotis rectimargo	0.5561	0.0246	0.5875	0.0236	0.5962	0.0233	0.6134	0.0163
Hemicephalis agenoria	0.3314	0.0268	0.3349	0.0302	0.3721	0.0331	0.3931	0.0236
Thysania zenobia	0.4102	0.0327	0.4329	0.0236	0.4623	0.0235	0.4971	0.0356
Chrysoglossa norburyi	0.5472	0.0225	0.5553	0.0253	0.5672	0.0237	0.5752	0.0205
Erbessa albilinea	0.6048	0.0365	0.6324	0.0336	0.6547	0.0136	0.6755	0.0174
Erbessa salvini	0.3562	0.0468	0.3634	0.0425	0.3867	0.0325	0.4143	0.0345
Nebulosa erymas	0.5432	0.0312	0.5647	0.0291	0.5699	0.0257	0.5935	0.0225
Tithraustes noctiluces	0.5438	0.0214	0.5624	0.0331	0.5912	0.0284	0.6086	0.0251
Polypoetes haruspex	0.5247	0.0216	0.5369	0.0234	0.5682	0.0273	0.5906	0.0202
Dioptis longipennis	0.5621	0.0281	0.5746	0.0212	0.5990	0.0187	0.6154	0.0175
Methionopsis ina	0.4721	0.0375	0.4835	0.0367	0.5014	0.0325	0.5102	0.0425
Neoxeniades luda	0.3742	0.0374	0.3852	0.0432	0.4176	0.0396	0.3975	0.0457
Saliana Burns	0.5042	0.0364	0.5356	0.0256	0.5494	0.0275	0.5731	0.0234
Saliana fusta	0.6480	0.0247	0.6597	0.0275	0.6968	0.0214	0.7346	0.0134
Talides Burns	0.5437	0.0256	0.5572	0.0247	0.5854	0.0173	0.6352	0.0176
Vettius conka	0.6417	0.0334	0.6782	0.0148	0.7332	0.0184	0.7544	0.0169
Aroma aroma	0.5437	0.0273	0.6035	0.0245	0.6204	0.0174	0.6461	0.0211

The performance of SRV-MSI is greater than all other approaches except for *Neoxeniades luda*, *Isostyla zetila*, *Atarnes sallei* and *Nascus Burns*

19.4.2 Image Retrieval Results

To test the performance of our SRV attribute-based approach for image retrieval with the proposed relevance feedback scheme, like for species identification in Sect. 19.4.2, we divided the entire image dataset into 10 folds. The parameters are determined using the same scheme as described in Sect. 19.4.1. We set the number of attributes to 300. In order to reduce the amount of work of submitting relevance feedback that are required by users, we propose to simulate the user interaction by launching queries and submitting feedback automatically by the system. The simulated process works in the following way: the system compares the ground-truth species labels of the retrieved images with the query, if the species label matches

Table 19.3 Comparison of the retrieval performance for the 50 species

Mean average precision

Species	BL-I	BL-II	SRV-IR	Species	BL-I	BL-II	SRV-IR
Ceroctenaamynta	0.4096	0.3872	0.4571	*Bardaximaperses*	0.2836	0.3064	0.3275
Eudocimamaterna	0.4538	0.4170	0.4764	*Dasylophiabasitincta*	0.3538	0.4152	0.4658
Eulepidotisfolium	0.3824	0.4115	0.4745	*Dasylophiamaxtla*	0.3628	0.3738	0.4145
Eulepidotisrectimargo	0.5572	0.5069	0.6130	*Nystaleacollaris*	0.3427	0.3735	0.3841
Hemicephalisagenoria	0.4187	0.3950	0.4712	*Tachudadiscreta*	0.2917	0.2978	0.3114
Thysaniazenobia	0.4104	0.3933	0.4705	*Atamessallei*	0.5832	0.6224	0.6778
Chrysoglossanorburyi	0.5856	0.5710	0.6786	*Dyscophellusphraxanor*	0.5324	0.5799	0.6128
Erbessaalbilinea	0.6045	0.5972	0.7153	*Tithrausteslambertae*	0.4846	0.4472	0.5315
Erbessasalvini	0.4587	0.4311	0.5478	*Entheusmatho*	0.4796	0.4925	0.5486
Nebulosaerymas	0.5219	0.5346	0.5857	*Hyalothyrusneleus*	0.6042	0.6584	0.6971
Tithraustesnoctiluces	0.5486	0.5148	0.5749	*NascusBurns*	0.2396	0.2846	0.3167
Polypoetesharuspex	0.5745	0.5237	0.5964	*Phocidesnigrescens*	0.5755	0.5942	0.6398
Dioptislongipennis	0.4816	0.4754	0.5048	*Quadruscontubernalis*	0.6492	0.7047	0.7168
Methionopsisina	0.3581	0.3847	0.3994	*Urbanusbelli*	0.5693	0.5480	0.5724
Neoxeniadesluda	0.3625	0.3827	0.4117	*MelanopygeBurns*	0.6454	0.6845	0.6992
SalianaBurns	0.5317	0.5485	0.5884	*Myscelusbelti*	0.6715	0.7047	0.7673
Salianafusta	0.6046	0.5917	0.6459	*Mysoriaambigua*	0.4917	0.4802	0.5746
TalidesBurns	0.5154	0.5308	0.5742	*Dicentrarustica*	0.3969	0.4105	0.4453
Vettiusconka	0.6296	0.6115	0.7135	*Farigiasagana*	0.2946	0.3072	0.3418
Aromaaroma	0.4537	0.4425	0.5289	*Hapigiodessigifredoma*	0.3634	0.3728	0.4051
Carystoidesescalantei	0.5046	0.4672	0.5274	*Malocampamatralis*	0.4746	0.4869	0.5537
Lirimirisguatemalensis	0.3234	0.3456	0.3753	*MeragisaJanzen*	0.5643	0.5756	0.6683
Isostylazetila	0.5924	0.5483	0.6175	*Naprepahoula*	0.3748	0.4245	0.4886
Oriciadomina	0.4641	0.4547	0.5044	*Pseudodryaspistacina*	0.2975	0.3174	0.3531
Scoturaleucophleps	0.5179	0.5357	0.5678	*Rifargiadissepta*	0.5648	0.5247	0.6190

the query, the system will mark the image as relevant, otherwise, the image is marked as irrelevant. By doing this, we assume the relevance feedback provided by the users will always by correct (i.e., users will only mark the relevant images as those from the same species category as the query). In each iteration, the retrieval precision is evaluated by the rank of the relevant images. Further statistical evaluation of the averaged precision for each species relies on standard image retrieval measure: *Mean average precision of top D retrieved images* over all the query images from a specific species category. Let D be the number of retrieved images and R be the relevant ones with size $|R|$. Given a query Q, the average precision is defined as $AP(Q) = \frac{1}{|R|} \sum_{i=1}^{|R|} \frac{i}{\text{Rank}(R_i)}$, and the mean average precision (MAP) is the averaged AP over all the testing images.

To demonstrate the effectiveness of our proposed retrieval framework, we use the following approaches as the baselines to compare the results: **Baseline-I**: The proposed image retrieval framework without relevance feedback scheme. **Baseline-II**: We reimplemented an insect image identification approach [11] and integrated it into our retrieval framework with five iterations of relevance feedback process. The features used are a combination of color, shape, and texture features, and there is no higher level image descriptor like our SRV attribute that has been used in the original approach. **SRV-IR**: Our proposed retrieval framework with relevance feedback scheme based on the SRV attributes.

We show the top twelve retrieved images in the application interface. However, the application can be adjusted to show more images upon request. Table 19.3 summarizes the mean averaged precision from the three approaches for all the fifty species. As we can observe, when RF scheme is applied (Baseline-II and SRV-IR), the mean averaged precision is increased compared to the retrieval without RF (Baseline-I), which demonstrates the effect of human interaction in improving the retrieval performance. When more retrieval iterations are involved in the searching process, and when more iterations of relevance feedback are provided, the system can find more relevant images matching user's search intention. In the two approaches that adopts relevance feedback scheme, our approach which uses SRV attribute-based image descriptor outperforms Baseline-II for all the species categories. The system response time for each individual query for a database of 1000 images is around 150 ms. For a database of 4000 images, the response time for each individual query is approximately 500 ms.

19.5 Conclusions

In summary, this chapter has introduced a novel insect species identification and retrieval system based on wing attributes in the moth image dataset. The purpose of the research is to design computer vision and pattern recognition approaches to conduct automated image analysis that can be used by the entomologists for insect studies. We have demonstrated the effectiveness of our system in species identification and image retrieval for fifty moth species.

Future research will include investigations on more effective feature and attributes as well as more advanced learning approaches which could address both the scalability and discrimination issues. Also, we will look into the scalability issue of the current system as the number of moths images increases. We will seek advanced image indexing strategies that involve modern technologies in big data and parallel computing.

Acknowledgment This work was supported in part by the National Science Foundation grants 0641076 and 0905671.

References

1. Carter D (1992) Butterflies and moths. Eyewitness handbooks
2. Kerr P, Fisher E, Buffington M (2008) Dome lighting for insect imaging under a microscope. Am Entomol 54:198–200
3. Buffington M, Gates M (2008) Advanced imaging techniques ii: using a compound microscope for photographing point-mount specimens. Am Entomol 54:222–224
4. Buffington M, Burks R, McNeil L (2005) Advanced techniques for imaging parasitic Hymenoptera (Insecta). Am Entomol 51:50–56
5. Wen C, Guyer DE, Li W (2009) Local feature-based identification and classification for orchard insects. Biosyst Eng 104(3):299–307
6. Francoy TM, Wittmann D, Drauschke M, Müller S, Steinhage V, Bezerra-Laure MAF, Jong DD, Goncalves LS (2008) Identification of africanized honey bees through wing morphometrics: two fast and efficient procedure. Apidologie 39(5):488–494
7. Yue J, Li Z, Liu L, Fu Z (2011) Content-based image retrieval using color and texture fused features. Math Comput Model 54:1121–1127
8. Bunte K, Biehl M, Jonkman M, Petkov N (2011) Learning effective color features for content based image retrieval in dermatology. Pattern Recogn 44:1892–1902
9. Singhai N, Shandilya S (2010) A survey on: content based image retrieval systems. Int J Comput Appl 4:22–26
10. Bhanu B, Li R, Heraty J, Murray E (2008) Automated classification of skippers based on parts representation. Am Entomol 228–231
11. Wang J, Lin C, Ji L, Liang A (2012) A new automatic identification system of insect images at the order level. Knowl-Based Syst 33:102–110
12. Divvala SK, Hoiem D, Hays JH, Efros AA, Hebert M (2009) An empirical study of context in object detection. In: IEEE conference on computer vision and pattern recognition, pp 1271–1278
13. Hanjalic A, Lienhart R, Ma WY, Smith JR (2008) The holy grail of multimedia information retrieval: so close or yet so far away? Proc IEEE 96(4):541–547
14. Pereira HM, Ferrier S, Walters M, Geller GN, Jongman RHG, Scholes RJ, Bruford MW, Brummitt N, Butchart SHM, Cardoso AC et al (2013) Essential biodiversity variables. Science 339(1):277–278
15. Bacon SJ, Bacher S, Aebi A (2012) Gaps in border controls are related to quarantine alien insect invasions in Europe. PLoS One 7(10). doi:10.1371/journal.pone.0047689
16. Kumschick S, Bacher S, Dawson W, Heikkilä J (2012) A conceptual framework for prioritization of invasive alien species for management according to their impact. NeoBiota 15(10):69–100

17. Steele PR, Pires JC (2011) Biodiversity assessment: State-of-the-art techniques in phylogenomics and species identification. Am J Bot 98(3):415–425
18. Qing Y, Liu QJ, Yang BJ, Chen HM, Tang J (2012) An insect imaging system to automatic rice light-trap pest identification. J Integr Agr 11:978–985
19. Ganchev T, Potamitis I, Fakotakis N (2007) Acoustic monitoring of singing insects. In: IEEE international conference on acoustics, speech and signal processing, vol 4
20. Meulemeester TD, Gerbaux P, Boulvin M, Coppée A, Rasmont P (2011) A simplified protocol for bumble bee species identification by cephalic secretion analysis. Int J Study Soc Arthropods 58(5):227–236
21. Joly A, Goëau H, Glotin H, Spampinato C, Bonnet P, Vellinga W, Planque R, Rauber A, Fisher R, Müller H (2014) Lifeclef 2014: multimedia life species identification challenges. Proc LifeCLEF 2014:229–249
22. Wang J, Ji L, Liang A, Yuan D (2011) The identification of butterfly families using content-based image retrieval. Biosyst Eng 111:24–32
23. Janzen DH, Hallwachs W (2009) Dynamic database for an inventory of the macrocaterpillar fauna, and its food plants and parasitoids, of area de conservacion guanacaste (acg), northwestern costa rica (nn-srnp-nnnnn voucher codes). http://janzen.sas.upenn.edu
24. Sun Y, Bhanu B (2012) Reflection symmetry-integrated image segmentation. IEEE Trans Pattern Anal Mach Intell 34(9):1827–1841
25. Duan K, Parikh D, Crandall D (2012) Discovering localized attributes for fine-grained recognition. In: IEEE conference on computer vision and pattern recognition, pp 3474–3481
26. Parikh D, Grauman K (2011) Interactively building a discriminative vocabulary of nameable attributes. In: IEEE conference on computer vision and pattern recognition, pp 1681–1688
27. Russell BC, Torralba A, Murphy KP, Freeman WT (2008) Labelme: a database and web-based tool for image annotation. Int J Comput Vis 77:157–173
28. Prasad VSN, Yegnanarayana B (2004) Finding axes of symmetry from potential fields. IEEE Trans Image Process 13(12):1559–1566
29. Haralick RM (1979) Statistical and structural approaches to texture. Proc IEEE 67:786–804
30. Lowe DG (2004) Distinctive image features from scale-invariant keypoints. Int J Comput Vis 60(2):91–110
31. Fu G, Shih F, Wang H (2011) A kernel-based parametric method for conditional density estimation. Pattern Recogn 44(2):284–294
32. Rubner Y, Tomasi C, Guibas LJ (2000) The earth mover's distance as a metric for image retrieval. Int J Comput Vis 40(2):99–121
33. Yin PY, Bhanu B, Chang KC (2008) Long-term cross-session relevance feedback using virtual features. IEEE Trans Knowl Data Eng 20(3):352–368
34. Dong A, Bhanu B (2005) Active concept learning in image databases. IEEE Trans Syst Man Cyber Part B 35:450–456
35. Sivic J, Russell B, Efros A, Zisserman A, Freeman W (2005) Discovering object categories in image collections. In: International conference on computer vision, pp 1543–1550

Index

A

Aβ plaques, 266, 276
Acoustic monitoring, 352
Acrosomal eversion, 183
Acrosome, 183
Actin, 183, 276
Actin-binding proteins, 266
Actin depolymerization, 266
Actin filaments, 240, 273
Actin-severing protein cofilin, 266
Action units (AUs), 331, 343
Active appearance models (AAM), 332
Active contour models, 105
Adaptive quadratic Voronoi tessellation
 (AQVT), 230, 232, 233
Addition, 305
Aerial hyphae, 244
Affective computing, 330
Affine transformation, 206
Agar, 240
Algorithm, 101, 104, 108, 111, 113, 114, 127,
 140, 148, 195, 229, 230, 243, 244, 247,
 271, 274, 287, 290, 298, 305, 308, 332
Algorithms, 193, 291, 310, 317, 319, 324, 326,
 330, 345
Alternative upstream effectors, 269
Alzheimer's disease (AD), 266, 268, 276
Amira, 152, 325
Angiogenesis, 89
Anisotropic diffusion, 197
Anomalous diffusion, 195
ANOVA analysis, 155
Apical growth, 202
Apoptosis, 152, 162
Apoptotic, 84, 86, 136
Apparent diffusion coefficients (ADC), 79, 84,
 88
Appearance, 292–294
Appearance/disappearance, 294, 298, 300

Appearance hyperedge, 299
Arabidopsis, 7, 9, 219
Arabidopsis hypocotyls, 194
Arabidopsis RIP1, 209
Arabidopsis SAM cells, 218, 229
Arabidopsis stomata, 194
Arabidopsis thaliana, 209, 217, 234, 240
Arabidopsis thaliana pavement cells, 241
Arabidopsis thaliana root meristem, 223
Arousal, 331, 345
Ascospores, 239
Asexual spores (conidia), 244
Astrocytes, 180
Attachment, 155
Attribute co-occurrence pattern detection, 364
Attribute detector, 362
AU 6, 331
AU 12, 331
Audio/visual emotion challenge, 345
Automated ischemic lesion detection schema,
 101
Automated processing, 152
Automated tracking algorithm, 276
Automation, 190
Average intensity (A), 126

B

Background, 355
Backward prediction, 296
Bank, 334
Basal hyphae, 244
Basal vegetative hyphae, 238
Baseline-I, 364, 367
Baseline-II, 364, 367
Bayes, 360
Bayes' theorem, 360
Bayesian network (BN), 62, 63
β-amyloid, 266
BioDBcore, 318

© Springer International Publishing Switzerland 2015
B. Bhanu and P. Talbot (eds.), *Video Bioinformatics*,
Computational Biology 22, DOI 10.1007/978-3-319-23724-4